Quarrying in Western Norway. An archaeological study of production and distribution in the Viking period and Middle Ages

Quarrying in Western Norway

An archaeological study of production and distribution in the Viking period and Middle Ages

Irene Baug

Archaeopress Archaeology

ARCHAEOPRESS PUBLISHING LTD
Summertown Pavillion
18-24 Middle Way
Oxford OX2 7LG

www.archaeopress.com

ISBN 978 1 78491 102 7
ISBN 978 1 78491 103 4 (e-Pdf)

Table of Contents

List of Figures

List of Tables

Acknowledgments

This study is has been financed by a scholarship from the Faculty of Humanities at the University of Bergen.

Several people have contributed to this work, first and foremost my supervisor Professor Ingvild Øye, who from the first scholarship application to the present volume has supported and contributed to my project. Ingvild's efficiency and professionalism is admirable, and her many hours of reading and commenting on my drafts (workdays, weekends and holidays) have been of the greatest importance. Thank you so much!

I am also grateful to the interdisciplinary project 'The Norwegian Millstone Landscape' (2009-2012) and to project manager Gurli Meyer and her colleagues, Tom Heldal and Tor Grenne, for allowing me to connect my studies to the project; this made an investigation of the distribution of the Hyllestad stones possible. Special gratitude goes to 'my' geologist Øystein J. Jansen for patiently sharing his knowledge in geology, and for being a good colleague and friend on our research trips and many hours spent studying museum assemblages, as well as for his contribution during my excavations in Ølve and Hatlestrand.

Grants from the Melzer Foundation and from the Faculty of Humanities enabled me to carry out the excavations in Hyllestad and Ølve/Hatlestrand. Several students voluntarily helped during the excavations: Halldis Hobæk, Anja Sætre, Marie Ødegaard, Camilla Jacobsen, Glenn Heine Orkelbog, Henriette Børslid Hop, Silje Foyn, Tina Granados, Åsne Helleve, and Kristoffer Nerhus. Thank you so much to you all! In addition Kristoffer Nerhus, Knut Johan Nerhus, Gunnar Nygård, and Audun Oppedal contributed greatly with their local knowledge of the quarry sites and surrounding areas. Kristoffer Nerhus and Knut Johan Nerhus were also of great help during the identification and mapping of quarries in Ølve/Hatlestrand. Special thanks also go to Atle Ove Martinussen for support and help during my identification and mapping of quarries in Ølve and Hatlestrand.

I would also like to extend special thanks to Arne Solli, who, time and again, patiently helped and answered my many questions concerning GIS and maps.

I am grateful to Helge Askvik and Morgan Ganerød for their help with dating the whetstones found together in cargoes with quernstones and millstones. Arild Marøy Hansen was generous in providing information on cargoes and allowing me to take samples from whetstones.

Thanks are also due to Torbjørn Løland and Astrid Waage at the Norwegian Millstone Centre for many fruitful discussions and cooperation.

Several other colleagues have helped contribute to a pleasant working environment for which I am most grateful. Special gratitude must go to my friends and colleagues at Bryggens Museum, where I have carried out my daily work. I would also like to express my thanks to my present and previous fellow PhD students at Bryggens Museum: Sigrid Samset Mygland, Janicke Larsen and Ole-Magne Nøttveit, and to Associate Professors Gitte Hansen and Alf Tore Hommedal at Bergen University Museum.

Justine Parer (Australia) and Gloria Thomas (England) proofread the preliminary manuscript and I thank them for their help.

Working on a research project sometimes makes for logistical problems at home. A big thank-you therefore goes to my parents for taking care of my children when workdays suddenly became too long and too busy.

Finally, my largest debt of gratitude goes to my husband Per Terje and our two sons Arin and Tristan. Per Terje - thanks for your help during the excavations, and for taking good care of the boys in those periods when I was stuck in quarries or my office! Thank you for all your support and for giving me the time needed to finish this study. And to our two fantastic boys, Arin and Tristan: thank you for just being there and reminding me of what is important in life.

This present work is based on research defended for my PhD on 11 October 2013. Professor Dagfinn Skre, University of Oslo (Norway) and Professor Eva Svensson, Karlstad University (Sweden) were opponents. I thank them for their comments and for many interesting discussions. The thesis was first published in 2013 and minor alterations have been made to this present edition. Proofreading and layout work for this edition has been financially supported by the Department of Archaeology, History, Cultural Studies and Religion, University of Bergen.

Bergen, October 2014

Irene Baug

1 Introduction

Over large swathes of history, grain and grain products have been an indispensable source of food, with bread and porridge as vital elements in diets. In the Middle Ages up to 80 per cent of consumption could derive from grain and grain products (Øye 2002:323). To make grain digestible, it had, however, to be crushed. Consequently, querns and quernstones, along with bakestones, flat stones used for baking bread over fireplaces, have played an important role in both prehistoric and historic households. The demand for these tools eventually formed the basis for large-scale production, which is the theme of this study.

The production of quernstones, millstones and bakestones is only rarely referred to in medieval sources, and until recently has gained little scholarly attention. The aim of this study is to illuminate different aspects related to the quarrying of these products, such as extent, age, distribution and the organisation of production and export of these objects. The study is based on analyses of two different quarry landscapes in western Norway: Hyllestad and Ølve and Hatlestrand in the counties of Sogn og Fjordane and Hordaland (Fig. 1.1). The areas are comprised of different geological resources making it possible to extract these products.

The Hyllestad quarries with their products gain the main focus. Besides quernstones and millstones the products included stone crosses, grave slabs, vessels and smoke vent stones. In Ølve and Hatlestrand tiles and building stones, as well as a few crosses, were extracted in addition to bakestones. In this project, the production and distribution of bakestones are emphasized. I have earlier (in my Master's thesis on the quarries in Hyllestad from 2001, published 2002) investigated four smaller areas. This study represented the first archaeological investigation into quernstone and millstone quarries in Scandinavia, and the first step towards knowledge of these quarries and this type of quarrying (Baug 2002). In this new project I follow up issues from my previous study, while additionally expanding the research area, as well as the objectives.

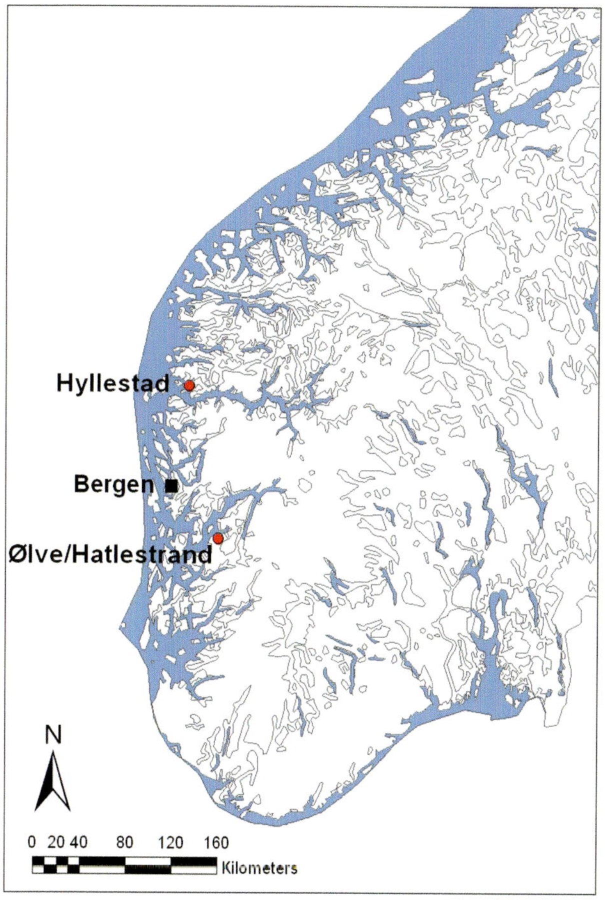

Figure 1.1: Location of the two quarry areas Hyllestad and Ølve/Hatlestrand (Map: Irene Baug, 2013)

1.1 Scope and objectives

Several different but interrelated aspects of quarrying are central to this study. An important goal is to achieve a clearer understanding of the quarries as parts of comprehensive and integrated systems on many levels, technologically, economically, socially and culturally. A particular objective is to gain an insight into the production, transportation, trade and exchange, as well as the organisation of these activities, both locally and regionally. The study therefore has a multi-scale approach, where the quarries and productions are studied on a micro level in relation to the wider local communities on the 'meso' level, and related to issues such as control and rights of use, as well as provenance studies and distribution in a wider geographical context - the macro level.

For Hyllestad a larger area has been investigated than in previous research; furthermore quarries both in the central and peripheral parts of the production landscape are taken into account. Additionally, it has been important to incorporate different quarry types based on different technologies, and extraction of different product types from those investigated before. I also look more closely at the societal contexts of both production and distribution of the objects through time. Incorporating new types of production site, such as the quarries in Ølve and Hatlestrand, represents a new empirical basis for the study of quarrying, with regard to production, distribution and the organisation of these activities. The production sites for bakestones are different and smaller in extent and production, as well as distribution, is generally less studied. A special challenge has been therefore to upgrade the empirical knowledge basis for both production and distribution, and to widen the perspectives. A comparison of the two different quarry landscapes, their products and geographical distribution may also open up a better understanding of trade, networks and actors involved on a broader basis. Seen together, the two quarry landscapes may illuminate the exploitation of different resources across a wider regional perspective.

Important issues to be assessed are socio-political conditions related to the exploitation of these resources, such as control of land and use rights related to production, as well as aspects concerning specialisation and professionalization. The two production areas represent two specialized quarry types, both with a limited range of products that seem to have been exported across shorter and longer distances. The productions most likely represented an important income for the people in control. Incorporating two different production areas, the analysis makes it possible to compare the two types of extraction and see to what extent they differ or concur in time, methods and organisation. A research question to consider, therefore, is: To what extent do the quarries indicate economic specialisation with activities controlled and organized by elites, or were they in the hands of local farmers on a seasonal basis? To what degree do the quarries indirectly reflect power and wealth?

I have earlier shown that the extensive production landscape in Hyllestad was located within several farms that were parts of larger medieval ecclesiastical estates. The estate structure may date to the Viking period, and should most likely be seen in connection with production (Baug 2002:90-105). Is it possible to trace similar conditions in Ølve and Hatlestrand? Other interesting questions concern the workers. Who worked in the quarries, and was the production based on seasonal activity carried out by the farmers, or people in the farmers' households, or by specialists?

The changing political, socio-economical and ideological conditions in the Viking period and Middle Ages are relevant when assessing production and resource exploitation. An important issue is related to what degree external conditions outside the local communities affected the exploitation and product types. The long-time perspective from the early Viking period into the early modern period makes it possible to illuminate stability or changes through time related to production processes and product types, as well as trade and exchange. The transition from the Viking period to the Middle Ages was characterized by far reaching changes, from more loosely organized provinces on a regional level to a more centralized kingdom, and with market-oriented trade run through towns. New institutions, churches and monasteries, were established, and trades and crafts professionalized. How did these changes affect the production of these products?

An important precondition for studying the quarries over a long-term perspective is to be able to date production, from beginning to end. When were the different areas used? Dating the extraction and products from different sites within the two quarry landscapes forms an essential part of the project, involving selected archaeological investigations within the two production areas. This has been necessary in order to assess the extent and time of the activity, and to place the production in societal contexts. The extent of the production landscapes, as well as the distribution of products, made it, however, difficult to cover all aspects and issues within the parameters of this project. A selection of quarries, product types and distribution areas has therefore been necessary. Do different quarry types represent different periods of production? To what degree were practical and functional aspects decisive for their location? And to what degree did socio-political aspects, such as property structure and control of land, have an influence on the location? Another central issue concerns how technological changes and possible innovations affected the range of products and distribution. The work also involves methodological challenges, as quarries represent a field that has rarely been investigated archaeologically. It was considered important to study the quarries and production without including large and destructive methods in the cultural landscapes. An important problem to come to grips with, relates to dating connected with this type of landscape.

Technological choices may involve more than the choice of carving techniques and tools. One of the aims of this study is to look into the technological know-how and organisation of the people who were directly or indirectly involved in the different production processes, including possible external entrepreneurs and actors who may have organized and also commissioned production of special products. Another important interconnected question concerns how the social and political situation in the Viking period and Middle Ages affected technology, and to what degree local agents or non-local elites affected the work and organisation in the quarries.

As for the distribution of products, it is necessary to identify and map their distribution both spatially, in larger geographical contexts, and chronologically. Do the products represent traded commodities or distribution through other means of transfer? What markets can be traced for the two production areas? Are there similarities or differences with regard to markets, transaction forms and possible trade networks for the two production areas and products produced?

During the last decade a renewed interest in production sites for quernstones and millstones can be seen, not least through the interdisciplinary project 'The Norwegian Millstone Landscape' -hereafter referred to as the Millstone project. It was funded by the Norwegian Research Council 2009-2012, and integrated the disciplines of geology, archaeology, history and cultural geography (www.millstone.no); my study has been connected to this project in the final stage. This has made it possible to investigate the distribution of quernstones and millstones more thoroughly, based on larger samples of materials within a wider geographical setting than would have been possible within the resources of a single project alone.

1.2 Geological conditions in Hyllestad

Stones suitable for grinding should not be too hard or too soft, and toughness and durability are important qualities. Both quernstones and millstones should have surface textures that neither became totally smooth, nor highly polished with use, which would result in gritty or discoloured flour, and would be too readily worn away or broken down. The quality of the stone was therefore dependent on their ability to maintain good grinding properties without frequent roughening of the grinding surfaces. In practice, a wide variety of rock types have been used for making both quernstones and millstones, but some were preferred to others and had higher value (Watts 2002:29). This is the case for the rock type in the Hyllestad quarries.

The geological condition for the production in Hyllestad was a specific variety of the rock type garnet mica schist. However, the mica schist in Hyllestad shows variations, and subtypes suitable for quernstones and millstones are recorded on the northern and eastern sides of the Åfjorden (Heldal and Bloxam 2008:15, 23; Grenne *et al.* 2008: 51). It was the combination of the soft schist together with the hard garnets that made the stone suitable for grinding. The hard garnets were worn down more slowly than the soft schist, meaning that the milling surface would remain a somewhat rougher surface for a longer period of use (Liebgott 1989:46; Carelli and Kresten 1997:115). At the same time the soft schist made it suitable and relatively easy to carve (Fig. 1.2).

The rock type targeted for quarrying in Hyllestad comprises coarse-grained aggregates of muscovite alternating with quartz-rich laminae, and with a varying amount of garnets and kyanite. The garnets also vary in size, and the extraction shows a preference for mica schist with garnets of 2-7 mm. There are instances of quarrying being ended when it reached garnet poor zones or zones where the garnets were too big. The presence of kyanite has generally been considered an identifying feature of Hyllestad stones. Variations in kyanite do not seem to have had an apparent value in terms of production (Grenne *et al.* 2008:51-52; Heldal and Bloxam 2008:15). The mica also varies in extent and texture, and two types of mica schist suitable for production are recognized. The most common form is a frequent occurrence of coarse aggregates of light mica, giving a characteristic silvery surface. This subtype was used when the products were carved directly from the bedrock. A second type consists of coarse-grained mica in more isolated corns. This type is also more quartz-rich, making the rock harder to carve, and it seems to have been exploited only in later periods when explosives were adopted (Grenne *et al.* 2008:52; Heldal and Bloxam 2008:15, 23).

FIGURE 1.2: THE GARNET MICA SCHIST FROM HYLLESTAD. THE DARK-RED GARNETS SEEN ARE C. 2-5MM IN SIZE (PHOTO: ØYSTEIN J. JANSEN)

Several quarries for extraction of querns and millstone are known in Norway. Characteristic of nearly all the Norwegian quarries is the exploitation of mica schist, however, with local variations. Five of the quarry areas, Hyllestad, Vågå, Selbu, Brønnøy, and Saltdal are larger than the others and thus considered particularly important production sites (Grenne *et al.* 2008:48; Fig. 1.3).

The geological survey of Norwegian quarries carried out within the Millstone project has made it possible to distinguish the rock type in Hyllestad in relation to the other quarries, and thus recognize the products from Hyllestad recovered within different archaeological contexts in Northern Europe. The rock type in the Hyllestad quarries is relatively easily identified, but in some cases more thorough geological analyses are necessary. Smaller geological differences, such as the size of the garnets, may also occur within one single quarry, making it difficult to trace the products back to the exact quarry where they were produced.

1.3 Geological conditions in Ølve/Hatlestrand

Rock types required for bakestones had to contain elements of soapstone, also referred to as steatite, making it possible to repeatedly heat and cool the objects without an imminent risk of fracturing. The geological precondition for the production of bakestones in the Ølve/Hatlestrand

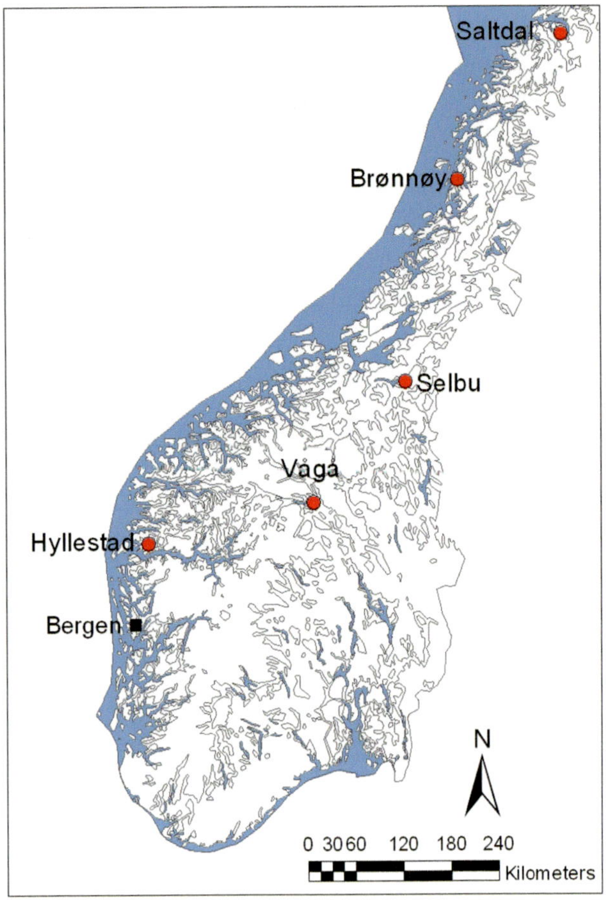

FIGURE 1.3: LOCATION OF MAJOR QUERNSTONE AND MILLSTONE QUARRIES IN NORWAY (MAP: IRENE BAUG, 2013)

FIGURE 1.4: FRAGMENTS OF TWO DIFFERENT BAKESTONES FROM BRYGGEN IN BERGEN (UPPER: BRM0/50627, DATED TO AD 1248-1332 BY MYGLAND (2003). LOWER: BRM0/50119), MOST LIKELY PRODUCED IN ØLVE/HATLESTRAND (10-CM SCALE) (PHOTOS: MARCIN GLADKI © MUSEUMSSENTERET I HORDALAND)

area has been a chlorite-rich talc-bearing green schist with some variation in the amounts of the different minerals (Fig. 1.4). The rock type has been referred to as schistose soapstone (Naterstad 1984), but the content of 'soapstone', that is the mineral talc, is not thought sufficiently high to justify the term soapstone (Jansen 2013:77). The term Ølve/Hatlestrand is mostly used when referring to the bakestone quarries in this area, as, in most cases, there is no need to distinguishing between the two communities.

The rock used for extracting bakestones is located in a certain zone of a larger greenstone complex, situated on the southern and eastern side of Lake Kvitebergsvatnet, as well as on the western side of the Kvinnheradsfjorden, a section of the Hardangerfjorden. Due to the content of talc, the rock is somewhat similar to soapstone but has a larger degree of schistosity, making it more suitable for the production of bakestones than soapstone vessels. The rock in the quarries is at the same time nearly lacking in the minerals quartz and feldspar that are common in the more normal greenstone complex in the area, and this has formed the basis for the quality of the stone in terms of softness and durability, as well as the stone's ability to stand repeated heating and cooling without fracturing (Naterstad 1984:161-162). The chlorite-rich talc-bearing green schist forms a certain stratigraphical level in the greenstone complex, and appears in a c. 5-6m thick layer, with a different rock type below and above. During production the layer of green schist had to be followed into the rock, with overhangs and underground quarries as a result.

The green schist in the Ølve/Hatlestrand area is rather more difficult to identify compared to the products from Hyllestad. A similar rock is also recovered in Øye in Melhus in the Sør-Trøndelag region, referred to as chlorite shale, and bakestones of chlorite-rich talc-bearing green schist are therefore likely to stem from one of these two quarry complexes. Bakestones were extracted in Øye, but seem to have been of minor importance, with building stones the main product. The quarry in Øye was most likely in use from the latter half of the 11th century until about AD 1200 (Lundberg 2007:86). A medieval quarry site for building stones also produced bakestones at the Ertenstein farm in Rennesøy, Rogaland (Fig. 1.5). The quarries at Ertenstein contain more mica and carbonate, and are thus easier to distinguish from Ølve/Hatlestrand. Additionally, traces from bakestone productions have been recorded in a few soapstone quarries, but where vessel production dominates (Jansen pers. comm. 2012). This production seems, however, to be of minor importance

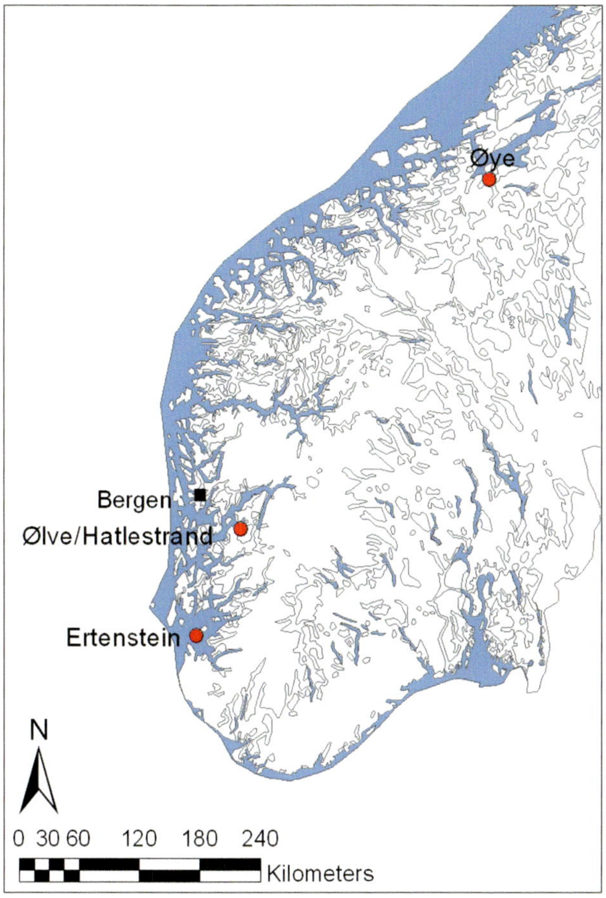

FIGURE 1.5: LOCATION OF MAJOR BAKESTONE QUARRIES (MAP: IRENE BAUG, 2013)

regarding bakestones compared to the quarries in Øye and Ølve/Hatletrand.

As the same resources and geological competence as Hyllestad have not been available for the bakestone quarries, investigation of the provenance of bakestones could not be carried out with the same thoroughness as for the quernstones and millstones within the scope of this project. As for distribution, the main focus therefore falls on these two product types.

1.4 The investigation areas

The quarries in Hyllestad and Ølve/Hatlestrand both represent long-abandoned production landscapes with large-scale extractions that developed over hundreds of years, leaving quarries, spoil heaps, remnants of roads, harbours, etc. as visible traces. There are similarities in the production techniques, and the quarrying in both areas has to a large degree affected the landscapes. Both sites also produced objects for larger markets. Bakestones as well as quernstones and millstones were important tools connected to the use of grain and grain products in daily diets. Querns and bakestones were used at a household level, while millstones represent a more specialized and professional activity, with milling on a larger scale. When using the terms quern and mill, they denote hand-turned and water-driven types respectively (Grenne *et al.* 2008:48). The Norwegian equivalent to quern (*kvern*) may also refer to minor water-driven constructions. I have used quernstone and quern to denote hand-driven querns and millstones, and mills for water-driven types, both smaller and larger mills. The many common features of the two production landscapes make it interesting to compare the social and socio-political backdrop connected to the production and distribution of these products. Both quarry landscapes are investigated in order to shed light on production and product types and to gain insights into temporal aspects of the activities, as well as the socio-economic and political contexts of which the quarries were part.

1.4.1 The Hyllestad quarries

The quarries are located by the Åfjorden, north of the outlet of the long Sognefjorden, roughly 90km north of Bergen. Innermost in Åfjorden, which is about 15km long, a small peninsula divides the fjord into two fjord arms, Sørefjorden to the north and Hyllestadfjorden to the south. The quarries are located on both the northern and eastern sides of Åfjorden, stretching from the shoreline into the mountains, about 200m a. sl. Altogether 367 quarries have been identified, spread over an area of about 20km^2 (Heldal and Bloxam 2008:47; Fig. 1.6), located at 15 different medieval farms. Most of the quarries are situated on the slopes above Åfjorden, less than 1km from the fjord. Not only did the rock type favour large-scale production in Hyllestad, but also the location, with its accessibility and proximity to harbours and transportation routes.

Visible traces of quarrying are found as carved bedrock and spoil heaps containing flakes, slabs, unfinished and destroyed products, located in front of the rocks and to a large degree covering them. The majority of the quarries is located in the 'outfield' (ON: *innan garðs*) or wasteland of historic farms, but some farms also have quarries in the 'infield' (ON: *utan garðs*) area close to the settlements. The vast production landscape and the large number of quarries have an industrial character. In some areas numerous quarries occur side-by-side and even on top of each other, and the density makes it difficult to recognize the original terrain. Large rocks have been quarried away, leaving severe changes to the landscape.

The extraction was carried out in open quarries, and two different extraction techniques have been recognized; with and without gunpowder. The oldest technique involved cutting the shape of the quernstones and millstones directly on the bedrock, then breaking the stone loose along its base. This is the quarry type investigated in this study. A detailed study of production techniques and tool marks was, however, outside the scope of this project and has been covered through the Millstone project, where technological aspects of carving techniques in Hyllestad are investigated (Løland in press). Parallels to the techniques used in Hyllestad can also be found in Ølve/Hatlestrand.

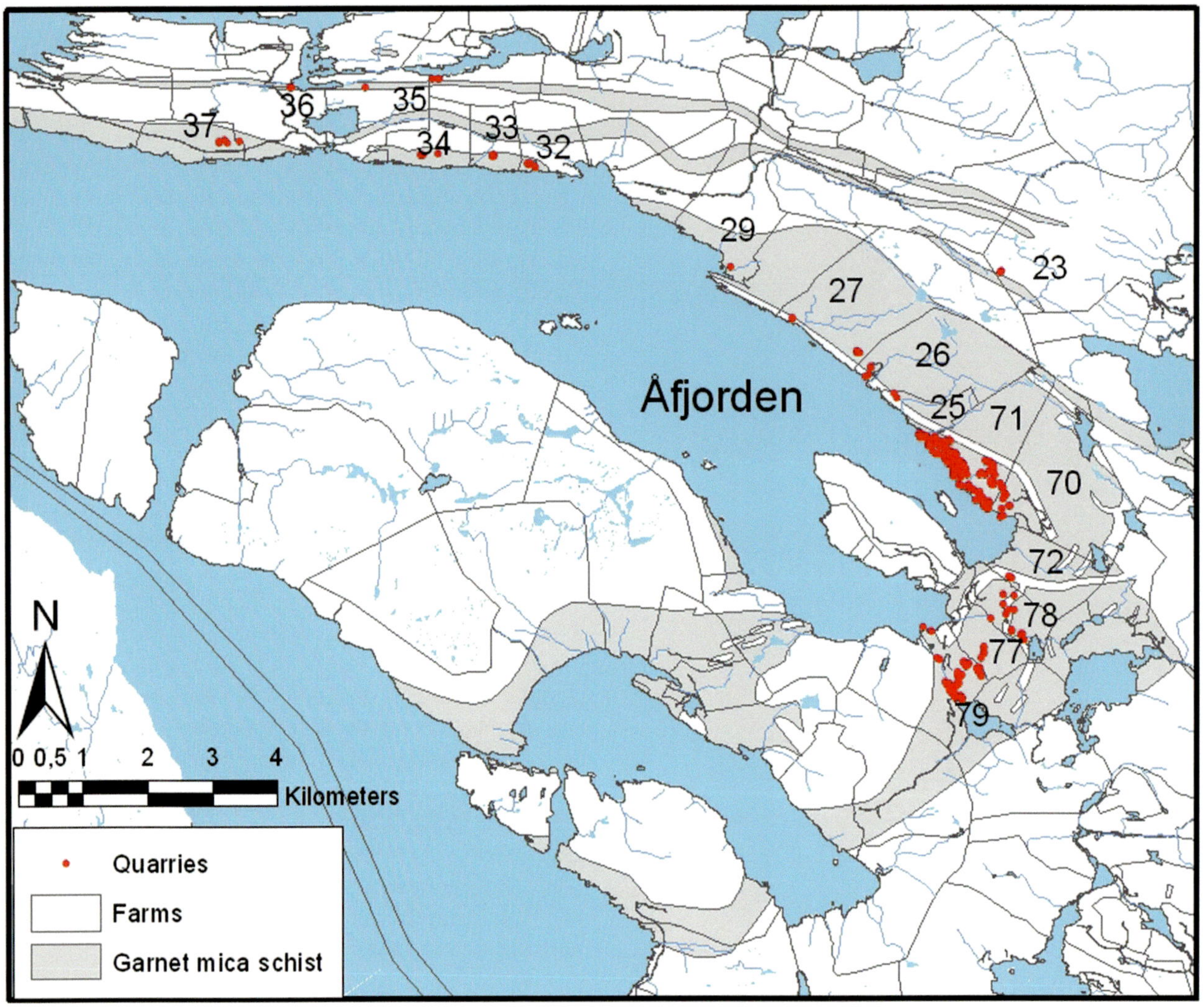

FIGURE 1.6: IDENTIFIED QUARRIES IN HYLLESTAD. FARM NUMBERS ARE SPECIFIED FOR THE FARMS WITH QUARRIES WITHIN THEIR BOUNDARIES (MAP: IRENE BAUG, 2013, NORGE DIGITALT. FKB-DATA, 1413 HYLLESTAD KOMMUNE, 32_1413EIENDOM_FLATE, 28.02.2009)

Sites investigated

In this project, altogether five different quarries within four different farms have been investigated in Hyllestad, two quarries at Rønset (no. 71) and one quarry at each of the Myklebust (no. 79), Sæsol (no. 78) and Sørbø (no. 32) farms. Sørbø is the only investigated farm located in the parish of Øn, whereas the first three farms mentioned are located in the parish of Hyllestad. Compared to my earlier study (Baug 2002), the research area has been expanded and new types of quarries and products, as well as work places, have been investigated.

Sørbø (no. 32)

Sørbø farm is located on the northern side of the Åfjorden and borders the farms of Ulvika (no. 31), to the east, and Indre Gil (no. 33), to the west. Sørbø was a part of Ulvika until the end of the 16th century. The farm was divided into two holdings in 1851 and today consists of several smaller units. The farm is hilly, but with wide boggy areas between the hills. Today large parts of the farm have been built over and comprise a rural centre, with a shop, nursing-home, and church (Førsund and Kellmer 1990:293-307), and the modern settlement development may have destroyed some of the quarries on the farm.

Rønset (no. 71)

The historical farm of Rønset is located on the northern side of the Sørefjorden, the small arm of an inlet furthest into the Åfjorden, bordering the Berge farm (no. 70), to the east, and the Leirpollen farm (no. 25) to the west. The latter is also the old border between the parishes of Hyllestad and Øn. The farm was divided into two holdings in 1610, and consists today of five units (Førsund and Kellmer 1993:219-229). Rønset has a shoreline of about 1.6km in length, along which four possible harbours have been identified, used for the distribution of quernstones and millstones. The settlement area and most of the infield of the two main holdings on the farm (holdings 1 and 3) are located on a flat stretch of land, just north of the road, about 50m.a.s.l. The buildings are still located as an agglomerated settlement. Most of the outer areas of the farm are located west of the settlement, in a hilly region that slopes southwards towards the fjord.

I have earlier investigated two quarry sites at Rønset, and the production has been dated to the Viking period (AD 730-965), perhaps with a start in the early Viking Age, continuing into the 16th or 17th centuries (Baug 2002:34-51). An archaeological investigation of the fossilized remains at Rønset indicates that the farm may have been established during Roman times, probably also with a settlement near the historically known settlement area (Foyn 2008). The farm thus seems to be older than the quarrying, and an interesting aspect to look into is how the farming activity at Rønset relates to the large-scale production of quernstones and millstones.

Sæsol (no. 78)

Sæsol farm is located along the steep hill east of the Sørefjorden. The farm has a shoreline approximately 200m in length, south of Stigedalen (small-holding 2 at Sørefjorden, no. 72), with good harbour conditions, but the settlement and the central infield of the farm is on a plateau about 180m a. sl. Sæsol borders the farms of Akse (no. 68) to the east, Sørefjorden (no. 72) to the north, and Hyllestad (no. 77) to the south. The farm was divided into two holdings in 1850, and consists today of five units (Førsund and Kellmer 1993:282-293). I investigated one quarry at the farm previously, indicating that the production may date back to the Viking Age, perhaps to the early Viking Age (AD 715-890) (Baug 2002:55-58). Farming activities in the form of arable and animal husbandry goes back to the 8th century AD at the earliest, and the farm may have been established contemporaneously with the quarrying activities (Sætre 2008).

Myklebust (no. 79)

Myklebust is situated on the eastern side of Hyllestadfjorden, the second arm of the Åfjorden, bordering the farm at Hyllestad (no. 77) to the north, where the border stretches up from the bay of Hyllestadvika, south of Kyrkjefjellet. To the east, the area borders Kleive farm (no. 80), extending to Lake Kleivevatnet. The first two farm holdings were separated in the 17th century, and the farm is now divided into several smaller units. Today the old farm has the largest concentration of settlements in Hyllestad, and the municipal buildings were established at Myklebust from the 1970s (Førsund and Kellmer 1993b:293-309). Consequently, some of the farm quarries may have been lost due to the recent development of the small centre.

In my earlier investigations at Myklebust, stone production was dated back to Viking times (AD 890-1015), and the extraction of millstones may also have begun this far back; output from the quarry continued until the late Middle Ages (AD 1315-1435) (Baug 2002:51-54). This project is revealing different quarry and product types from those studied previously.

1.4.2 The quarries in Ølve/Hatlestrand

The quarries are located at Ølve and Hatlestrand in the municipality of Kvinnherad, on the western side of the Hardangerfjorden, c. 100km southeast of Bergen. Altogether 71 quarries have been identified so far, located at nine historical farms: four in the parish of Ølve and five in Hatlestrand. Most of the quarries however are in Ølve, south and east of Lake Kvitebergsvatnet. The production landscape reflects large-scale production.

Unlike Hyllestad, where only open quarries are found, production at Ølve/Hatlestrand was carried out in open quarries, overhangs, and underground sites. The spoil heaps are located in front of, and partly covering, the rocks in the open quarries, as well as underneath and inside the overhangs and underground workings. Today many of the underground quarries have collapsed, leaving huge stone blocks in front of (or inside) the quarries, completely or partly blocking the access. However it is still possible to enter several of the quarries, and the sizes vary from only a few metres to c. 30m deep. The proportions of the quarries, together with the spoil heaps, bear witness to large-scale activity, and, like at Hyllestad, the production landscape and many of the quarries at Ølve/Hatlestrand have an industrial character.

As at Hyllestad, the output production of Ølve/Hatlestrand was diverse. Not only bakestones were produced. According to priest and writer Peder Claussøn Friis (early 17th century), thin oblong slabs 2 *alen* (= 1.20m) in length and 1.5 *alen* (= 94cm) broad were removed for drying corn and oats (Friis 1632:72). Tiles, building stones, and church crosses were produced at some quarries. Many of the sites produced different items at different periods, while others may have produced different objects at the same time. In this project, the production of bakestones is the main focus, but other objects are also considered so as to give an idea of activity as a whole.

Written evidence from the 16th century onwards gives information on the production of bakestones and roof tiles (NRJ IV:474; Friis 1632:71-74; Hoff and Lidén 2000:156). In the late 19th century the quarries are also thought to have supplied building material for Danish castles, as well as cornerstones and blocks for window- and door-frames for parish churches, and for the manor house at Rosendal (Haukenæs 1888:68, 135). Grave crosses were also extracted from some of the quarries. In the 20th century it was believed that soapstone vessels were extracted from the quarries as well (Vaage 1972:125–128), but this kind of production has not been verified.

These quarries have not been the object of detailed archaeological investigations before this current research. In 1931, four quarry sites at Ølve were surveyed by Kristen Lindøe: Bakkhidlaren, Tuftahidlaren, Lambagarshidlaren and Kvitebergana. Test pits inside the underground quarries of Bakkhidlaren and Lambagarshidlaren revealed production waste. The fourth site mentioned,

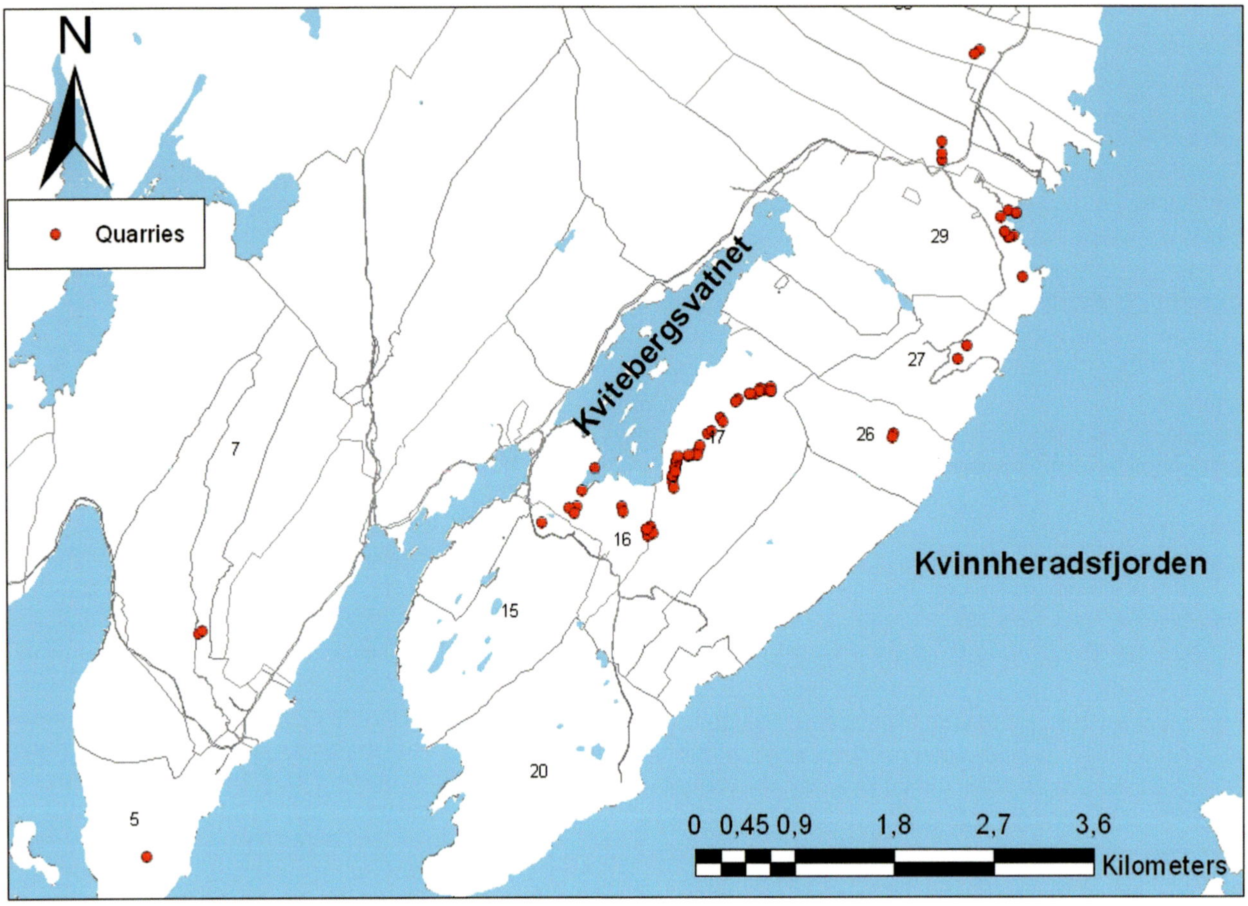

Figure 1.7: Identified quarries in Ølve and Hatlestrand. Farm numbers are specified for farms with quarries within their boundaries (Map: Irene Baug, 2013, Norge Digitalt FKB-data, 1224 Kvinnherad commune, 32_1224eiendom_flate 21.11.2007)

Kvitebergana, is not known today, but it was allegedly located not far from Kvitafjell. No quarries or spoil were identified here (Top Ark, University Museum of Bergen). In the 1980s, archaeologist Birthe Weber conducted small-scale investigations at a work site in Kvitafjell and in the Bakkhidlaren quarry, finding a pickaxe and an iron fitting (Top Ark, University Museum of Bergen), but unfortunately the documentation is unavailable. Most knowledge of the quarries has thus come only indirectly through geological surveys and finds of bakestones in urban contexts from around AD 1050-1100 onwards (Naterstad 1984; Weber 1984:159). Therefore my investigations of the quarries in this current project represent a new empirical basis for the study of quarrying.

The sites

The archaeological investigations focused on two different farms, Fugleberg (no. 17) and Netteland (no. 29), where four quarries and one site for the finishing of bakestones were investigated.

Fugleberg (no. 17)

Fugleberg farm is in the north-eastern part of the Ølve community, east and southeast of the 3km-long Lake Kvitebergsvatnet. Fugleberg lies next to the farms of Tufta (no. 16) to the west, and Skarvatun (no. 18) to the east. Fugleberg was divided into five holdings in the 18th century and today consists of six units. From the 18th century, tiles for private use were extracted from the farm's quarries, however there is no record of production of bakestones here in this period (Vaage 1987:172-179).

Netteland (no. 29)

Netteland (Hatlestrand), on the western side of Kvinnheradsfjorden, part of Hardangerfjorden (Kvinnherad), borders on three farms: Ask (no. 27) to the south, and Atramadal (no. 23) and Hjortland (no. 30) to the north-west. The farm was divided into four holdings in the 18th century and now consists of seven units. Netteland has a long shoreline with good harbours on the bays of Netlandsvågen and Hellebergsvika in the Kvinnheradsfjorden. In the 1600s Netlandsvågen was used as a harbour for timber exports to Scotland (Vaage 1987:230-242). Many of the quarries at the farm, such as the investigated quarry Båtahidlaren, are located within a fairly short distance to Netlandsvågen, and most likely large amounts of stone products were exported there.

Overall, this study, therefore, takes in three levels of research: at the micro level, with a focus on the quarries and productions of Hyllestad and Ølve/Hatlestrand; at the 'meso' level, where social and socio-political relations within the local and regional communities are studied in relation to organisational aspects; and at the macro level, investigating distribution and trade from the supra-regional perspective.

2 State of Research

Despite their vital role in producing daily bread, querns, millstones and bakestones represent an understudied field of research. Mining and quarrying were only a few years ago described as a 'new' subject in Norwegian archaeology (Stylegar and Landmark 1999). In this chapter, I present the state of research relating to such quarries and their products relevant to this present study, including also relevant research outside of Norway. As knowledge of how the products were used is important in terms of understanding the quarries and their wider context, I also give a brief overview of the use and development of the tools employed, based on studies of the technological development of different types of mills (see, e.g., Norman 1936; Ek 1962; Lund 2001; Fisher 2004).

2.1 Quernstones and millstones within Scandinavian research

Quarry sites for the production of quernstones and millstones form some of the largest 'cultural' landscapes in Norway. Several larger and smaller quarry landscapes have been identified, but their productions have only rarely been looked into archaeologically. The most comprehensive investigation so far has been my own study of selected quarries from Hyllestad, in connection with the present author's Master's thesis in 2001 (Baug 2002).

The first explorations at Hyllestad were by the historian Ottar Rønneseth. In an article first published in Germany in 1968, and translated into Norwegian in 1977, he looked at quarries and production marks at the farms at Rønset, where most of the quarries were identified (Rønneseth 1968; 1977). Based on the formation of the workings, with their rocks and spoil heaps, Rønneseth discussed the possible production techniques and outputs of the quarries, and also posing questions about their dating. However, the limited resources available prevented firm conclusions and answers. This scholar was the first to acknowledge that quernstones and millstones from Hyllestad had a wide distribution and played an important role outside the local community.

An important precondition for my research has been the geological mapping carried out in Hyllestad from 2005-2007, whereby the geologist Tom Heldal, from the Geological Survey of Norway (NGU), with the archaeologist Elizabeth Bloxam mapped the entire quarry landscape of Hyllestad (Heldal and Bloxam 2008). Several different quarry types and extraction techniques have been identified, including quarries and the production of quern stones, millstones and slabs. Detailed maps and a database of the quarries, their geology and remains of roads and infrastructure within the production landscape are now available. This survey has provided an important starting point for the archaeological investigations as part of my own study, providing an overview of the quarry landscape, including the different types of quarries, their extraction techniques and the artefacts they produced. All digital mapping data for the Hyllestad quarries used in this project also derive from the registrations carried out by Heldal and Bloxam.

As well as Hyllestad, the millstone quarries of Selbu (Sør-Trøndelag) are so far the most widely known and discussed production sites in Norway. Based on oral tradition from the 18th and 19th centuries, the Norwegian engineer P.O. Rollseth has discussed questions concerning the technologies, means of production, and working processes related to the organisation of production, as well its distribution (Rollseth 1947). As the activities described here are several hundred years younger than the oldest production from Hyllestad, his documentation gives insight to the later stages of millstone production, and although not directly relevant to my project, it can offer vital information about technology and the tools used in the quarries. The millstones from Selbu were formed by carving and hewing, using picks, hammers and chisels; even though the work was done from slabs, and not directly from the bedrock, similar tools may have been used at Hyllestad.

Far more millstones than quernstones were produced at Selbu. While direct carving from bedrock was done at most of the Hyllestad quarries, the Selbu stones were extracted from slabs that had first been wedged, chiselled or blasted from the bedrock. The oldest documented quarrying at Selbu goes back at least to the 17th century (Friis 1632:97), but so far archaeological investigations of structural remains connected to the quarries have not proven activity before the 18th century (Alsvik *et al.* 1981). At Selbu, NGU conducted a survey similar to that done at Hyllestad. In a *Geological Survey of Norway Special Publication*, the two production landscapes with their different geologies, quarry types, extraction techniques and histories, were compared, revealing both similarities and differences, and in this way provided new perspectives on changes in extraction techniques over time (Grenne *et al.* 2008). Hyllestad and Selbu represent the two largest areas for querns and millstone production in Norway, but differ both in duration and production practice.

So far, little research has been carried out on the other Norwegian quarry landscapes. Two small-scale archaeological rescue excavations were carried out in a quarry in Saltdal (Nordland) in 2006 and 2009, and dating production to c. AD 900-1190 and AD 1040-1270 (Helberg

2010; 2010/2011). The Tolstad and Vågå quarries in eastern Norway are the earliest recorded in documents (the 1400s onwards), providing information about the disposal and ownership of land and quarries (Teigum 2011). Both areas revealed important comparative material for the Hyllestad quarries. They operated, at least for a while, simultaneously with Hyllestad, and their products may have competed in the same markets as Hyllestad. Lately, local interest in the production sites at Saltdal, as well as Vågå, has flourished (see, e.g., Brekken 1980; Trones 2003; Titland 2003), but has so far not resulted in more scholarly investigations of the quarries.

In the 1990s, questions concerning the distribution of stone products gained attention, but within limited areas and only to a small extent in Scandinavia (Elfwendahl and Kresten 1993; Schön 1995; Hansen 1992; 1997; Carelli and Kresten 1997; Hvoslef 2009). The first study of the distribution of Hyllestad products in wider geographical contexts was conducted by the Swedish archaeologist Peter Carelli and the geologist Peter Kresten, in an article in *Acta Archaeologica* in 1997, and represents the first interdisciplinary study of distribution of such objects in Scandinavia. They showed that quernstones and millstones from Hyllestad were distributed and traded over large areas, and mainly within the borders of medieval eastern Denmark. By investigating the archaeological finds geologically, as well as chronologically, they were able to identify provenance and throw new light on trade and trade networks in early towns from the Viking period and the early Middle Ages. The article clearly demonstrates the importance of provenance studies related to distribution. At the time, however, Carelli and Kresten lacked information on the geology of most of the other Norwegian quarries. Several quarries besides Hyllestad, such as Saltdal, Vågå and Brønnøy, could therefore be the sites of origin. New ways of looking at the export of Hyllestad stones, involving new geological information, both in Norway and abroad, has been important for my study. The German historian, Meinrad Pohl (2011; 2012 [2011]), used transportation costs to explain the different quernstone regions indicated by Carelli and Kresten (1997), providing new insights into distribution. This aspect will also be further considered here for the distribution of Hyllestad stones.

Marine archaeology can also provide helpful evidence for the study of quernstone and millstone distribution, as well as other aspects related to the history of quarries. Three different shipwrecks containing stones from Hyllestad were investigated by Arild Marøy Hansen (Bergen Maritime Museum), as well as certain harbours in Hyllestad used for transportation (Hansen 1992; 1997). Over recent decades, more intensive surveying has increased the number of wrecks found with sunken cargoes of quernstones and millstones; by 2011 as many as 18 such cargoes have been discovered along the Norwegian coast (Nymoen 2011). According to marine archaeologist Pål Nymoen (Norwegian Maritime Museum), the stone cargoes are only rarely large enough to represent the main commodity on-board. The weights of the cargoes indicate that the ships most likely would also have been loaded with other goods, and Nymoen discusses the stones' possible function as ballast. The cargoes have only exceptionally been dated (Nymoen 2012:177), nevertheless the finds are important for any discussion of the distribution of products from Hyllestad.

A renewed interest in quarries, querns, and millstones has been seen over recent years; the Hyllestad sites, for example, have witnessed several local initiatives (cf. Thue 2000; Førsund 2005; 2006a; 2006b; 2007; Waage 2005; Nordstrand 2007), as well as an annual seminar held in Hyllestad since 1999 (Baug 2000b; 2005a; 2005b; 2006a; 2006b; 2007).

The Millstone project has provided a new scientific basis for interdisciplinary research on a national level. Important objectives have included attempts to identify geological, archaeological, technological and historical elements of the quarry landscapes of Norway, and to investigate their duration and development (e.g. Baug and Løland 2011; Heldal and Meyer 2011; Prøsch-Danielsen and Soltvedt 2011; Baug and Jansen 2014; Prøsch-Danielsen and Heldal 2014; Grenne *et al.* 2014a; Grenne *et al.* 2014b). In this present work I draw upon published results from the project, as well as a few preliminary analyses, although many avenues of research are still in progress.

It has to be remembered that only a few quarries and production sites have been subjected to thorough archaeological investigations; consequently the overall number of excavated sites is so limited that even one small-scale excavation could change long-held suppositions.

Several quernstone quarries are also known from Sweden (Welin 1942; Pettersson 1977; Beiron 2006; Hockensmith 2009:190-193), but have not been involved in the archaeological research. Production of quernstones and millstones from Malung in Sweden, from the 18th century, is well known through the ethnological study of Täpp John-Erik Pettersson (Pettersson 1977). Even though the study mainly concerns more recent production, it does provide an overview of the majority of the Scandinavian querns and millstone quarries, irrespective of size and age. Based on more recent written sources, Pettersson sheds light on production, distribution and ownership, thus touching upon many aspects included in this present study.

2.2 Quernstone and millstone research in areas of Western and Central Europe

Compared to Scandinavian archaeology, quernstones and millstone production has been more fully researched both in France and Germany, and often based on broad interdisciplinary collaboration. Over recent decades, a larger scholarly interest in quarries has also resulted in several research projects and publications. In this work reference is made to some artefacts that may have been circulating in some of the same markets as the Hyllestad products, and these aspects will be further investigated here.

Germany is in the forefront when it comes to archaeological investigations of quarrying and mining. Thematically the studies have a wide range, from metallurgy to different types of quarries. Already by the end of the 19th century and the beginning of the 20th, quarries from the Roman period, including querns and millstone quarries, attracted scholarly interest, especially related to technological aspects and production processes. Nevertheless, few systematic investigations were carried out, and not until the post-war period did the quarries achieve greater archaeological attention. Even though several studies on querns and millstone quarries have been published, relatively few archaeologists have engaged themselves in the subject (Mangartz 1998:1-4). Attention can be drawn to two different German quarry landscapes producing stones for markets in Northern Europe at the same time as Hyllestad. One is the volcanic mountain area of Hohe Buche, Andernach, and the other is the quarry region of Mayen: both are in the Rhine area in the district of Mayen-Koblenz. The new German research interest tends to approach quarries and production contextually, within larger societal and cultural spheres, and thus correspond to my own approach.

In 1992, a large research project began on the quarries of Hohe Buche (Mangartz 1998), and resulted in a detailed mapping of quarries and production remains and techniques. The focus however was on Roman productions, and activities in the Middle Ages gained less attention. Quernstone and millstone production at Hohe Buche was on a small scale compared to the Mayen quarries, which have attracted most interest. Large-scale production and distribution started in the pre-Roman Iron Age and continued until recent times (Mangartz 2008). The quarries in the Rhine area can be characterized as some of the major preindustrial production sites in Central Europe. Recent investigations have given a broad and nuanced picture of the activity, especially connected to the research project 'Vulkanpark-Forschungen' (Mangartz 2000; 2008; Oesterwind 2000; Pohl 2012). Production and distribution in the Rhine area provide comparative material to the Hyllestad quarries – both with regard to ownership and organisation of production, as well as distribution of stones, issues that will be further followed up in this study.

Quernstones and millstones of basalt lava from the Rhine area were exported over large parts of Northern Europe, and this distribution has been studied by different researchers (Parkhouse 1976; 1997; Schön 1995; Pohl 2011; 2012; Coulter 2011). In Hedeby and Schleswig in Germany, the German archaeologist Volkmar Schön (1995) identified querns of garnet mica schist, presumably from Hyllestad, along with basalt lava stones. In the Middle Ages, the quarries of the Rhine area seem to have been competing production sites with the Hyllestad sites (Carelli and Kresten 1997:22).

Over the last 30 years, several excavations of quarries have been carried out in France, including investigations of distribution and trading routes. Pre-Roman and Roman quarries were central to the research (Belmont and Hockensmith 2006:2; Hockensmith 2009) but recently medieval production has become an important field of research. The French archaeologist Alain Belmont, one of the leading researchers of millstone quarries in Europe, has conducted several archaeological investigations of millstone production sites within different areas in France, dating from the Middle Ages to the end of the 18th century. His studies are based on interdisciplinary cooperation, drawing upon archaeological, historical, geological, and anthropological information (Belmont 2006a; 2006b). Belmont's studies clearly demonstrate the importance of interdisciplinary research, and face many of the same problems found at Hyllestad. Belmont's material, however, was distributed over different areas from the Hyllestad stones, and are thus not of direct relevance the present work.

Several publications on Swiss quarries have also appeared over the last few years (Hockensmith 2009:194-196, with references), with some of the quarries seeming to have similar production techniques as in Hyllestad (Andersen et al. 1999; Anderson 2006). They represent useful comparative material for studies of carving techniques, but as this is not one of the objectives of this study it is not followed up here.

Several quarries for quernstones and millstones are also identified within the British Isles, especially Roman quarries (Hockensmith 2009:151-165, with references). However no thorough investigations of production sites have been carried out. Many quarries are small, and seem mainly to have produced stones for local markets (Tucker 1982; 1984; Wright 1982; Jobey 1986). Additionally, a substantial part of millstone production, both in Scotland and England, especially in the 18th and 19th centuries, was based on imports of smaller pieces of stones that were dressed and carefully fitted together to form complete millstones by different Scottish and English factories (Tucker 1982; 1984). This activity is of less relevance for Norwegian quarrying.

In parallel with an increased Norwegian interest in the topic, a new interest and fresh research into quernstone and millstone quarries can be seen in several European countries. Since 1995 this trend has resulted in four international conferences focusing on querns and millstone quarries: two in France (Barboff *et al.* 2003; Belmont and Mangartz 2006); one in Rome (Williams and Peacock 2011); and one in Bergen and partly in Hyllestad (2011) (Selsing 2014). These seminars have contributed towards internationalising Norwegian research. Methodological challenges and experiences within different subjects have been highlighted and discussed, which have contributed to a wider knowledge of quarries and their products in different geographical areas and at different times.

2.3 Bakestones

While quernstone and millstone quarries are located in many areas of Europe, production of bakestones has so far only been identified in a few locations in Norway and Shetland, and the study of their production has been marginal. In Norway small-scale production of bakestones occurred in different soapstone quarries in the Middle Ages, but the quarries of Ølve/Hatlestrand (Kvinnherad), Øye (Sør-Trøndelag) (Lundberg 2007), and perhaps also the quarries of Rennesøy (Rogaland), are the only production sites so far known where extraction of bakestones seems to have had any importance.

The first publication of a production site for bakestone appeared in 1984, following a survey carried out in Ølve (1983) by archaeologist Birthe Weber and geologist Johan Naterstad (Naterstad 1984; Weber 1984). Their main objective was to locate and map the quarries and to identify production marks and technologies, as well as to investigate the geology over the larger production area. This pioneer work was important for bringing this type of production and resource exploitation into the archaeological discourse.

Weber has also focused on the distribution of bakestones in medieval urban contexts, such as Oslo and Trondheim, and as far west as Shetland and the Faroe Islands. Her focus has been on chronology and mineral identification of the bakestones, connecting them to quarries near Lake Kvitebergsvatnet in Ølve (Weber 1984; 1989; 1990; 1999; Naterstad 1989). Until the 1980s, production of bakestones at Øye, in the community of Melhus (Sør-Trøndelag), was unknown, which may cause a problem for the provenance analyses of Weber and Naterstad. They do, however, open up the possibility that extraction of bakestones may have occurred as a by-product of some of the building stone quarries of Sør-Trøndelag (Naterstad 1984:23; Weber 1989:9, 19). Their analyses, mainly related to rock type, carving traces and size of quarries, have a narrower scope than this present study.

Today, a greater extraction of bakestones is documented from the quarries of Øye, Sør-Trøndelag, a site with a geology rather similar to Ølve and Hatlestrand. The Øye quarries were investigated in 2007 for a Master's project (Lundberg 2007), and small-scale excavation at one quarry revealed that building stones were the main product. The production of both building stones and bakestones is dated to the Middle Ages (AD 1040-1255 and AD 1170-1285). As there are several similarities between the two quarry sites, Øye stands out as providing interesting comparative material. Similar carving traces are to be found in quarries in both Øye and Ølve/Hatlestrand, and the same types of tools and carving techniques may be expected at both quarry sites.

A study of bakestones from the Middle Ages into the early modern period from parts of the Bryggen site (Bergen) has been conducted by Sigrun Solbakken Tengesdal, as a Master's project (Tengesdal 2010). Here, the use of bakestones was documented within a period from c. AD 1120/30 to c. 1700. Many of these bakestones most likely came from the quarries of Ølve and Hatlestrand, but not totally (Tom Heldal and Øystein Jansen pers. comm. 2010). The study provides information for the use of bakestones within urban contexts.

Norwegian bakestones have been found at several locations in Shetland, but also at a few sites on the Faeroe Islands and Iceland (Hamilton 1956:183; Weber 1999; Forster 2004:168 with reference to Buttler 1984; Forster 2009:65). So far, Shetland is the only place outside Norway where production of bakestones has been documented. A few quarries have been investigated and bakestones were extracted together with soapstone vessels (Buttler and Forster 2009; Forster 2009:65; Turner 2009:2; Turner *et al.* 2009). Generally, bakestones have gained little scholarly attention, being mainly mentioned in publications from excavations (e.g. Hamilton 1956; Weber 1999). Distribution and possible transaction methods of soapstone artefacts in Shetland have been taken up in a PhD study by Amanda Kate Forster (2004). Forster sheds light upon the early trade of Norwegian soapstone goods in Viking times and the Middle Ages, where also bakestones and their distribution in the North Atlantic islands are examined.

This brief survey demonstrates that research on bakestones has been modest, which should be seen in relation to distribution and use within a rather limited geographical area. Nevertheless, large quantities of bakestones have been identified in medieval urban contexts in Norway, and to some degree in the North Atlantic islands. Investigation of the trade and transaction forms, and the production of the objects, have, however, rarely been taken into account.

2.4 Different forms of querns and mills

Stones from the Hyllestad quarries were used in different types of querns and mills through the centuries, representing remains from earlier technologies of food production. The oldest type of quern is the so-called saddle quern, comprising a hollow shaped/concave lower stone, on which the operator, with both hands, pushed and pulled a smaller and rounded or oval upper stone. The ground flour fell from the sides and lower section (Cederlund 1964:536; Watts 2002:25). This type of quern is not found in the Hyllestad quarries. In the early Iron Age a new type of quern appeared: the rotary quern. The oldest type of rotary quern, also called the beehive quern, was hand operated and consisted of two circular stones (quernstones) of small diameter (Fig. 2.1). The earliest known examples of rotary querns come from Catalonia in Spain, dated to the 6th and 5th centuries BC (Watts 2002:27-28). Nearly at the same time, during the pre-Roman Iron Age (c. 500-0 BC), the first rotary querns appear in Scandinavia and the British Isles (Cederlund 1964:536; Watts 2002:28). The origin of the rotary quern is still unknown. These earliest quernstones vary in shape and size, and in Norway they

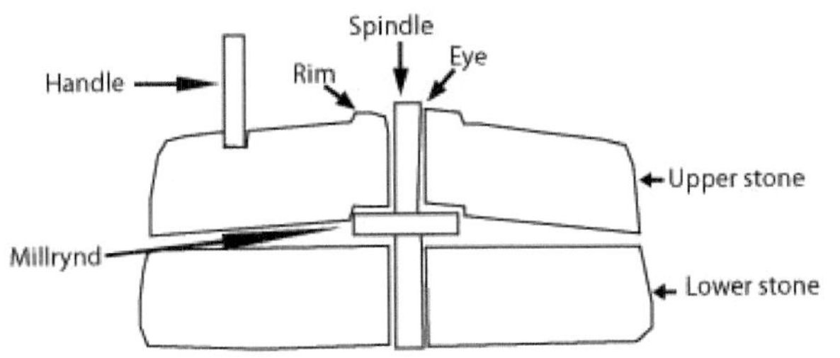

FIGURE 2.1: CROSS SECTION OF A HAND QUERN AND TERMINOLOGY RELATED TO THE DIFFERENT ELEMENTS

stem from different types of rock, normally extracted from loose boulders within local communities where they were found (Løland in prep.). In all likelihood they were produced to meet local needs, and no exchange can be documented this early. Recent studies indicate that the rotary quern in Scandinavia was only for ritual use at special places this early (Zachrisson 2004:153-154; in press; Bergström 2007:179), and did not become common equipment for producing flour for daily bread until the late Iron Age.

Mills driven by water-power represent an important breakthrough in milling technology, and a development from grinding within households to milling on a larger scale, often carried out by professional millers. Two different mill types are known: the horizontally placed waterwheel with vertical axle (Norw. *kvernkall*) and a vertically placed waterwheel with horizontal axle (Berge 1979:7; Fig. 2.2), the latter often operated on a communal or institutional basis. It is, however, not known which of the two types represent the oldest watermill. The possibility that the two different types were the result of topographical conditions with different possibilities for exploiting water power, rather than a different degree of technological knowledge, should also be considered (Fisher 2004:28-32, 108).

From the Viking era onwards, quernstones became a more standardized product, and the size normally varied from 35 to 60cm in diameter, and from about 10 to 20cm in thickness, although both smaller and thicker quernstones are identified - as they are also at Hyllestad. Millstones are generally larger than hand-driven quernstones. From 1700-1800s onwards, stones measuring approximately 60-130cm in diameter are found in watermills in both Norway and Sweden. In Sweden, these stones are documented at mills with horizontally-placed wheels (Ek 1962:10; Berge 1979:17). In Hyllestad, millstones of c. 140cm in diameter are known from modern-period quarries. Nevertheless, there is an overlap between the largest quernstones and the smallest millstones, in some cases making it difficult to decide whether a stone was meant for a mill or a quern. Stones measuring 50-60cm in diameter may have been used for both hand-querns and watermills. This is supported by observations from Vågå and Heidal (Oppland), where quernstones for hand-querns of 55cm in diameter and a few millstones of 50cm in diameter are found in watermills with horizontal wheels (Gurli Meyer pers. comm. 2013).

When and where watermills were first introduced is still an open question. The oldest known watermills with vertical wheel are documented in Italy and Switzerland, dated to the first century AD, while the oldest mill with a horizontal wheel is registered in Tunisia, dated to the late 3rd or early 4th century AD; written sources indicate that the technology appeared even earlier, around 240 BC. During Roman times the technique spread to most parts of the Empire (Watts 2002:47; Fisher 2004:19, 21). It is still uncertain when this technology reached Scandinavia.

FIGURE 2.2: LEFT: MILL WITH HORIZONTAL WATERWHEEL WITH VERTICAL AXLE. RIGHT: MILL WITH VERTICAL WATER WHEEL WITH HORIZONTAL AXLE (SOURCE: BERGE 1979:7)

Written evidence indicates an introduction in the early Middle Ages; in Denmark in the 12th century; and in Sweden and Norway in the early 13th century (DN XII nr 860 1209; RN bd. I no. 216, 336; Ek 1964:542, Helle 1982:430; Fisher 2004:35). Archaeological investigations, however, indicate a date for mills with vertical wheels to the late Viking period (Kristensen *et al.* 1997:209-210; Fisher 2004:107; Jessen in press). This concurs with results from my previous investigations in Hyllestad (Baug 2002:38, 53).

2.5 Bakestones and their use

Bakestones from Ølve and Hatlestrand are most often shaped as round or oval plates of different sizes, from about 25-50cm in diameter, even though larger stones are known. The bakestones are normally about 1cm thick, even though thinner (0.2cm) and thicker (3cm) stones occur (Tengesdal 2010:23-24). Normally, one or both sides of the plates are covered with carved, long and thin, parallel grooves (Fig. 2.3). The grooves may have been a result of the production technology and a means of making the plates thinner, but they may also have had a function related to baking.

So far no stone baking plates have been found dated to the Viking period in Norway. Iron frying pans, however, seem to have been commonly used in this period and appear among Viking grave goods. The frying pan consisted of rounded and barely concave plates, up until 28cm in diameter, with a long shaft (Petersen 1951:417-421). It is possible that this utensil was also used for baking bread. From the Middle Ages there are clear indications that iron frying pans were used for this purpose, and first and foremost to have been common in Norway, but also recorded in north-western Sweden and Iceland (Olaus Magnus: 595; Campbell 1951:11, 16-17; Granlund 1956:307-308).

Figure 2.3: Complete and unused bakestone found in the quarries in Ølve/Hatlestrand (Photo: Atle Ove Martinussen)

A description of bakestones used for baking *flatbrød* (Eng. *flatbread*) from Hardanger in 1632 (Friis 1632:71-74), indicates that such objects were commonly used in Norway as late as the 17th century, and were produced in the Ølve/Hatlestrand area. At that time bakestones were exported over large parts of Norway. According to Fredrik Grøn it is possible that not only bakestones, but also the baking of flatbread, may have originated along the western coast of Norway (Grøn 1939:93; 1942:62). As well as bread, chemical analyses of remains on bakestones show that food containing vegetable fats, oily crops, such as beans and peas, dairy fats, most likely from cattle, were cooked on the stones (Reiersen 1999:74-75). There are, however, also indications that iron baking plates were used in the Middle Ages. The Swedish Archbishop, Olaus Magnus, gives (1524) a thorough description of the baking of flatbread in Scandinavia, but only with reference to iron baking plates (Olaus Magnus: 595; Campbell 1950:14). It is possible, of course, that there were regional variations in forms of baking plates.

2.6 An overall assessment

The modest extent of the distribution and use of bakestones has limited research interests. While quernstones and millstones have seen increased attention recently in Scandinavia and Europe as a whole, this is not the case with bakestones. Yet, as this overview demonstrates, a broader and more analytical approach to the quarrying and trade and exchange of these different products is still underrepresented compared to the technical and descriptive interest generally in quarries and their products in the United States and Europe, as summarized by Hockensmith (2009). The most thorough works on quarries are those studies by Alain Belmon and Fritz Mangartz (Belmont 2006a; 2006b; Mangartz 2008). Although they mainly have a technical slant, the organisation of production and distribution are also covered. A narrower, empirical approach characterizes research on the distribution of both bakestones and quern/millstones, where the spatial distribution and dating of the products are emphasized. Only occasionally are the modes of trade and exchanges discussed and analyzed (cf. Carelli and Kresten 1997; Forster 2004; Pohl 2011; 2012).

Empirical and technical approaches constitute a necessary first step for the study of quarries, production, and distribution. The results of these works form a necessary basis for further studies and a precondition for approaching quarries and their products in wider societal contexts. In this study, the aim is to widen the perspectives with regard to both production and distribution, and to add a more interpretive approach. By taking different theoretical perspectives into consideration, the present author also hopes to arrive at a better understanding of production, distribution, and organisation on different levels.

3 Theoretical Perspectives

How can we understand and interpret physical remains - the fragments of human activities? What causes changes and stability in the physical remains left by humans, and how can we make meaningful interpretations from these fragments? Any approach to Prehistory and the Middle Ages requires an understanding of what factors, processes and forces have impacted on people's relations to their surroundings in time and space.

Humans shape and transform the landscape they live in, and their landscapes, therefore, constitute testimonies, although fragmentary, to the lives and activities of past generations – to the people who once lived there. The remains from many of their activities are often still visible in the landscape long after the people who created them have gone, as we will see from the two quarry landscapes highlighted in this study.

Landscapes, however, are continually changing and should therefore be studied as processes rather than fixed 'objects' (Ingold 2000:189,198, 200-201). The relation between humans and landscape may thus be seen as a continuous 'structuration process', where the landscape is both a medium for, and a result of, human action (Tilley 1994:23).

My objective in this work is to reach an understanding of the interactions happening between the quarries at Ølve/Hatlestrand and Hyllestad, their wider, local communities, and beyond them to general Scandinavian society in the Viking and Middle Ages. As far as possible, I also try to make the agents visible, as indicated at the level of social groups. On the micro level, my focus is the work in the quarries, their technologies and productions, and, indirectly, also the agents in the background - the stonecutters and their skills. The special mineral deposits constituted important resources, and the utilization of these landscapes and resources, and how these factors have contributed to shaping the landscape, are also emphasized in this account. From the beginning of the Viking period onwards, the rise in serial productions of commodities, aimed at wider populations, can be seen in Scandinavia. Earlier studies have shown that to understand important socio-economic development, production may be more essential than transactions. It was not necessarily the transaction forms that changed in the early Viking period, but the forms of production (Callmer 1995; Moreland 2000a; 2000b; Skre 2007:450-451). Within this study, the quarry sites and their inherent activities conducted have centre stage.

Landscape cannot be understood in economic terms alone (Gerrard 2005 [2003]:228), and the socio-political organisation of activities is also important. In addition to economic, technological and functional approaches to production areas, contemporarily cultural norms and values provide an important framework for this present study, as far as they can be recognized. People's cognition and relation to their physical surroundings have also to be understood within the wider social, socio-political and cultural contexts, of which the quarries and their activities formed part. Consequently, perspectives and levels of analysis must include social aspects of the landscape. Control and disposal of resources are central aspects in this respect, and the agents influencing the production landscapes will also be taken into account.

Distribution of different products within larger geographical contexts is looked into. From the Viking period onwards commodities became more widely distributed than previously (Skre 2007:451), such as soapstone vessels and quernstones, and both production areas produced for other, larger markets. The contexts in which the objects were found may also shed light on the agents involved in the distribution, as well as the users of different products. Due to large-scale production for larger markets, output was also influenced by conditions beyond the local community. To what degree institutional and structural changes in society had influence upon production, distribution and trade in different products, therefore also requires investigation.

This work also focuses on choices and actions related to structural forces in the society, both social and physical, and the structures' impact upon humans. There is no uniform explanation for the notional structure, and it may be used on several levels. Structure may denote manifest forms in a landscape as well as societal organisation, and it can be used to explain both relations between parts, as well as denoting the totality (see, e.g., Sabo 2005:38-39). In this study, physical structures, such as quarries and spoil heaps, as well as societal structures, are all examined.

People adapt to their physical surroundings, but at the same time they are able to make choices and exploit their physical surroundings to their own advantage. People are not only passive observers of the landscape and its resources; they also change the landscape, and transform it actively to adapt it to their own needs (Gansum *et al.* 1997:18).

This present work discusses stability, change and development in production sites and their outputs, and also how changes in trading-systems, as well as structural and socio-political conditions, influenced these activities. To what degree it is possible for agents to influence their surroundings varies, related to their social position and power. In addition to options there are always restrictions that limit activities. To what degree individuals are able to change these structures is a debated issue. Structures may represent too strong an influence upon humans, and to such a degree that it becomes an obstacle to change and development. A production system, with its organisation and established trading routes and contact network, may thus have limited their options, and the amount of capital invested in prevailing systems may have consolidated the state of affairs and made development and change of these structures difficult.

Possibilities and limitations in the local landscape need further investigation, such as access to resources, and to what degree the exploitation of these was a result of local involvement, choices and solutions. At the same time, I investigate to what degree external factors and social conditions have influenced the development of the production landscapes and the activities there, as well as the distribution and contact networks related to production. Important aspects to be discussed include the social and socio-political conditions related to the quarries, in terms of production, technological aspects, and distribution, including trade and exchange.

This work draws on several theoretical approaches, especially elements in Anthony Giddens' (2004) [1984] structuration theory, concerning the relation between human action, social structures, and change or stability in the long-term perspective. The French historian Fernand Braudel, and his three-tiered division of structures (1980) [1969], is important when discussing processes of change on different temporal levels. The British geographer Robert Dodghson, and his work *Society in Time and Space. A Geographical Perspectives on Societal Change* (1998) have also inspired my study. While Giddens emphasizes social structure, Dodgshon, as a cultural geographer, underlines the physical structures in the landscape, such as settlements and traffic arteries. His expansion of the structuration theory is therefore especially relevant within the archaeological discourse: not only social structures, but also physical structures have an impact on people and their actions.

3.1 Social dimensions of production landscapes

The landscape with its physical conditions and resources has significant influence upon its use and exploitation. Humans adapt to the landscape and find solutions that give acceptable results, according to given efforts and means; use and utilization of the landscape represent social activities (Sabo 2005:79-80). Thus quarries and their productions reflect past actions through everyday practices, and also decisions, an aspect that cannot be separated from power and influence, from the direct or indirect control of who does what, where and when (Pred 1985:339). Everyday practices involve routines, which again demand rules to structuralize activities. In this way actions create structures and norms related to access to resources, which in turn may have a regulating effect on the people involved - who could do what, where and when.

Production is embedded in political, social and economic systems (Costin 1991:2), and in this section I examine to what degree socio-political aspects, such as control and disposal over land and resources, were relevant for this type of production and extraction. Were there any restrictions related to access to the special geological resources? Do the quarries represent an economic specialisation and activities controlled and organized by elites, or were the quarries located on common land utilized by people living in the area or nearby farms? According to Dodgshon the density of decision-makers and weight of decision-making tends to thin out towards peripheral and marginal areas (Dodgshon 1998:197). To what degree the intensification of the exploitation of so-called outfield resources in the Viking period, such as quarrying, was result of local initiative or the result of initiatives taken by elites in an attempt to increase production through an organized activity based on the work of tenant farmers, and perhaps slaves, in the early part of the period, thus also need to be considered. Giddens' structuration theory represents a theoretical framework for understanding the relation between human actions and social structures, also as applied to early societies. His theory is an important analytical tool for identifying different phenomena that constituted important structures in connection with the quarries.

Giddens defines 'structure' as 'rules and resources recursively implicated in social reproduction', i.e. production and reproduction of social systems. He deals with two kinds of resources: 'Authoritative resources' referring to non-material resources connected to the ability to co-ordinate and control human agents and their actions (Giddens 2004:xxxi, 33). In my study this may be related to control of activities in quarries, such as production and trade or exchange, as well as control of those who conducted these activities, for example stonecutters and distributors. Giddens' so called 'allocative resources' refer to control of objects or commodities - material products, comprising raw material, production and production technology, as well as the goods produced (Giddens 2004: xxxi, 33, 258). In the present study this relates to control of quarries, extracted products, as well as technologies used. As a consequence, the exercise of power is a central feature that concerns both material and immaterial resources (Cassel 1993:11). The different types of resources are also connected to each other and in such a way that the agents controlling the quarries and the products would probably also be in control of the people active in production and distribution. When using the notion 'rule', I relate it to aspects at different levels, including property rights and legal relations concerning both production and transaction forms. There may also have been more local norms structuring routines, like the use of extraction techniques and the way the work in each quarry would have been organized.

An important element in Giddens' structuration theory is what he denotes as 'the duality of structure', meaning that the structural properties of social systems are both the medium and outcome of the practices they organize. Structure is not to be equated with constraint, but represents both possibilities and restrictions; influence between human action and structure is mutual (Giddens 2004:25). In other words, social structures are both a premise for, and an outcome of actions, in mutual relation to the human agents and their actions. Actions create and affect structures, but these structures formed by human beings cause new choices and actions.

Giddens' theory thus aims to explain relations between human actions and social structures. It has, however, been criticized for using a too strict definition of the notion 'structure'. Dodgshon has criticized Giddens for reducing structure to a social form, while other structures, such as physical structures, including material culture and landscapes, are excluded (Dodgshon 1998:29). Contrary to Giddens, Dodgshon chooses to emphasize the physical structures in the landscape, and his work thus represents an elaboration of the structuration theory that is of relevance for archaeologists working in the landscape.

Giddens' perspectives can be used to understand the relation between actions and physical structures. His theory has also been used in several Scandinavian archaeological analyses over the last decade (cf. Sabo 2005; Sindbæk 2005; Hansen 2005 [2004]; Zehetner 2007). In the same way as there is a mutual relation between actions and social structures, the physical structures are also in a mutual relation to human actions. Humans produce material culture out of certain intentions, which again affect their actions and practices. This interaction between action and structure, ideas and materiality, is important in the archaeological discourse (Olsen 1997:210-211). The notion structure can thus be used on several levels, and within this study both the physical structures in the landscape, as well as social structures, are important.

Quarries represent both physical and social structures. They have a geological precondition and a physical shape as a result of long-term production, formed by repetitive and routine actions of various individuals and groups, and at the same time they are parts of a larger societal structure - a socio-political organisation of production, distribution and disposal of resources. The actions created new structures, both material (quarries and products) and social (organisation of the work), and these structures inter-react and cause new actions. In this way man produces structures, but when the structures are first established they impact on further human thought and action.

Actions take place within a physical setting. The agents behind them affect and create the physical setting, such as the surroundings they live in. However the physical setting also influences and 'structurates' actions. Related to the quarry landscapes in Hyllestad and Ølve/Hatlestrand they can be seen as results of the production that took place there. The landscapes and the geological conditions represented both possibilities and restrictions that stonecutters and landowners had to relate to – they were constrained by a situation they were only partly capable of changing.

People seldom participate on equal levels in the production and reproduction of these structures. The person who wanted change must have had the possibility or power to carry it out, and to such a degree that it would influence others and their actions (Giddens 2004:14-16). Giddens uses the term 'agency', which refers to people's capability of making decisive changes through their actions (Giddens 2004:9). This represents the entrepreneur perspective put forward by the social anthropologist Fredrik Barth. People who take the initiative and attempt to gain profit in some form, and by doing so act in a way that influences and controls other people and resources, are by Barth denoted as 'entrepreneurs'. Goods obtained through entrepreneurial activities are not restricted to material forms alone, but may also be intangible factors such as power (Barth 1972:5-6, 8). To what degree individuals would have the possibility to change and influence both physical setting and social structure is, however, dependent on where the different agents relate in a social and spatial context.

Transferring these perspectives to the stratified society in the Viking period and early-medieval Norway, it is likely that people from the upper levels of society were the ones who had the resources, and thus were able to execute power and make changes, and to be the entrepreneurs. A landowner thus had far greater influence than a subordinate farmer or slave in the early phase of the period. Landowners may have possessed the appropriate allocative and authoritative resources to be able to take decisions that could affect and change their surroundings.

To be able to understand the societal organisation in the Viking period and the Middle Ages, in and around the quarry sites, it is important to include all available sources, both archaeological and written evidence. The totality of the landscape is important, and an interpretation of the relations between different components in landscape is critical to its understanding. Humans have affected the whole landscape, and the totality is thus more important than single structures (Gansum *et al.* 1997:17). According to Dodgshon, social organisation makes a physical imprint on the landscape (Dodgshon 1997:88-89). Such physical imprints or investments he calls the 'built environment'. Built environments are created from certain social situations, and consolidate in this way the social organisation prevailing at time of the creation of the built environment (Dodgshon 1997:139-140; Sabo 2005:41).

This concept also has a certain transfer value to the prehistoric and historic landscapes that the quarries were part of. Physical structures, like burial mounds from the Viking period and farm and estate structures from the Middle Ages and early modern period, represent manifest structures of power and disposal, and thus give information relating to the prevailing societal structure at the time of the quarrying. The national Land law of 1274 refers to the ON expression *haugsóðalsmaðr* (L VII, 16), which has been interpreted as a descendent of the person buried in the mound (Taranger 1915:107). According to a document from 1316, people needed to clarify their ownership of allodium land back to pagan times: to *haughs ok til heiðni* (NgL III, 121) for cases where land and inheritance were pivotal, implying that people could legalize their right to the land by referring to mound and family. Evidence thus points towards a situation where burial mounds were regarded as symbolic expressions of an allodium privilege to ancestral farms, implying that only people with such rights were permitted to demonstrate this through burial mounds. Over recent decades several scholars have pointed to a possible connection between the occurrence of prehistoric burial monuments and land rights (cf. Zachrisson 1994; Skre 1998; Iversen 1999 [1997]; 2008 [2004]). Consequently, monumental graves have a possible connection to the allodial rights to land in the Middle Ages, as well as the previous period (Iversen 2008:66), which also constitutes an important premise for this study. An important question is whether it is possible to identify a connection between quarries and the activities there, and later property rights over production landscapes. Allodium and property rights were not only connected to infield areas. Allodium also applied on certain valuable outfield resources (Robberstad 1967:494; Iversen 2008:67), and this may have been the case for quarries. Relations between superior landowners and a subordinated labour force in the late Iron Age and the Middle Ages may in this way have been decisive factors in how production in the quarries and the distribution of the different products were organized. If the production areas of Hyllestad and Ølve/Hatlestrand were structured and changed, not only as a result of local initiatives, then the regional and strategic needs of the landlords need to be discussed. Were the workers in the quarries in a position of decision-making, or was production a result of choices taken by societal elites, in or outside the local community?

The transition between the late Iron Age and the Middle Ages in Norway is characterized by institutional and structural changes, with the emergence of new political, religious and economic institutions and socio-political elites. Changing power groups may have had further consequences for property structure and the physical organisation of the landscape (Iversen 2008:54). These are all processes that possibly influenced both resource utilization and trade. An issue to be looked into is whether it is possible to substantiate a connection between external societal conditions and the activities in and around the quarries. How societal changes, such as the emergence of new elites and institutions ('King and Church'), along with urbanisation and an increase in commercial trade, influenced production and distribution in the quarries, should therefore also be studied. Here, I focus on different agents and their opportunities to influence the exploitation and change of their physical surroundings. Do agents, as Giddens claims, influence and change structures through power and resources? External structural and societal changes, and altered disposal and power relations, therefore have to be considered in terms of production and distribution, stability and changes to the quarries.

Since uses of land and resource utilization were entangled in social interactions between different social groups, this should also be assessed in any analysis concerning quarries and the resource utilization. During the Viking period and the Middle Ages large changes in property ownership and disposal of land took place in Norway; around 1300 AD ecclesiastical institutions became the largest landowners with about 40% of lands in their position (Øye 2002:221, with references). To what extent such institutions were also involved in the production and distribution studied here, should also be discussed. The study of entrepreneurs and their choices and actions in relation to the wider community is important, since these are all factors involving the entrepreneur, but which he also has the possibility of changing (Barth 1972:7). If the social structures are, as Giddens points out, both a medium and an outcome of the actions and practices they organize (Giddens 2004:25), the quarries should be studied and interpreted according to the historical and societal situation in the Viking Age and the Middle Ages of which they formed part.

Whether the quarrying was carried out by specialists, or was a seasonal activity conducted by local farmers and their servants will be looked into. Specialisation is, by Cathy Costin, explained as a means of production where people produce for profit or commercial return, and is distinguished from the domestic mode of production, where people primarily produce for their own subsistence (Costin 1991:3-4). Production in the quarries may have represented a primary occupation only on a seasonal basis, and where agriculture constituted the main activity for the rest of the year, a situation termed 'proto-industrialisation' by the historian Franklin Mendels (1972). According to Mendels, this form of semi-professional activity existed before industrialisation in rural societies, where agriculture was always the main activity.

The distinction between so-called 'independent specialists', who produce goods for unspecified demand, and 'attached specialists', producing for a social elite or a governing institution (Brumfield and Earle 1987:5), may also be important concepts to assess in terms of the organisation of production as well as trade. The possibility that some product types extracted from the quarries represent commissioned work for special receivers, rather than a serial production based on routine commercial exchange, or trade for a wider market, will also be discussed. Attached and independent specialists thus usually produce different types of goods, high-value or utilitarian products respectively (Costin 1991:11). The distinction put forward by Brumfield and Earle between attached and independent specialists may, however, not cover all aspects of production. It is possible that actors involved in the production of quernstones and millstones (and bakestones) were also attached specialists obliged to produce for their landowners. This may again be a question related to power, and agents with allocative and authoritative resources, in this respect people with control and disposal over quarries and products, production and distribution may have been those who ordered or authorized production of various commodities - regardless of whether they were high-value objects or utilitarian goods.

3.2 Technology

A study of quarries and production also involves technological aspects, such as extraction techniques and production processes. Several scholars have stressed the importance of treating technology as an integrated part of society as a whole (e.g. Lemonnier 1992; Pfaffenberg 1992; Miller 2007). The last decades have seen a change in the understanding of technology, from being regarded as a more-or-less purely functional element within processual archaeology to now emphasising its social role. Technology is complex, and should not be separated from other aspects of society and which may lead to a fragmentation of knowledge. Like any other social product, technology cannot be reduced to a single dimension (Lemonnier 1992:118). It is within this theoretical framework that the technological aspects of the quarries are approached in this study.

Technical procedures involve routines and routine practices, and the remains of quarries are the results of routine and repetitive actions over long periods of time. Giddens maintains that routine is a basic element of daily social activity (Giddens 2004:xxiii, 60-64; 376). Repetitive activities carried out at quarries can be explained as a standardisation of production, and according to Giddens, the foundation of those social and material structures being influenced, created or maintained. When the quarries with their technology and organisation were first established, they had further influence upon human actions.

Technology includes production and techniques as well as the specialized knowledge and organisation of the people involved in them (Miller 2007:31). This wide definition of technology fits the perspectives discussed above. Technological aspects related to quarries should not be seen as in some way separate, where extraction processes and changes in quarry types can be deduced from function alone. The production and technologies involved should also be seen within a social setting related to the local and regional community on three levels: the fundamental social and material setting at the micro level (the quarry landscapes); the meso level within the local and regional community; and on the macro level within the wider social, political and ideological context. A further important issue, then, is to what degree the social and political changes in the Viking period and Middle Ages affected the technology in the quarries.

A survey of carving traces and an interpretation of techniques in several of the largest quernstone and millstone quarries in Norway is currently being carried out (Løland in prep.). Parallels from extractions from quernstone quarries can most likely be found in bakestone quarries. Experimental archaeology at several quernstone quarries in Norway, including Hyllestad, has shed light on technical and functional aspects of production. Such experimental techniques can be especially useful for creating and testing hypotheses about past production (Miller 2007:35), and may further indicate how products were extracted. In this study I mainly draw upon these results, benefitting from the work of others, and link them to quarries and products, focusing on production processes as well as the organisation of the production, in what Pfaffenberg terms as a 'sociotechnical system' (Pfaffenberg 1992:497). To make sociotechnical systems work, agents use existing resources but modify them to function within the systems. In this sense Pfaffenberg's system may be seen in relation to Giddens' structuration theory and his concept of the 'duality of structure', whereby people construct their social world by using available resources and structures, and where these activities in turn modify the structures (Pfaffenberg 1992:500). In this way an interaction and mutual influence between human action and structures, both social and material, is important also for the study of technology.

Another approach for studying production, work processes, and the actors involved is the *chaîne opératoire*, a theoretical approach developed by the archaeologist and ethnologist Leroi-Gourhan (1993) [1964], emphasising functional, symbolic and social aspects of technology. What is important is how different operational steps in the production link together to transfer raw materials from its natural to a manufactured state. Gaining insights into the totality of operational sequences constituting any technological system is problematic within archaeology (Barndon 2002:8). This approach has therefore not been the focus of this present study.

The creation of new artefacts, or technical procedures, represents a certain break of routine behaviour, or at least requires an acceptance that other products and behaviour is possible and acceptable (Lemonnier 1993:22). The stonecutters most likely had the opportunity to choose between different mineral resources and technical options, such as extractions from below ground or open quarries at Ølve and Hatlestrand, and deep or shallow quarries in Hyllestad. Another option may have been to extract products directly from the bedrock, or from slabs already removed from the rock face. And whether or not to fashion the 'eyes' of querns or millstones before removing the stones from the bedrock was also something the stonecutters had to decide. Changes in technology may not have replaced the existing processes, but rather modified or supplemented them, a view that follows the *chaîne opératoire* approach (Barndon 2002:10). Whether different quarry types and technologies, as well as the production of new products in the quarries, represented innovations or adaptions, and to what degree these were result of external factors and influence, or if choices made at the local or regional level influenced technological development, will all have to be discussed.

Studies on quarrying in the late Iron Age and Middle Ages have so far mainly focused on functional aspects relating to production (cf. Skjølsvold 1961; Østerås 2002; Gardelin 2006; Lundberg 2007), whereas other technologies, such as iron smelting, or aspects of ritual and magic are discussed also in connection with technological studies based on ethno-archaeological material (cf. Rijal 1998; Barndon 2004; Håland 2004). Some technologies, such as iron smelting, may have been more ritualized than others, making such an approach relevant. Work processes in connection with quarrying may also have been ritualized, and differences between different products may have existed - such as the production of crosses being more ritualized than the extraction of quernstones and bakestones. However, gaining insight into the totality of operational sequences constituting a technological system is problematic based on the archaeological material (Barndon 2002:8; Tveiten 2012) and is outside the scope of this study.

Those involved in production faced several problems, and different actors faced different problems, representing varied technological and hierarchical levels. Some difficulties were technical, such as how to work and what tools to use, and related to stonecutters and others working in the quarries. A possible hierarchy among ordinary workers, skilled workers and specialists may have existed at the quarry sites. Actors controlling production had to decide when to work in the quarries - seasonal or all-year, and they may also have had to relate to fluctuations in demand for their products, shifting trade systems, as well as issues related to the organisation and access to resources and production. Such economic and socio-political aspects could have been as important and problematic as the technical challenges (Miller 2007:15-16). Technological choices were, then, based on several factors, not only efficiency, but also social, economic and political relations.

3.3 Trade, routes and 'routinisation'

Production should be studied in relation to distribution and consumption of the products (Costin 1991:2-3), and the scope of trade and exchange and for different products in different areas is another issue to consider. This may shed light on the conditions that formed the basis of trade and distribution. Was it an exchange where products went from hand to hand through networks of personal contacts, or was it a more organized and 'routinised' trade? Do some of the products represent commissioned objects produced for specific groups? These questions are central when discussing stability and change, both spatially and temporally, with regard to both production and distribution of the different products involved.

Archaeological studies of exchange in pre-modern societies have been based on models of trade and exchange, and from the latter half of the 20th century to a large degree influenced by Karl Polanyi and the so-called substantivist school (cf. Brumfiel and Earle 1987; Hodges 1989; Christophersen 1989), in contrast to those termed as 'formalists'. Recent studies, on the other hand, have stressed the plurality of transaction forms existing in Viking times and the Middle Ages, and criticized the strict classification by the early substantivists of redistribution, reciprocity and market oriented trade. They argue rather that different exchange systems most likely existed side-by-side, with no sharp distinction between personal related gift-giving and commercial exchange with a plurality of transaction forms (Moreland 2000a; 2000b; Gustin 2004; Sindbæk 2005; Skre 2007:448-449; Skre 2008). The recent discussion of material exchange in pre-modern societies is part of a general trend less focused on models and an understanding of pre-modern economies as dynamic and complex. The models may represent tools important for the understanding of pre-modern societies, but instead of being regarded as ideal types, the models should be compared to, and modified by, the empirical results (Skre 2007:448-449). Economic action is contextual and influenced by social relations, and the historical models do not capture the dynamic complexity of pre-modern economies.

How foreign artefacts found in archaeological contexts were distributed thus needs to be discussed with an open mind, bearing in mind that several forms of exchange may have existed. A distribution map of products from the quarries may reflect different levels of interaction between different areas over time. Large numbers of quernstones, millstones and bakestones in a town, or market place, may indicate the existence of more-or-less regular trade and commercial exchanges, but the products in themselves do not allow us to reconstruct the organisation of the trade and trading routes with any certainty. At the same time, the presence of a few similar products in a town, market place, or within other contexts outside the quarry areas, does reflect some form of contact or exchange, although not necessarily in terms of trade. The objects may have been brought there as personal belongings by itinerant people, for use at the site or for further distribution. Hence, a contextual approach is emphasized, where both the context and the volumes of relevant artefacts must be considered.

Again, Giddens' structuration theory serves as a useful perspective for understanding trade with products from the quarries; it has previously been used successfully for the study of trade and trade routes in Viking Scandinavia (Sindbæk 2005). Additionally a trade route may be seen as social practice and as an expression of 'routinisation' where exchange or trade between specific regions had become routine. A route is thus not only a geographical construct, but a social institution and historical process: a repetitive praxis, a routine (Sindbæk 2005:32-33). Motivated actions of individuals led to the organisation of exchange and trade, and a routinised transaction implies that the transportation of commodities repeatedly took place along specific routes, and that exchanges and trade were carried out in an organized manner at specific locations (Sindbæk 2005:261-262, 268). Trade or exchange thus entails interaction between different areas and involves different zones with a great variation in span or scale, which Giddens calls 'regionalisation'. According to Giddens, regions existing over large time spans would depend upon a high degree of institutionalisation (Giddens 2004:122, 376). This means that tradition and routines would be important for maintaining and legitimizing social systems.

The products studied here indicate regional transactions as well as long-distance trade for some objects, and both forms may be the result of repetitive actions that led to routine, organized transactions. Nevertheless, trade or exchange within local areas may have had a different cultural meaning compared to long-distance trade and have changed over time.

Long-distance trade extending beyond the local and regional contexts would involve agencies with other, often unfamiliar, social norms and rules compared to exchange or trade within the local or regional community. Commercial exchange/trade with outsiders therefore represents a larger degree of uncertainty, and often involved elements of mistrust and conflict (cf. Dahl 1998). Protective elements to secure peaceful transactions must therefore have been important (Gustin 2004:166-174; Skre 2007:450-451). Agents with the appropriate resources, both authoritative and allocative, may have been those able to secure markets, trading routes and networks, and thereby make the transactions workable.

The establishment of routines and traditions may in this way have been important - both for local and long-distance trade. According to Giddens, routinisation is vital to the psychological mechanisms where trust and security are sustained across the daily activities of social life (Giddens 2004:xxiii). Routinisation consists of the formation of habits, conventions and institutions which could, potentially, make demanding practices, such as long-distance trade, systematic and therefore manageable (Sindbæk 2005:40). Routines would make it easier to interact with strangers from different cultural, perhaps also different language areas. Routinisation of transactions may have further led to long-lasting trading networks and organized commercial trade. When assuming that routines and traditions are vital for maintaining and legitimizing social systems, the time/space dimensions of distribution patterns for products from Hyllestad and Ølve/Hatlestrand are essential for understanding the organisation of the distribution of various products.

Production is closely tied to exchange, and different forms of production may correspond to different types of exchange. The ability to identify the physical source or production centre of an artefact is, then, essential for determining the presence and extent of the exchange (Earle 1982:4, 8). In this case, objects from quarries with good opportunities for provenance, represent an important source. Both quarry landscapes represented in this study involve serial production. This presupposes a considerable number of customers and is usually aimed at a specific market and exploited for profit (Skre 2007:450). However, the quarries were also production sites for more specialized and exclusive objects, such as stone crosses, grave slabs and building stones, which never reached markets as ordinary commodities. They were evidently subject to different forms of transaction - production on demand. The different products from the two quarry areas may thus have represented alternative commodity spheres. While quernstones and bakestones were everyday products meant for the general population, stone crosses, grave slabs and building stones may have been items of high status, aimed at, and commissioned within, the upper strata of society.

3.4 Structural durability and change: which processes define stability and which cause change?

All structures are exposed to change, and the long-term perspective of this study makes it possible to shed light on the questions of stability or changes through time. The production sites at Hyllestad and Ølve/Hatlestrand generated different products over a long period of time, and represent activities based on trade surplus and transactions of goods. Both quarry landscapes reveal changes and development in quarry types, and the type of products produced and exported, but at the same time continuity in terms of certain product types, technologies, and markets.

All structures – their emergence, change and possible end – are linked to date. Braudel's model and analysis of structural changes in societies of differing durations (1980), emphasizes the connection between individual actions and larger societal structures, where structure is perceived as a more-or-less fixed connection between humans and their landscapes (Braudel 1980:31; Iversen 2008:51). Braudel differentiates between three temporal levels of change, where structures with a long time-span interact with structures lasting only a short period; changes to one of these levels, to a certain degree, also influence the other temporal levels. The widest temporal scale is *la longue durée*, referring to geographical or environmental long-lasting structures, such as climate, vegetation or agricultural practices, where changes occur slowly, and structures may last for centuries. 'Conjuncture', on the meso level, is characterized by social and economic changes with more durability, for example a generation - sometimes longer. This may be represented by the rise and fall of social, religious, political and economy-based institutions, as well as economic fluctuations. Of special interest for this study is the property structure in the quarry areas, such as the presence of single farms and estates, where actions and changes on the meso level may have long lasting effects. The third level, *l'histoire événementelle*, refers to events in the short term, often based on the activities of a single individual. This may represent short-term political events, but may also have had long-lasting consequences. Braudel's three-tiered division of structures is thus a useful analytical tool for this present analysis, where the investigated sources stretch over a period of c. 700 to 1200 years; several levels interact and relations between human actions, society and landscape are central. Changes in property relations within the quarry landscapes occurred from the Viking period to the Middle Ages. The property structures may in many cases have been sustained even though land and quarries changed ownership.

The production landscapes make it possible to study both changes and more durable trends. The landscape, with its geological resources, represents long-term structures that constitute a basis for a large-scale production over centuries. The geological conditions represent both possibilities and obstacles for stoneworkers. The resources were not inexhaustible and production was not static. Quarries were abandoned and production moved to new areas when supplies of good stone ran out. Development and changes in quarry types, technological processes and products also occurred, and indicate both slower and more rapid changes. Both contemporary and later documentary sources reveal changes, institutional and structural, which affected the quarries and disposal of resources in the period of study.

In their initial phase the quarries represented a time with new possibilities, and the quarries may be seen as areas for innovation, representing new forms of production and with new products and new ideas. The actors, or entrepreneurs, to use Barth's terminology, were exploiting a new niche, with a new flow of goods (Barth 1972:9). The initial phase of the quarrying may thus be characterized as innovation in resource utilization, and a change in how the resources were expanded.

New structures have to be adapted to prevailing conditions, and the more adapted a structure becomes the more difficult is it to change, because capital and resources involved could be at risk from possible change. The initial phase is, then, normally followed by routinised and repetitive actions with heavy investment in labour and resources. This view is also supported by Barth, who claims that when the initial choice concerning an enterprise is made, this may act as a restriction on the entrepreneurs' freedom of choice (Barth 1972:9-10). This may imply that when the quarries with their markets, and perhaps trading networks, were first established they may have been more difficult to change.

Dodgshon also emphasizes those processes that sustain societal structures and hinder change. Influenced by the structuration theory he adds a geographical perspective, stressing the physical structures in the landscape, settlements, production systems and transport factors. By focusing on geographical variations in a time/space perspective, his approach represents an important aspect of my study when it comes to change and stability related to production and trade as well as organisation. He argues that an understanding of change should begin with a clearer conception of the circumstances that make changes difficult (Dodgshon 1998:15-16), a situation he denotes as 'inertia'. Compared to Giddens, Dodgshon is more sceptical of an individual's ability to influence structures, claiming that even though people can affect social and physical factors, structures are generally stronger and an obstacle to change. When analysing change, then, attention must be directed also on what constrains change, as well as the forces that promote it (Dodgshon 1998:162).

3.4.1 Innovation and change in the centre and periphery

Innovation and change occur, according to Dodgshon, more easily in areas with fewer investments, that is, in the periphery of the built environment. In such areas actors have more flexibility in terms of alternatives, or to use Dodgshon's term, there are greater resources of 'unused freedom' in the periphery compared to core areas (Dodgshon 1998:180). Greater resources are invested in the central areas, and people involved in these investments are those who benefit from the prevailing societal system and wish to maintain it. These agents are thus less flexible in their actions and less susceptible to change and innovation compared to those within the more peripheral areas (Dodgshon 1998:188). Dodgshon seems to share his thinking here with Giddens. The latter sees those who occupy production centres as those with 'control over the resources which allow them to maintain differentiations between themselves and those in peripheral regions' (Giddens 2004:131). The binary concept centre/periphery - central/marginal - may in this way include both geographical areas and different social groups. A centre may constitute the resourceful elements within a society, that is, people with power and opportunity to control others, while the periphery comprises those who were controlled. How we use the concepts 'centre/periphery' thus depends on perspective and the level of analysis, and in this study both geographical space and different social strata will be considered.

The quarries are analysed on different levels, with regard to the local environment as well as in relation to larger societal contexts. On a local level they are studied according to the possible estates, or single farms, they belonged to. Both quarry landscapes represent typical 'outfield' resources, and from a solely agricultural perspective they are located at the periphery of farms. A farm may be seen as a small society, where the 'infield', especially the area around the settlement nucleus, is more intensively used for arable farming compared to the outfield. Farming comprises both organisation and utilization of land, where, for instance, the fencing off of fields demands an investment in the landscape that also represents a stabilising factor, creating a stronger form of inertia compared to more peripheral parts of the farm (Zehetner 2007). However, in some cases the work in the quarries may have been more intensive than the agricultural activity, and to what degree the quarries may be seen as central or peripheral should be discussed in relation to their contemporary sources.

4 Methodological Approaches

In this chapter I present the terminology and methodology used in the investigation of production, the products extracted, and product distribution. To shed light on the different research questions, several sources and methodological approaches are needed. The three different levels referred to previously, micro-, meso- and macro-level, are essential in terms of the methodologies; change or stability in activities connected to the quarries are investigated and analysed on different spatial levels.

On the micro level I focus on quarry production. Archaeological investigations are important to help clarify the chronology of production, as well as the technologies and product ranges extracted, to reveal changes over time. On the meso level it is important to study the socio-economic contexts with regard to control over land and resources in the local communities around the quarries. On the macro-level the distribution of different products is mapped across a wide area, including markets and contact networks connected to the two quarry landscapes.

4.1 The quarries: definitions

The quarries of Hyllestad and Ølve/Hatlestrand represent different extraction processes. In this study a quarry is defined as a production unit consisting of a source of bedrock from which products are extracted; the areas nearby are used for spoil heaps, debris, and production waste. Some areas, however, developed more or less as coherent production sites, where the dense concentration of spoil heaps, and the rocks from which the products were extracted, make it difficult to distinguish between the different quarries. It is therefore also hard to estimate the number of quarries located in the area. In the most intensively exploited areas quarrying activities have caused radical changes to the landscape and the original topography is barely recognisable. Such clusters of quarries are hereafter referred to as quarry complexes. Each quarry complex may thus comprise several quarries, possibly used in different periods. When using the term quarry landscape or production landscape, the references are to the overall production areas of Hyllestad and Ølve/Hatlestrand where extractions took place.

Several different types of quarries are identified in Hyllestad and Ølve/Hatlestrand. The term open quarry refers to those sites where only surface exploitation took place, i.e. where extractions were made directly from surface rock. This type of quarry is identified in both Hyllestad and Ølve/Hatlestrand. An underground quarry involves below surface extraction, where production results in caves or overhangs. This type of quarry is only identified in Ølve/Hatlestrand.

In a few cases, in both quarry landscapes, workplaces are identified in connection with quarrying. A workplace is characterized as a location where secondary production takes place, such as the finishing of extracted objects (i.e. bakestones or quernstones), or the shaping of objects from worked slabs extracted from the bedrock. Secondary production was undertaken at special places, either within a quarry site or at nearby locations, and not inside the quarries themselves.

4.1.1 Classification of the different quarry types at Hyllestad

Two different techniques were used at Hyllestad: with and without gunpowder. Gunpowder was used for the most recent production, and only millstones seem to have been produced using it (Rønneseth 1977:61), with stones extracted from blocks already blasted out of the rock. According to Rønneseth, some quarries where gunpowder was used at Hyllestad were abandoned in 1751 (Rønneseth 1977:20-21), thus giving a *terminus post quem* for this kind of production. Blasting technology must have been introduced in the first half of the 18th century, and in some quarries production continued until the 1930s. Large amounts of stone blocks in the shape of talus are clear evidence of this recent production, and this kind of production is easily distinguished from the older techniques. Production using gunpowder represents only a minor part of the total output when compared with older quarries. Only 18 quarries of this type have been identified, at five different farms (Heldal and Bloxam 2008:57) and mainly concentrated in an area on the north-eastern side of the Åfjorden, and at a higher altitude than the older quarries. The location also coincides with a harder rock type, with more quartz and smaller amounts of finer flakes of mica compared to the earlier hand-worked quarries. This type of rock is much harder and seems to have been exploited only in later periods when blasting became the primary extraction technique. This selective way of quarrying may have been a consequence of the harder mica schist being more suitable for blasting. Harder mica schist and more complete garnets also led to better grinding properties and durability of millstones (Grenne *et al.* 2008:52, 62; Heldal and Bloxam 2008:15, 23). This production technique, however, is outside the scope of this present study.

Only the older quarry types, prior to the use of gunpowder, are included, extractions where quernstones and millstones were normally shaped and cut directly from the bedrock. The quarries can be further divided into different subtypes, with the typology based on the geological surveys of Tom Heldal (Heldal and Bloxam 2008).

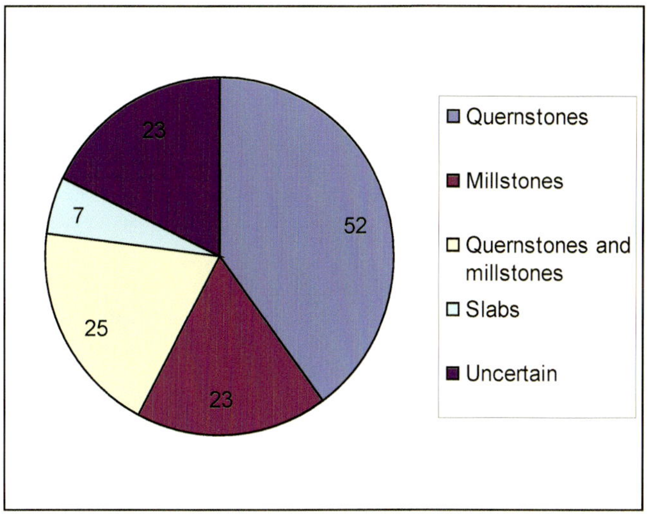

FIGURE 4.1: PRODUCTS EXTRACTED FROM SHALLOW QUARRIES. THE FIGURES SHOW THE NUMBERS OF QUARRIES BASED ON SURFACE RECOVERY (HELDAL AND BLOXAM 2008: FIG. 6-3)

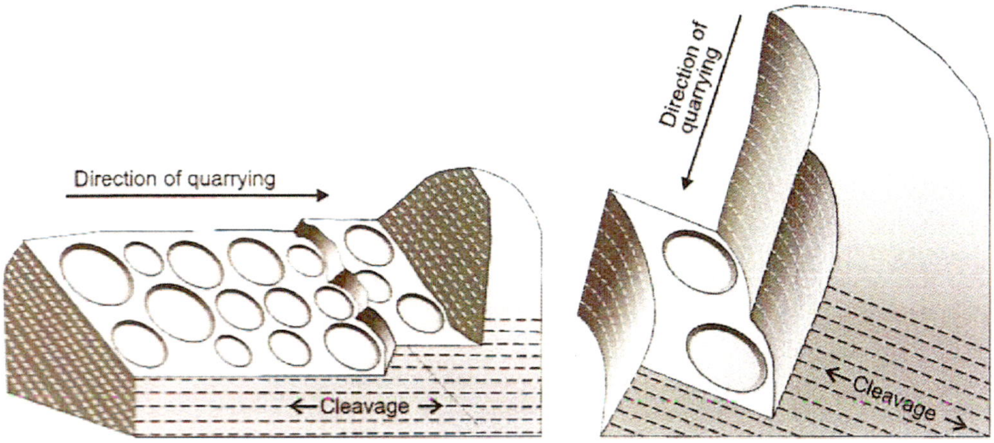

FIGURE 4.2: SCHEMATIC FIGURE OF THE TWO SUBTYPES OF BEDROCK QUARRIES: SHALLOW AND DEEP (GRENNE ET AL. 2008: FIG. 18)

Shallow quarries, with stones cut along the cleavage plane, are the most common quarry type at Hyllestad, and altogether 130 quarries have been identified so far. During production the stone cutters obviously dealt with the mica schist by exploiting one layer at a time. The stones were hewn close to each other, leaving circular traces on the rock. Both querns and larger millstones were extracted from this type of quarry (Fig. 4.1). As the quarries could have several production phases the spoil heaps were haphazard, indicating that they may have been repeatedly moved, or over-layered. The spoil heaps often cover the original rock, or at least large sections of it, making it difficult to identify the extent of the quarry without further excavation. Querns, millstones, and in a few cases slabs, were extracted from such quarries. The majority of quarries, however, reveal only querns for hand-working, which may indicate old age (Heldal and Bloxam 2008:47-50). Whether the shallow quarry type actually represents the oldest production phase at Hyllestad can only be substantiated by further archaeological excavations.

The second quarry subtype is characterized by deeper extraction in taller steps, here denoted as deep quarries. Quernstones and millstones were quarried in columns, one under the other, leaving tall, carved walls, sometimes with step-like shape. Instead of following the cleavage plane, vertical or sub-vertical slices of mica schist were exploited (Fig. 4.2). This type is also common at Hyllestad, represented at 45 identified quarries (Fig. 4.3). Large spoil heaps are usually found at these quarries, located in a semicircle around the production rock, with the latest sections of spoil right in front of the rock. Although different products (querns, millstones, slabs) were taken from such quarries, extraction from deep quarries is more associated with millstones (Heldal and Bloxam 2008:51-53). Only further excavations can shed further light on this question.

A third common quarry type at Hyllestad is a combination of deep and shallow working (16 quarries). Several extraction phases may explain this quarry type. A quarry

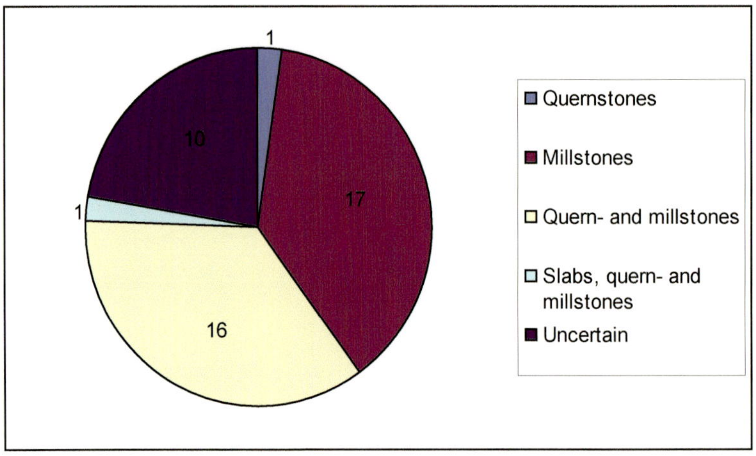

FIGURE 4.3: PRODUCTS EXTRACTED FROM DEEP QUARRIES. THE FIGURES SHOW THE NUMBERS OF QUARRIES BASED ON SURFACE RECOVERY (HELDAL AND BLOXAM 2008: FIG. 6-7)

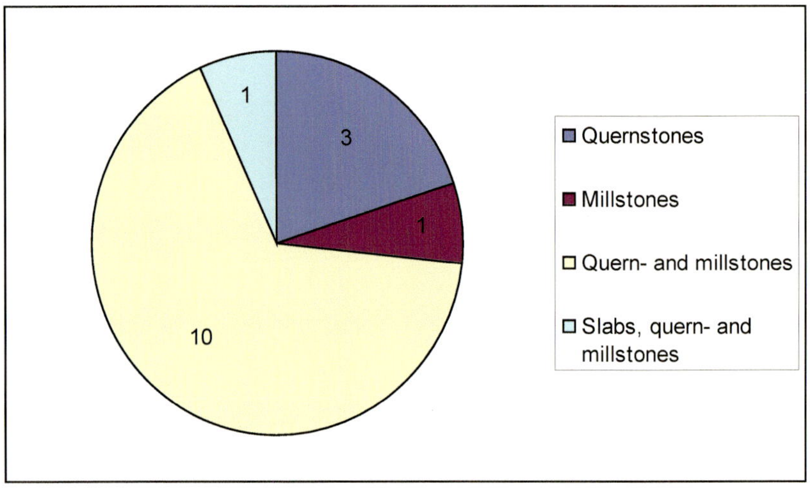

FIGURE 4.4: PRODUCTS EXTRACTED FROM COMBINED SHALLOW AND DEEP QUARRIES. THE FIGURES INDICATE THE NUMBERS OF QUARRIES BASED ON SURFACE RECOVERY (HELDAL AND BLOXAM 2008: FIG. 6-10)

may have started as a shallow one at first, and in a later period deeper production was continued. The majority of these quarries show remains from the extraction of both quernstones and millstones (Heldal and Bloxam 2008:54-55; Fig. 4.4).

Most of the quarries at Hyllestad can be classified within these three types. However a few exhibit different extraction techniques. Quarries in scree deposits feature loose, angular boulders are found at Rønset, where three quarries of this type have been identified. The quarries appear as circular depressions in scree surrounded by piles of unusable talus blocks, broken querns and waste from carving. The extraction has been explained as *ad hoc* activity, most likely for domestic use rather than standardized manufacture (Grenne *et al.* 2008:57; Heldal and Bloxam 2008:47). Another uncommon quarry type is the so-called boulder quarry, found at the Gil (Inner Gill, no. 33) and Borsholmen farms (no. 37), and seven in total. Here large, loose boulders from natural rock falls have been exploited. As many as ten quernstones could be extracted from one block. With the exception of one millstone, only querns were produced at these quarries (Heldal and Bloxam 2008:56). The lack of pronounced spoil heaps in relation to these quarry types makes it difficult to date the quarries.

One other quarry type, where raw blocks were split from the rock by wedges, has been found at Rønset. The date of this is uncertain, but Heldal and Bloxam suggest that the quarry may represent a transition phase between the quarries with extraction directly from the rock and quarries using gunpowder (Heldal and Bloxam 2008:56). Here also, the lack of obvious spoil heaps makes them difficult to date.

Some quarries are harder to identify and classify than others. They can appear as hollows in the terrain, often with a spoil heap located around the hollow. All accessible rocks are extracted, making them difficult to classify. They are, however, evidence of the earliest production based on hand-working (Heldal and Bloxam 2008:57; Fig. 4.5).

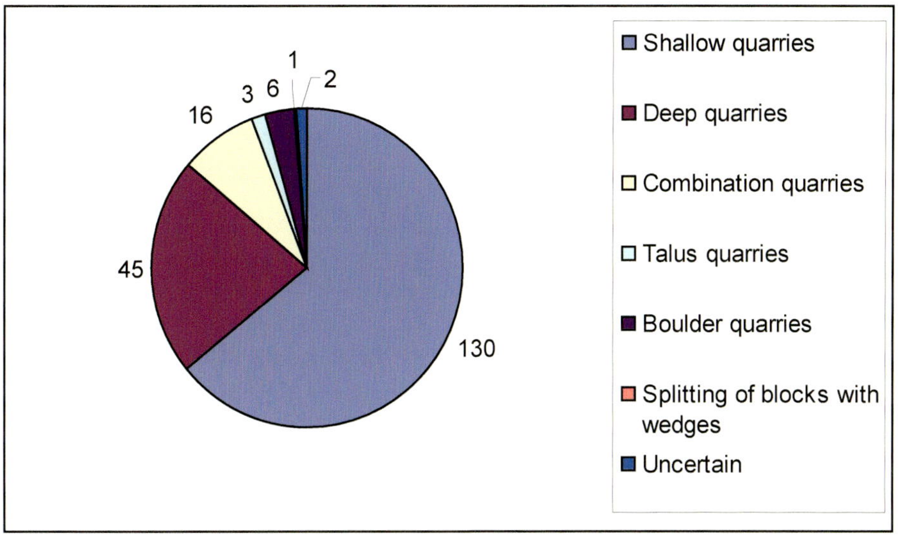

FIGURE 4.5: DIFFERENT TYPES OF HAND-WORKED QUARRIES. (THE 72 PROSPECTING QUARRIES ARE NOT INCLUDED.) (SOURCE: HELDAL AND BLOXAM 2008)

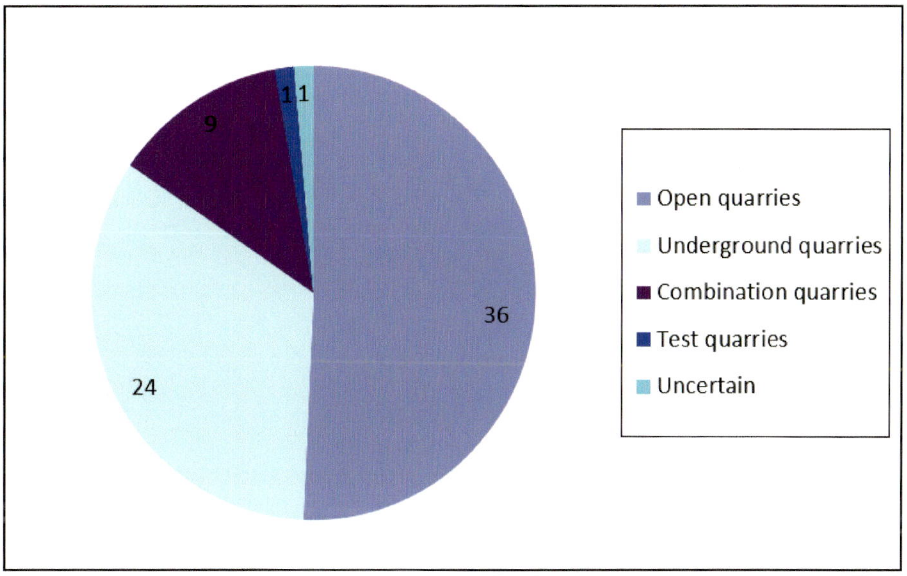

FIGURE 4.6: QUARRY TYPES IDENTIFIED AT ØLVE/HATLESTRAND

Additionally, 72 trial extractions were identified. Production is restricted to the rock surface, and only a few stones, often only between one and ten, had been extracted. Many of these quarries are located in areas of poor rock, which may explain why they were abandoned. In several cases test extractions were done and without removal of stones from the rock (Heldal and Bloxam 2008:59).

4.1.2 Classification of different quarry types at Ølve/Hatlestrand

Production at Ølve/Hatlestrand was carried out at both *underground quarries* and *open quarries* by means of surface exploitation. At a few quarries a combination of extraction from underground quarries and from the surface is found (Fig. 4.6).

Underground exploitation has left overhangs as well as large caves, from only a few metres to around 30m deep. The entrances to the caves are at the sides, not from the tops of the quarries, as can also be seen at many underground quarries in Germany. A few of the largest underground quarries have several caves, some with entrances from the surface, while others can only be accessed through other caves. In several caves huge stone blocks have collapsed, today lying both inside and in front of the quarries, in some cases blocking their access.

As at Hyllestad, the Ølve/Hatlestrand quarries were carved directly from the bedrock by cutting the outlines of the products into the rock and then breaking them loose. There are still obvious traces of hewing: furrows, circles, depression marks, hewed channels, etc. The cleavage plane was followed during production and horizontal slices of the talc-bearing green schist exploited. All identified Ølve/Hatlestrand quarries may be characterized as deep quarries, where bakestones and other products (building stones and tiles) were extracted, leaving high walls. These carved walls are often curved, and of different form and size depending on the products manufactured. The curves are close to each other, with only a sharp edge in between.

Regardless of what was produced, the products in most cases seem to have been extracted from the top downwards. Extraction from the roof in underground quarries was also the norm, identified in all quarries by the intact roof. Common traces in the roofs are rectangular or oval marks, some with straight, others with rounded corners. Visible traces of carved channels seem to result from loosening the products from the rock. Carving of channels seems to have been more common in the production of building stones than bakestones and tiles, thus giving some indications of product types extracted.

In most cases, bakestones seem to have been carved one by one from the bedrock, but in some quarries traces of extraction of larger blocks are found. Some of these blocks may possibly have been split into bakestones and tiles. Identification of products extracted at the different quarries was, however, difficult and based only on production marks in the bedrocks and on surface recovery. In most cases hewing traces do not give clear indication of what products were produced at the different quarries. Later extractions can also obliterate traces of earlier production, especially the removal of bakestones. Any classification of the quarries therefore based on product types extracted is problematic.

4.2 Methodological approaches on the micro level: the archaeological investigations

Because of the extent of the two quarry landscapes, comprising 367 and 71 quarries, only selected quarries could be studied archaeologically. Analyses of the spatial extent of quarries within the entire production landscape have been important for this selection. At Ølve/Hatlestrand the identification and mapping of quarries were carried out in collaboration with the geologist Øystein J. Jansen from the University Museum of Bergen. Together with the geological mapping and classification of the Hyllestad quarries by Heldal and Bloxam (Heldal and Bloxam 2008) these surveys formed the basis for choice of research areas for archaeological investigations.

It was important to investigate and date a representative sample of quarries within the two production landscapes, as well as to analyse the spatial extent of the quarry landscapes to assess each quarry archaeologically. For the purposes of this research it was relevant to investigate those quarries with varying distances from harbours and transportation routes, and according to central and more marginal production areas, with regard to size, location and stone quality, as well as quarries related to different farms and landowners. Different types of quarries have thus been identified within both production landscapes, and have also been included with regard to technological differences as well as range of objects produced.

Heldal's geological survey made possible a fuller and more systematic classification of different quarry types at Hyllestad, with regard to technology and mineralogical composition, prior to the archaeological investigations. This made it easier to select different types of quarries for the archaeological survey. The two main quarry types at Hyllestad, shallow and deep, were investigated in an attempt to arrive at relative and absolute chronologies with regard to these two types, and to assess the product types extracted from the different quarries. At Ølve/Hatlestrand quarries of varying size and with different carving traces were selected for investigation so as to shed light on the different products, and to date the extraction.

In the underground quarries of Ølve/Hatlestrand large amounts of production waste were located both inside and outside the quarries. Due to restrictions from the Directorate for Cultural Heritage with regard to the number of trenches and test pits, only external spoil heaps were chosen for investigation. While these are considered to represent several periods of extraction, the internal debris most likely represents the remains from the final and finishing phases. An important aim for the archaeological investigations was to shed light on both the beginning and end of the extraction process, and the spoil heaps in front of the quarries were therefore considered best suited for excavation. Additionally, a workplace within the quarry landscape at Ølve and a rock shelter at Hyllestad that may have been used in connection with quarrying were investigated. Based on these selection criteria, the overall aim has been to elucidate activities within different areas of the two quarry landscapes in time and space.

The archaeological investigations relating to the project were carried out during three field seasons (summer and autumn of 2006, 2008 and 2009), and added to my previous investigations of Hyllestad in 2001, when four quarry sites producing quernstones and millstones were investigated (Baug 2002).

As the quarries at Ølve have only been slightly investigated before, neither the beginning nor the end of activities there is known. Identification and dating of the production and product types was therefore necessary for studying the quarries in wider temporal and societal contexts. Nine trial trenches were dug into the spoil heaps in selected quarries. Where possible, the trenches were placed so that one end exposed the carved rock and the other went into the nearest spoil heap, so as to identify the relation between the carved rock and the heap. At some quarries the siting of the investigated spoil heaps prevented this and other orientations were chosen. All trenches were 1.5m wide, 2.6 - 5m long, and from about 20cm to approximately 2m deep. The sections of the trenches have been drawn to a scale of 1:20 or 1:10. At the Ølve workplace, as well as below the

s = Stone
qs = Quernstone
ms = Millstone
B = Bedrock
⊕ = stone cross
① = Number of layers
● = Cample for C14-datings and objects found
┬─┬─┬ = Carved edge. The short lines indicate the lower level
─■─■─ = Natural crack in the rock

FIGURE 4.7: EXPLANATION OF SYMBOLS USED IN THE DRAWINGS IN CHAPTERS 5 AND 6

Rønset rock shelter, test pits were made and drawn to a scale of 1:10. All deposits contain large amounts of stones, but only the largest and most striking were drawn (Fig. 4.7). In eight of the trenches/test pits, samples for C14-dating, mostly charcoal, were taken. In four trenches no datable material was found. Stratigraphical analyses of the different layers in the trenches and test pits combined with C14-dating indicate the date of production and products, in a few cases both the beginning and end of the activity. The methodology has proven to be well suited for investigating and dating quarries (Baug 2002), however it is difficult to date the oldest productions, as early quarries are often covered with waste from more recent extractions. The earliest dating only provides therefore an *ante quem* date for the oldest products.

4.2.1 Dating and interpreting material in the trenches and test pits

Radiocarbon dating of organic material, mainly charcoal, combined with stratigraphical analyses, is the main method for dating the quarries. Conventional dates have been analysed and calibrated at the Laboratory for Radiologic Dating, Norwegian University of Science and Technology (NTNU) Trondheim (T-). Samples containing only small amounts of charcoal for accelerator mass spectrometry (AMS) radiocarbon analysis were calibrated by the University of Uppsala, Sweden (TUa-).

Charcoal is found either as small and scattered pieces in the trenches and test pits or as bigger collections of both small and larger pieces of charcoal and ashes, the latter most likely representing traces of fireplaces. In a few cases the fragments are found so close together, that it is likely they originally came from the same piece of charcoal. The amounts and scattering of the charcoal indicate that it derives from activities related to the quarries investigated, and was most likely deposited at the same time as the debris in which it is found. Larger amounts of charcoal and ash most likely represent deliberate actions of workers in the quarries, i.e. the dumping of deposits from nearby fireplaces, while smaller and more scattered pieces of charcoal may have entered the spoil accidentally. Charcoals may also have been airborne from fires in or around the quarry sites.

A necessary and frequently repeated activity related to quarrying must have been the sharpening of tools - a daily activity. The presence of smithies near, or on the quarry sites is therefore also likely. In one of the excavated trenches a piece of slag was found, probably from a nearby smithy. Slag was also found at the investigated Ølve workplace, as well as below the Rønset rock shelter (see Chapters 5 and 6), and substantiating such a hypothesis. It is also possible that fireplaces near the quarry sites were used for preparing food.

The contextual conditions of charcoal in the trenches are not always ideal, and the relation between date and context has therefore received much attention. The stratigraphical layers have in some cases been disturbed and destroyed the relationship of the samples to the context. Large-scale and long-lasting production makes it likely that spoil heaps were moved to make space for further quarrying, and a spoil heap may thus contain waste from several production phases. Another methodological problem was the occasional lack of compact deposits in many of the spoil heaps. The large number of stones can make depressions, which, together with flowing water, cause spoil heaps to collapse, and further erosion may lead to stratigraphic disturbances in some of the heaps. Smaller samples could easily be moved from their original contexts by such physical processes (Pendergast 2000:237). The majority of the C14-dating samples from the quarries were small fragmentary charcoals for AMS radiocarbon analysis. Larger collections of charcoal and ash, on the other hand, are more likely to represent primary deposition. Consequently, it is important to compare the different samples to evaluate their relation to the relative chronology identified in the layers.

An extensive accumulation of charcoal containing several small fragments may contain charcoal with a mix of younger and older material, perhaps also representing different types of activity (Pendergast 2000:237). Only in a few cases did the charcoal samples collected from the spoil heaps constitute only one piece of charcoal; the normal situation was samples of several small fragments. The dating of such amalgamations of charcoal would involve dating a representative part of the charcoal from an area, and this may be more helpful than dating one single piece of charcoal, which, theoretically, may come from a different period. The same arguments have also been used with regard to C14-datings from agricultural contexts (Zehetner 2007:62). All the samples were taken from deposits interpreted as production waste. The risk that the samples from the spoil heaps reflect different activities is not great. There are few traces of human activity after production ended at the quarry sites. Although many of the areas were used as pastures and grazing land in recent times, activities that would leave behind charcoal traces are not documented. Altogether, the possibility that the charcoal derives from other activities than those related to quarrying is regarded as rather small.

In a few cases the lack of charcoal made it necessary to date other organic material, such as hazelnuts, although there is a greater uncertainty relating such material to quarrying. In all the areas where hazelnuts were found in the trenches, hazels still grow today, and small residues may have been transported into the spoil heaps by natural processes and are thus unsuitable for dating quarry activities. This is further discussed in the analyses of the quarries in Chapters 5 and 6.

In some of the trenches it was difficult to delimit and separate the different deposits. The content of the layers was normally more or less the same: large amounts of stones, sand and gravel. Some deposits may also have similar colour nuances, making it harder to recognize layers. In some cases one identified layer may therefore contain deposits from several different production phases, even though this is not observable in the section of the trench. One C14-dating may therefore not necessarily date the whole layer in which it was found. To try and solve such problems several C14 samples were taken from the same layer at different levels. Several samples for C14-dating, both from the search trenches and test pits, but also from each layer in a trench/pit, should make it possible to reveal a chronology and to consider whether or not the absolute datings given by the samples fit into a relative chronology, substantiating the C14-dating. For some of the samples there is a larger degree of uncertainty in terms of the relation to quarrying, and these are discussed in more detail in the following analysis (Chapters 5 and 6).

Another problem related to charcoal as dating material is the possibility that the charcoal derives from wood that has been dead for quite a long period before being burnt (Gulliksen 1979:70); such dating may be misleading related to the production phases. This may be particularly relevant for charcoal from pine, and in a few cases in earlier investigation at Hyllestad quarries, pine charcoal gave an unlikely high reading (Baug 2002:41-42, 57, 59). Identification of the wood species is therefore also important for assessing such problems; comparison with other dating results from the same trench may also help to resolve chronological discrepancies. Another potential problem when considering C14-analyses is the pollution of the samples by organic material of a younger date (Gulliksen 1979:70). This is usually recognized at the laboratory and removed there.

When collecting samples for C14-dating, charcoal fragments within a limited area were therefore preferred as much as possible. The limited and random presence of datable material in the trenches made it occasionally difficult to collect samples directly from the sections after the trenches had been excavated. Charcoal and other datable material normally had to be collected during excavations, and then the section was marked to show where the samples were removed, horizontally and vertically. Charcoal and other material that had dropped down from the profile were not collected. In cases where an amalgamation of fragments comprises one sample, fragments found close together were collected, preferably within a horizontal radius not greater than 10cm, normally around 5cm, and with a maximum of 1-3cm vertical difference.

The age given by radiological dating is not absolute, but has a standard deviation of 68% (1σ). This represents a 68% chance that the correct age is to be found within the given intervals. Consequently, in three out of ten samples the correct age is to be found outside the given interval (Gulliksen 1979:70). Occasionally, it may therefore be necessary to expand the standard deviation and use the 95.4% (2σ) or 99.7% (3σ) confidence for calibrated dates (Bowman 1990:49). In this present study the expansion of the standard deviation is based on the online OxCal program[1].

As long as the potential sources for errors are identified and considered, radiocarbon dating can also be appropriate for quarry dating. However, to support C14-analyses from quarries it is important to compare the results with indirect dating methods of the different products from the quarries from other contexts. The distributed stones are dated according to the cultural layers in which they are found, giving a *terminus ante quem* dating for the objects, meaning that the oldest stones found, and the quarries, are older or at least simultaneous with the cultural deposits within which they are located. Exact production dates cannot be retrieved in this way, but the results are important in order to assess excavation methods and the relevance of datable material from the spoil heaps, as well as dating the distribution and use of the objects in other contexts.

4.3 Methodological approaches at the meso level: retrospective analyses

Studies of the local communities associated with the quarries, including archaeological as well as documentary evidence, were undertaken to gain insights into the landholding and property structures at the time of quarrying. The objective is to shed light on aspects such as disposal and control over mineral resources.

Written evidence from the Middle Ages and early modern period, mainly cadastres and diplomas, were analysed by using a retrospective approach, but related to archaeological structures in the areas, especially grave mounds. To assess changes and stability in the property structures within the two regions studied, the spatial distribution of prehistoric burial mounds was studied and analysed in relation to known property structures from later periods. This can help indicate the situation in the Middle Ages, and in relation to the later Iron Age, and traces signs of change or stability over time.

[1] *Oxford Radiocarbon Accelerator Unit*, University of Oxford (https://c14.arch.ox.ac.uk/oxcal/OxCal.html

To approach these questions, medieval property structures and estates, as well as documentary sources from the Middle Ages and early modern period, are analysed to shed light on prehistoric and medieval estates and property structures. Research indicates that the absence of burial mounds within a medieval estate, coupled with a higher frequency of freeholders, may indicate that the estate dates back to Viking times, if not earlier. The nucleus of an old estate was often surrounded by farms without burial mounds, indicating a subordinate status where the surrounding farms were held by dependant farmers, while farms with burial mounds can be related to the presence of freeholders (Iversen 1999). The use of prehistoric burial mounds as symbols and indicators of property rights thus forms a basis for comparison with later conditions. This method, however, still requires a critical evaluation of sources. The number of prehistoric burial markers most likely decreased over the centuries, meaning that a reconstruction of prehistoric freeholds based on the graves would only represent a minimum figure (Iversen 2008:89). This method has earlier proven successful for analyses concerning estates in the later Iron Age and the Middle Ages in western Norway. The approach may reveal tendencies in the areas studied, but it should not be used on a like-for-like level (Iversen 1999; 2008). In my study, the oldest structures, with or without burial mounds, are emphasized, including tendencies in the recorded evidence from the Middle Ages and early modern period.

The concepts of property and ownership rights, and the disposal of land and resources, do not however represent stable features, but ones that have changed over time. By combining different sources, such as medieval property structures, documentary sources from the early modern period and archaeological evidence, both stability and changes in the two areas in the period can be studied. A more detailed discussion of this method and the use and combination of different sources is presented in the analysis of the property structures in Chapter 7.

The mapping of the known and preserved burial mounds at Hyllestad and Ølve/Hatlestrand has been done by using previous archaeological registrations and investigations in the two areas (Fett 1954; 1956; Dommasnes 1976; Fasteland 1983; Hatleskog 1997; Iversen 1999). Additionally, information in the topographical archive at Bergen University Museum, as well as the Directorate for Cultural Heritage's index of cultural monuments, Askeladden, is included.

4.4 Methodological approaches on the macro level: provenance studies and distribution

An investigation of the spatial distribution of different products on a local, regional and supra regional level sheds light on networks and commercial contacts related to the quarries and gives indications of the degree of contacts and trade with different areas and the scale and importance of activity in the quarry landscapes. Activities in both research areas can also be dated through the distributed products found in different contexts outside the quarry areas.

The special character of the rock types exploited means that it is possible to trace the distributed products back to both production areas at Hyllestad and Ølve/Hatlestrand. However, because of the similarities and differences with regard to the geology within each quarry landscape, it is more difficult to trace an object back to the exact quarry where it was made. The provenance study is therefore limited and takes in the entire production landscapes, without being able to give closer identification of potential quarries as sources for different objects.

Spatial analyses of the distribution patterns for products from the two quarry areas are only possible through an understanding of the geological composition within the quarries and collaboration with geologists. Cooperation with the Millstone project made it possible to investigate the distribution of stones in wider contexts, which is not possible for bakestones. The rock type at the Hyllestad quarries is relatively easily identified, and it has in most cases been possible for the geologists to conduct provenance studies without more thorough geological analyses. When in doubt, geochemical analyses were undertaken on a few stones.

Detailed information with regard to the distribution of quernstones and millstones, such as exact numbers and percentages of finds in relation to geology and provenance at the different sites, will be published within the Millstone project.[2] In this study therefore I intend only to give a brief overview of these preliminary results, within the context of the trade in Hyllestad stones. An assemblage where the number of Hyllestad stones exceeds 40 objects is regarded as a rough indication of a large and organized trade, while numbers below 40 may indicate a more random distribution, but this has to be discussed contextually. It is important to study the amount of Hyllestad stones in relation to the total amount of stones from the different contexts, as well as the timeline of each site. A domination of Hyllestad stones at a site may indicate a large-scale and routinised distribution even though the number of finds may be somewhat low. This will be further clarified and discussed in Chapters 7 and 10.

Outside the quarries, the dating of cultural deposits and stones found is based on C14-dating, dendrochronology, or connected to archaeologically documented finds, such as runic inscriptions, pottery and coins. Therefore information with regard to context and date of the distributed quernstones and millstones varies, and for a few of the investigated sites no such information is found. Nevertheless they are included in the analysis as they give important information of contact areas.

[2] Hauken and Andersen in press, Baug and Jansen in prep. (a-d); Baug *et al.* in prep. (a-b); Grenne *et al.* in prep. (a-b); Meyer *et al.* in prep.; Jansen *et al.* in prep.

The bakestones found in Norway comprise an extensive archaeological record, but with regard to this distribution I mostly rely on the work of others (cf. Weber 1984; 1989; 1990; 1999; Forster 2004; Tengesdal 2010).

5 Archaeological Investigations of Quarries at Hyllestad

To gain a better understanding of the quarries in terms of chronology, dimension, production techniques and product types, archaeological investigations were carried out at selected sites within the Hyllestad production landscape. The aim was to expand the research area and quarry types compared to my previous investigations, to cover quarries with different extraction techniques, different product types and different locations within the production landscape as a whole. Additionally, a rock shelter was investigated that might have been used in connection with quarrying. In this chapter the analyses and results from the archaeological investigations are presented and discussed as an empirical basis for further analysis.

5.1 Aims and methods

The main goal of the archaeological investigations was to assess technological issues and date the extractions and products at different quarries within this extensive production area, if possible to clarify both the beginning and end of production. To obtain the best representation through limited archaeological surveys, the investigation areas were selected on the advice of geologists, but also based on a spatial model with emphasis on the location of the individual quarries within the entire production area. As pointed out in the preceding chapter, several different quarry types are known in Hyllestad (Heldal and Bloxam 2008), varying considerably in size, from c. 2 to 40m in length. An issue to be considered is if the quarry types represent different periods in the production, or if different product types extracted were decisive for the types of quarries. To approach this question, different quarries have been incorporated with regard to technology and product types as well as different geological conditions. The sites are located on the outskirts of various historical farms within the quarry area, and with different distances to transport routes and harbours. Because of restricted excavation permits from the Directorate for Cultural Heritage, only one trench per quarry was excavated, with the exception of a site at Myklebust, where two trenches were opened because of the size of the quarry. The aim was to establish a chronology between different spoil heaps, and to shed light on working processes and different production phases.

Altogether five quarries and one rock shelter were selected, located at four different farms: two quarries and a rock shelter at Rønset (no. 71) and one quarry at the farms of Myklebust (no. 79), Sørbø (nr. 32) and Sæsol (nr. 78), and six trenches and two test pits were opened for archaeological analyses (Fig. 5.1). Sørbø, then, is the only new farm compared to my investigations in 1998-

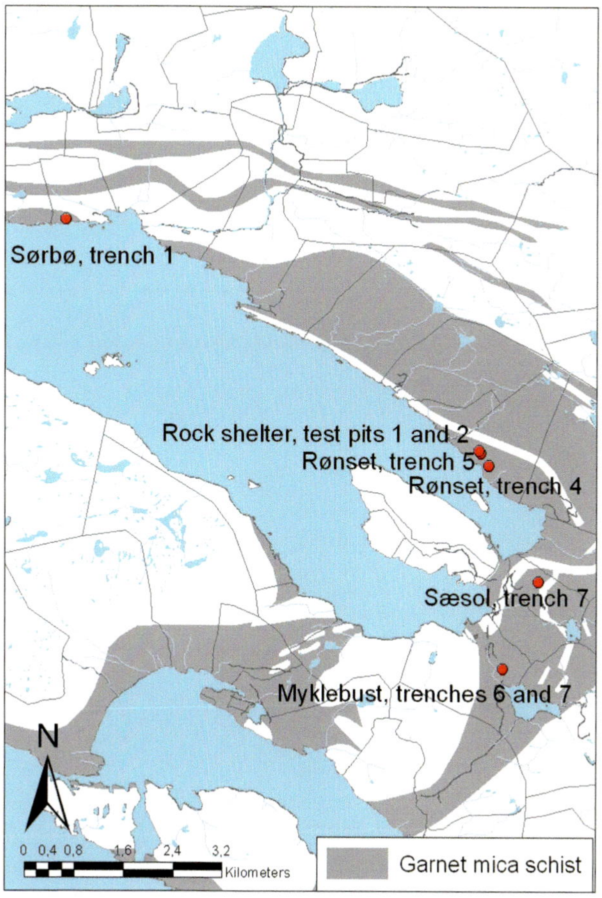

FIGURE 5.1: MAP SHOWING THE LOCATION OF THE EXCAVATIONS (MAP: IRENE BAUG, 2013, NORGE DIGITALT. FKB-DATA, 1413 HYLLESTAD KOMMUNE, 32_1413EIENDOM_FLATE, 28.02.2009)

1999 (Baug 2002). In order to make the comparison to my earlier investigations easier, the numbers given to the trenches connect to my previous investigation, and do not refer to the amount of trenches excavated at each farm.

The largest concentration of quarries is located at the Rønset (no. 71) and Myklebust (no. 79) farms, representing the main areas for production at Hyllestad. Whereas shallow quarries are found within the whole production area of Hyllestad, the sites at Rønset and Myklebust, with one exception, are the only farms where deep quarries and a combination of quarry types have been located so far (Heldal and Bloxam 2008:51-55; Grenne *et al.* 2008:58) (Figures. 5.2 and 5.3). Millstone production was also concentrated at these two farms (Heldal and Bloxam 2008:45), which makes them especially interesting research areas. Both Rønset and Myklebust have remains of large-scale, varied and long-lasting production. The discovery of different quarry types with extraction of products other

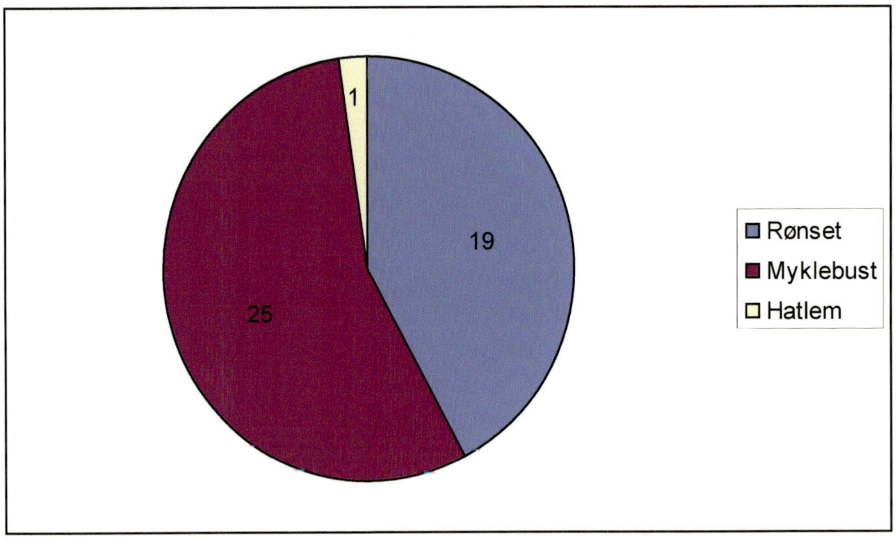

Figure 5.2: Distribution of Hyllestad deep quarries. N=45 (Source: Heldal and Bloxam 2008: fig. 6-8)

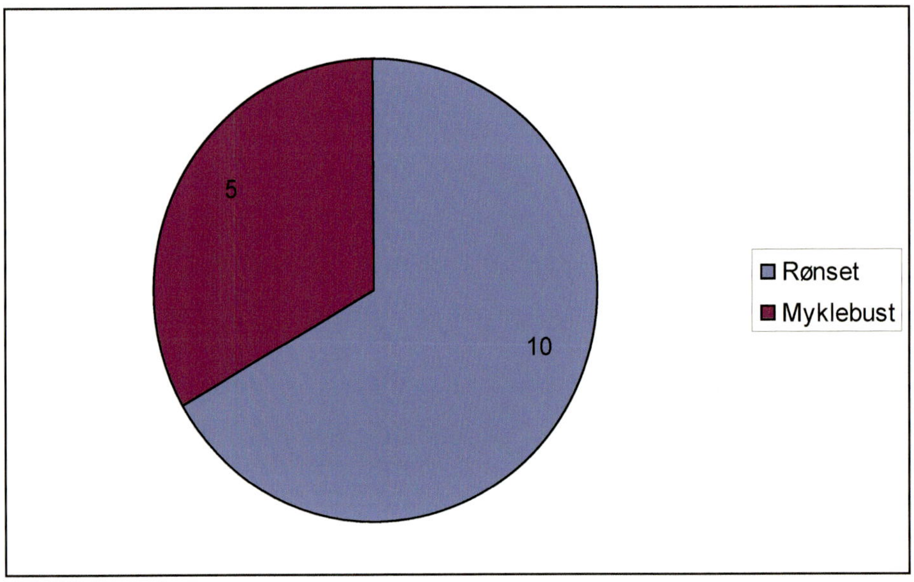

Figure 5.3: Distribution of Hyllestad combination quarries. N=15 (Source: Heldal and Bloxam 2008: 6-11)

than quernstones and millstones also made them interesting for further investigation. Here numerous quarries occur in close proximity - even on top of each other. In some areas rocks have totally disappeared following extensive extraction, and large amounts of production waste cover up most of the original rocks where production took place, and in the areas around the quarries, making it difficult to identify the original terrain.

It was also relevant to investigate different types of quarries, and those quarries where there were greater distances between farm interiors and their possible harbours. The research area was consequently extended to provide new information and uncover possible differences and similarities in the production landscape. In this way more representative information could be gained on production at different quarry types and other sites within the farm areas.

The more marginal production areas are located northwest of Rønset, as well as on the eastern side of the Åfjorden. Here the quarries differ, compared to the core areas, being generally smaller and more scattered, and only shallow quarries have been identified (Fig. 5.4). According to the survey by Heldal and Bloxam, more than 50% of the identified quarries only produced quernstones, signifying that the main product in the marginal areas were stones for hand querns, even though some remains from millstone production have also been found (Heldal and Bloxam 2008:45). An investigation into the more marginal quarries therefore also constitutes an important comparison with the more central areas of Rønset and Myklebust in terms

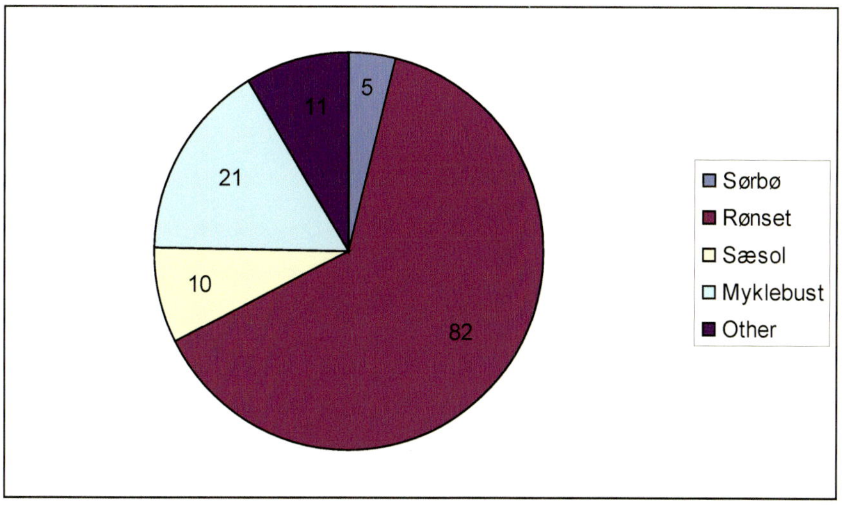

FIGURE 5.4: DISTRIBUTION OF HYLLESTAD SHALLOW QUARRIES. N=129 (SOURCE: HELDAL AND BLOXAM 2008: FIG. 6-4)

of chronology, technology, quarry type, product type and quarry size.

A further object of this study was to investigate quarries at different locations within the marginal production areas, so as to shed light on possible differences in chronology, product types and size of quarry. Consequently, one quarry at Sørbø and one at Sæsol were selected as case studies: Sørbø farm represents an area not previously investigated.

5.2 Archaeological investigations at Rønset (no. 71)

Rønset farm, with its 129 quarries and 36 identified trial extractions, has the densest concentration of quarries, including production of millstones on a large scale (Fig. 5.5). Quarries here that were entirely, or partly utilized for the manufacture of slabs have also been identified. The topography at Rønset is slightly sloping and hilly, with quarries and spoil heaps all over the outlying fields. A few quarries were also located at the farm's historic infield area, but all of these were disturbed and covered by more recent farming activities and buildings.

Two Rønset quarry sites were previously investigated. One is within the border zone between the farm's outfield and infield areas (Rønset trench 1), roughly 200m northwest of the farm buildings at holdings 1 and 3, while the other is located further away on the small Otringsneset peninsula, c. 30m from the fjord (Otringsneset, trenches 1 and 2) (Baug 2002:32). Production at the quarry investigated northwest of the farmhouses has been dated to the end of the Viking period/ beginning of the Middle Ages (990-1160 AD), and manufacture seems to have continued up until the 16th or 17th centuries (Baug 2002:38-44). The working can be characterized as a combination of deep and shallow quarry. At its northern end there is a high, carved wall, and to the east a rock wall with visible circles of extracted quernstones and millstones. The deep quarry is covered with spoil that may derive from extractions in the shallow quarry, which would indicate that the deep quarry is the oldest (Fig. 5.6). It is, however, difficult to determine where the spoil heap comes from. Several production phases and quarries may be represented in the spoil, and this is also supported by the dating from the spoil heap that spans a long period of time. Since the 3m-tall carved wall is completely covered with debris, the possibility that the deep quarry at this site is older than the combined quarry should be considered.

The same phenomenon can be seen at Otringsneset. A deep quarry was worked in an older, shallow quarry. At the upper part of the rock wall, the exploitation follows the cleavage plane, and circular extraction marks are still visible. The lower part of the same rock has been exploited vertically, leaving a carved wall with grooves from hewing. These traces may, however, represent a younger phase of a limited range compared to production in the shallow quarry. Only quernstones were produced, and the activity is dated widely to the Viking period (AD 730-965) and seems to have ended in the 13th century, AD 1180-1285 (Baug 2002:34-51). The long time span supports the assumption of two production phases at the quarry. What characterizes both the earlier investigated quarries is that the spoil heaps cover older quarries, which makes it difficult to date the oldest production. Extraction at the sites must, however, predate the oldest dating from the spoil heaps.

Archaeological investigation of the Rønset infield area was carried out by Silje Foyn (2008), dating the earliest traces of agriculture and probably cultivation to the early Bronze Age, with continuity and intensification of the arable land through the Iron Age and into the Middle Ages. Not all the farm zones were used simultaneously. The investigation indicates a shift to more regular fields in the early Iron Age, most likely in the Roman period, which again indicates a division of the farm's infield and outfield area during this period. Hence the farm seems to

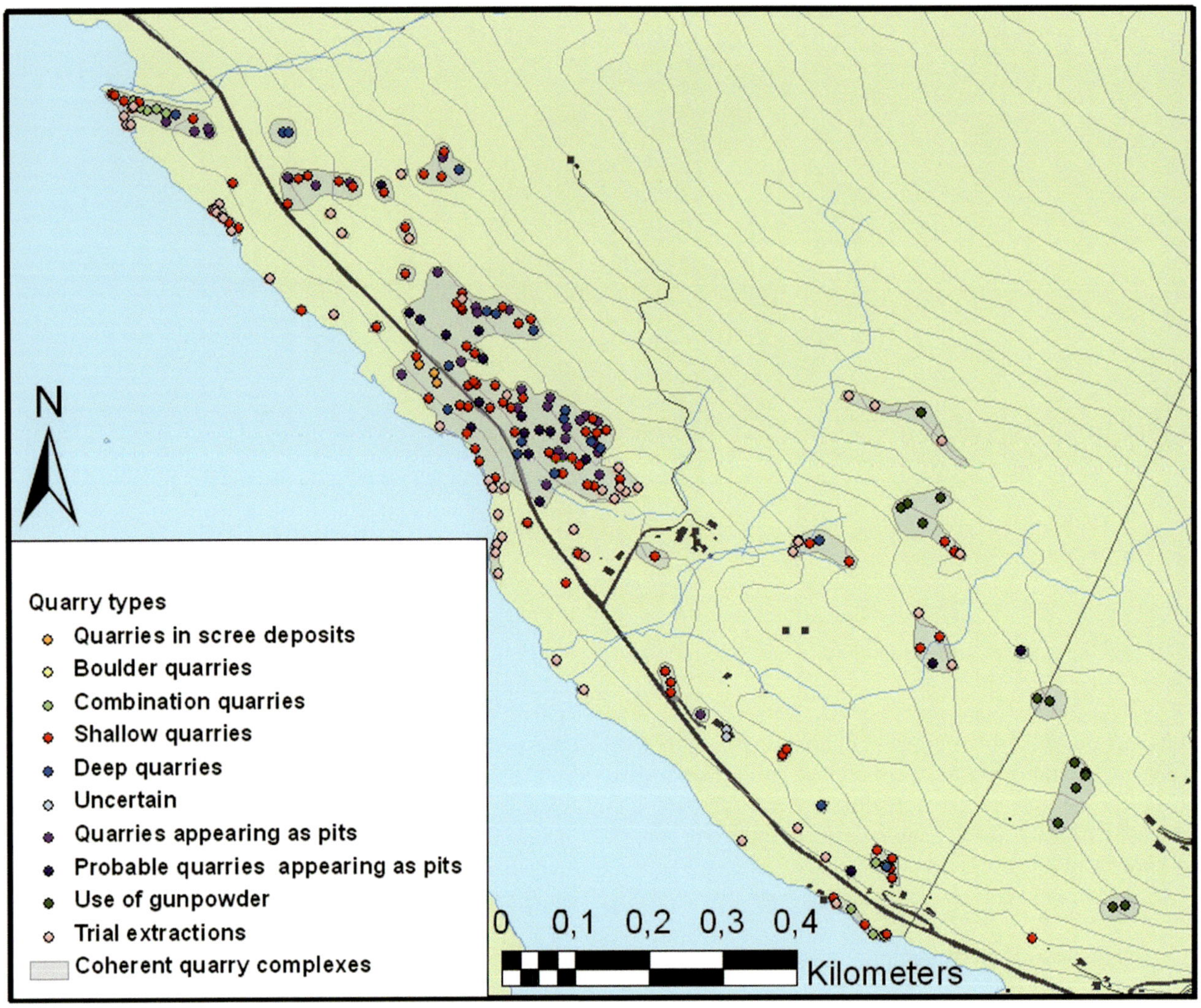

FIGURE 5.5: MAP OF THE DIFFERENT TYPES OF QUARRIES AT RØNSET (MAP: IRENE BAUG, 2013) (MAP DATA FOR THE HYLLESTAD QUARRIES FROM A SURVEY BY TOM HELDAL, NGU)

have been established during Roman times, probably also with a settlement near the known settlement area (Foyn 2008:70-85), but not necessarily a farm in the same sense as the historically known farm.

In this present project, two small-scale excavations were carried out at two different quarries that differ slightly from the two sites previously investigated at Rønset in terms of quarry type, location, and output.

5.2.1 Rønset trench 4: quarry site for slab production

The present author's earlier investigations concentrated on querns and millstone production. Consequently, it was important to also include other types of production for comparison, both chronologically and technologically. A quarry producing slabs therefore was of great potential interest. The investigated quarry is located in the outfield area,[3] about 180m east of the farm houses of holding 3. At present the site is overgrown with birch and hazel dominating the vegetation; the area is used now for grazing cattle in summer. An old forest track is located 15m south of the quarry, but the working site itself seems to have been undisturbed by human activity after production ended.

At the quarry site, just below the moss, a 2.8m-long, 56cm-wide and 6-10cm-thick slab was found (Fig. 5.7), with hewing traces that clearly show the slab had been extracted from the rock.

About 12m west of the investigated quarry another smaller production site was located, with visible traces of extraction of rectangular slabs. Marks on the rock show that a 70-72cm-wide slab had been extracted at some time. The length is somewhat uncertain, since it is partly covered with soil and vegetation, but about 1.4m is visible above the ground (Fig. 5.8). A loose slab was also found at the site, 43cm wide and 1.2m long, most likely representing a fragment of an extracted slab. This clearly indicates that products other than quernstones and millstones were produced in this area.

[3] UTM WGS84, zone 32: east 300938 north 6790092, c. 45m.a.s.l.

Figure 5.6: The carved rock wall at Rønset trench 1 (Photo: Kim Søderstrøm and Jørgen Magnus © Riksantikvaren)

Figure 5.7: Slab from Rønset trench 4 (Photo: Irene Baug, 2006)

Figure 5.8: Quarry with slab extraction (Photo: Irene Baug, 2006)

The investigated site is a deep quarry with a c. 3m-tall carved wall to the east. The south end of the rock continues into a more pronounced oval spoil heap, about 1m high, oriented in a northeast-southwest direction. The area where the trench was dug is flat and c. 10 x 8 m. The west end of this area borders a fissured rock wall without production marks, most likely representing a natural fissure in the rock. Between this rock and the spoil heap there is a 2.5m-wide opening, probably where the products (and possibly also spoil) were transported out of the quarry. North of the flat area the terrain slants upwards about 23m, and the whole area is covered with spoil. In some places carved rock with production marks is visible through the spoil.

Rønset trench 4

As the spoil heap at the southern end of the quarry most likely derives from production in the rock wall at the eastern end of the site, it was decided to make a trench in this heap to study the relation between the carved rock wall, the heap itself, and the possibly older worked-out area of the quarry. The trench was located about 1.8m from, and parallel to the carved rock wall, crossing what was interpreted as a previously worked part of the quarry, and continues into the spoil heap to the south.

Trench 4 was 1.5m wide, 3.5m long and from 20cm to 1.2m deep (Fig. 5.9). Both at the northern and southern ends of the trench, carved bedrock was reached at the base; at the north end, slanting bedrock with production marks was uncovered. The western end of this rock is naturally fissured, but also shows traces of quarrying, probably representing an attempt to take advantage of the natural fault during production. At the southern end of the trench carved rock was visible in the profiles to the east and south. Characteristic grooves from production were visible on the rock, but leave no indications of extracted products. It is, however, obvious that the spoil covers an older part of the quarry that continues underneath the spoil heap. Consequently, the excavation only covered the youngest phase of production, whereas spoil from the oldest production most likely was dumped in the sloping area south of the quarry. Water in the trench made it difficult to continue excavating, and the work had to be stopped before the base level was reached in most parts of the trench.

Profile interpretation: Rønset trench 4

Rønset trench 4 consists of three layers including topsoil (layer 1) (Fig. 5.10), and the content of the trench is interpreted as spoil, most likely from production of the carved rock wall to the east.

Layer 2 was a red-brown layer consisting of humus and sand, as well as small roots. The most dominant elements in the layer were flakes, 20-30cm in length. Several larger slabs were also represented, measuring 1m in length and about 10-20cm thick. Remains from carving on many of the slabs show that straight-sided products were made

FIGURE 5.9: RØNSET TRENCH 4 DURING EXCAVATION (PHOTO: IRENE BAUG, 2006)

there. The slabs tend to be more or less quadrangular, or rectangular in shape. Two samples for C14-dating were taken from the layer. The uppermost sample, TUa-6709, was a hazelnut located about 10cm below the top soil, and was dated to after AD 1960, and so cannot be related to the production in the quarry. As mentioned in Chapter 4, the contextual relations in the spoil heaps are not always ideal for C14-sampling. Large amounts of stones, slabs and possible hollows in the spoil may have made it possible for smaller material, e.g. a hazelnut, to enter the spoil. This dating should therefore not be taken into further consideration. The second sample of charcoal from layer 2, TUa-6710, is dated to 700 ± 35, calibrated to AD 1285-1300, indicating a production in the high Middle Ages. Layer 3 had a grey colour with touch of yellow, consisting of sand and small flakes with a size of 3-4cm, most likely the result of crumbling from the bedrock uncovered at the south end of the trench. One charcoal sample for C14-dating was taken from layer 3, TUa-6711, dated to 810 ± 35, calibrated to AD 1220-1280, which also indicates an activity in the high Middle Ages (Tab. 5.1; Fig. 5.11). Even though both TUa-6710 and TUa-6711 are from pine charcoal, it is likely that the dating is indicative of quarry production, as they coincide chronologically (Tab. 5.1,

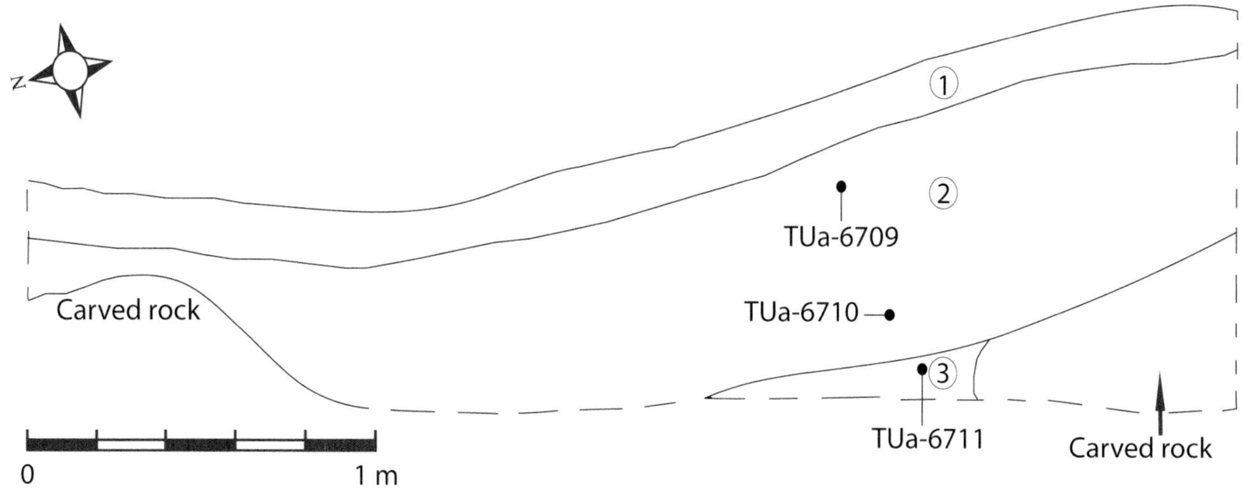

1. Topsoil.
2. Red-brown layer of humus, sand and flakes. Some small roots. Flakes 20-20cm long and 0.5-3cm thick dominate the deposit. Larger slabs c. 1m long and 10-12cm thick are also present. Some sections of the layer contained only stones (with no humus or sand), often with small gaps in between. Many of the stones (flakes and slabs) have disintegrated.
3. A grey sand layer with traces of yellow – most likely from weathering of the carved rock. Some smaller flakes (c. 3-4cm) are also present.

Figure 5.10: Section of Rønset trench 4 (Drawing: Irene Baug, 2006)

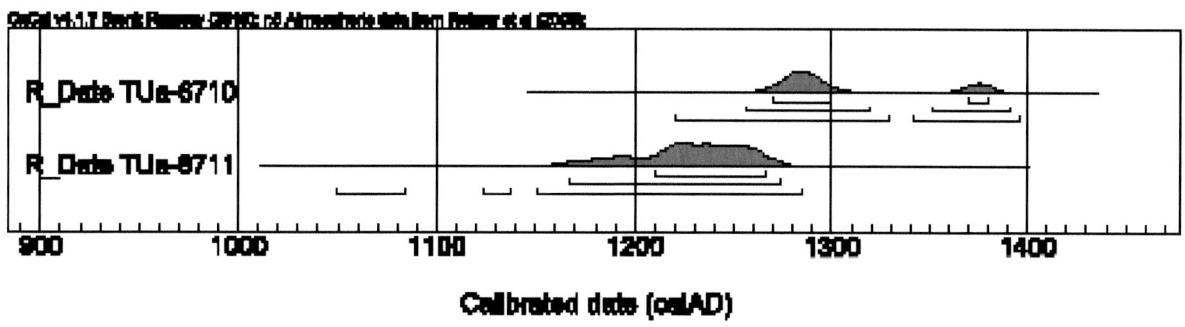

Figure 5.11: Multiple plot diagram of C14-analyses from Rønset trench 4

Fig. 5.11). Dating production to the high Middle Ages also coincides well with the earlier investigated quarries at the farm (Baug 2002:59).

Table 5.1: C14-analyses from Rønset trench 4

Lab. Ref.	Context	Age BP	Calibrated Age	Material
TUa-6709	Rønset trench 4, layer 2	73.8 ± 0.6% activity	After AD 1960	hazelnut
TUa-6710	Rønset trench 4, layer 2	700 ± 35	AD 1285-1300	pine charcoal
TUa-6711	Rønset trench 4, layer 3	810 ± 35	AD 1220-1280	pine charcoal

As no fragments of querns or millstones were found in the quarry, and together with the shape of flakes and slabs found in the trench, it seems that other product types were manufactured here; finds in the spoil, both from the trench and from surface recovery, indicate slab extraction. Whether they were meant for raw blocks for other artefact types, or represent finished products, still remain open questions. There was no clear indication of destroyed products in the trench and it is therefore possible that slabs may have been transported to other areas for finishing.

Finds of exported objects from Hyllestad indicate that both grave slabs and stone crosses are possible options. A destroyed cross has been found at the farm, c. 100m south of the quarry, located in a talus below the road and not far from Aurgota, a possible harbour used for onward shipment (Baug 2002:67-68). The find may therefore be an indication of the extraction of stone crosses at Rønset,

FIGURE 5.12: PATH TOWARDS RØNSET TRENCH 5

FIGURE 5.13: THE 6M-HIGH CARVED ROCK WALL NORTH OF RØNSET TRENCH 5 (PHOTO: IRENE BAUG, 2008)

an activity that may have been carried out at or near the investigated quarry.

5.2.2 Rønset trench 5: quarry site for quernstone and millstone production

A large quarry located in the outlying field, c. 390m northwest of the houses of holding 3, was chosen for archaeological research.[4] This is a deep quarry, and as mentioned above, most of the deep quarries seem to have had produced millstones, often in combination with quernstones. Earlier investigations, both at Rønset and Myklebust, provided rather early dates for the extraction of millstones, indicating that production had already begun in the late Viking period and in the transition to the Middle Ages (AD 990-1160 and AD 890-1050) (Baug 2002:34-38, 51-53, 60) - an unexpected early date.

The quarry is located in an overgrown area of alder and rowan, and where ramsons cover the ground and production waste. The area is rather steep and slopes towards the sea. Because of the steep terrain, large amounts of production waste from the investigated quarry must have ended up further down the hill, covering older quarries. Some 57m southeast of the quarry the ruin of a small house, earlier used as a barn for calves (A. Rønneseth pers. comm. 2008) is located. From the shed an old path is visible, going westwards towards the investigated quarry (Fig. 5.12). The area has been used as grazing land for cattle otherwise the quarry site seems undisturbed by human activities since production ended.

The quarry is in a 42m-long run of four sites (I-IV), all deep quarries with high, carved walls, and several spoil heaps. Quarry I is located furthest west. It is 8m wide and consists of a carved rock wall to the north. At the south-eastern end of the quarry is a 7m-long heap – most likely a rock covered with production waste. Quarry II is 6.5m wide, and the northeast wall defines the southwest wall of quarry III. Southeast of quarry III is a 17.5m-wide area with spoil heaps and marking the site of quarry IV.

The investigated quarry (III) has a 6m-high carved rock to the north (Fig. 5.13). The top of the rock is nearly flat, and circular extraction marks are visible over an area 3.3m wide. The quarry is about 6.4m wide with uncarved rock to the west and east, most likely due to the natural cracks. A large spoil heap is located 7.3m in front of the rock to the south. The area between the heap and the rock is flat,

[4] UTM WGS84, zone 32, east 300810 and north 6790280, 67m.a.s.l.

FIGURE 5.14: RØNSET TRENCH 5 DURING EXCAVATION; THE STONE WALL IS VISIBLE AT THE SOUTHERN END (PHOTO: IRENE BAUG, 2008)

and most likely represents a previously-worked area of the quarry.

Rønset trench 5

A 3.5m-long and 1.5m-wide trench was excavated at the northern end of the spoil heap, 5.2m south of the carved rock wall (Fig. 5.14). The trench covers parts of the flat area in front of this rock, and continues into the spoil heap to the south. It was opened to establish the chronology between the heap and the previous extracted area, and in this way could shed light on the work in the quarry and possible production phases. The trench was between 60cm and 2.3m deep, but because of the danger of collapsing walls the excavation had to cease before the base of the spoil heap was reached.

At the southern end of the trench a stone wall, also with traces of quernstones, was discovered (Fig. 5.15). The wall is located on the northern side of the spoil heap, probably a supporting wall to prevent spoil from falling into the production area, indicating a planned handling of the quarry spoil. The wall slants and leans towards the heap, suggesting that it was built when production was still in progress. As the spoil heap grew an even taller wall was needed to keep production waste away from the extraction area. The wall was covered with a layer (30-40cm) of soil and production waste, either from the end phase of extraction or caused by a later collapse of the spoil heap. The wall measured more than 2.1m in height when the excavation ended, but its foundations were not reached.

Profile interpretation: Rønset trench 5

Trench 5 consisted of seven layers, including topsoil (layer 1) (Fig. 5.16). Layer 2 was a dark-brown layer of humus and flakes, 2-30cm in size. There were also flakes (20-30cm) and slabs (50-60cm). The layer contained many small roots, and it is possible that the layer consists of deposits that might have come through the topsoil. Layer 3 is a brown layer consisting of humus and stone flakes, some small (2-3cm) and some larger (20-30cm). Some small roots were also found within this layer as well as larger stones and slabs, some up to 50cm in size. Two millstones were identified. One millstone had a diameter of over 1m; the other was located in the profile but not excavated, so the total size is unrecorded. The layer is interpreted as having collapsed from the spoil heap after quarry production had ceased, or spoil from another quarry, possibly quarry IV, after the production at quarry III had stopped. The dating of the layer thus gives a *terminus post quem* for production in quarry III. Two samples were taken from layer 3. The upper sample, TUa-7928, was taken from c. 40cm below the topsoil, and dated to 1030 ± 35, calibrated to AD 990-1025. The second sample, TUa-7931, was taken c. 65cm below sample TUa-7928, dated to 870 ± 35, calibrated to AD 1165-1225, and indicating a reverse stratigraphy. This supports the theory of collapse from the spoil heap, or perhaps the movement of nearby spoil related to later extraction. Sample TUa-7928 is dated to the late Viking period and consisted of pine charcoal. This tree species needs care in connection with C14-analyses because of the possibility of high age before being burning (Gulliksen 1979:70). As sample TUa-7928 indicates a somewhat

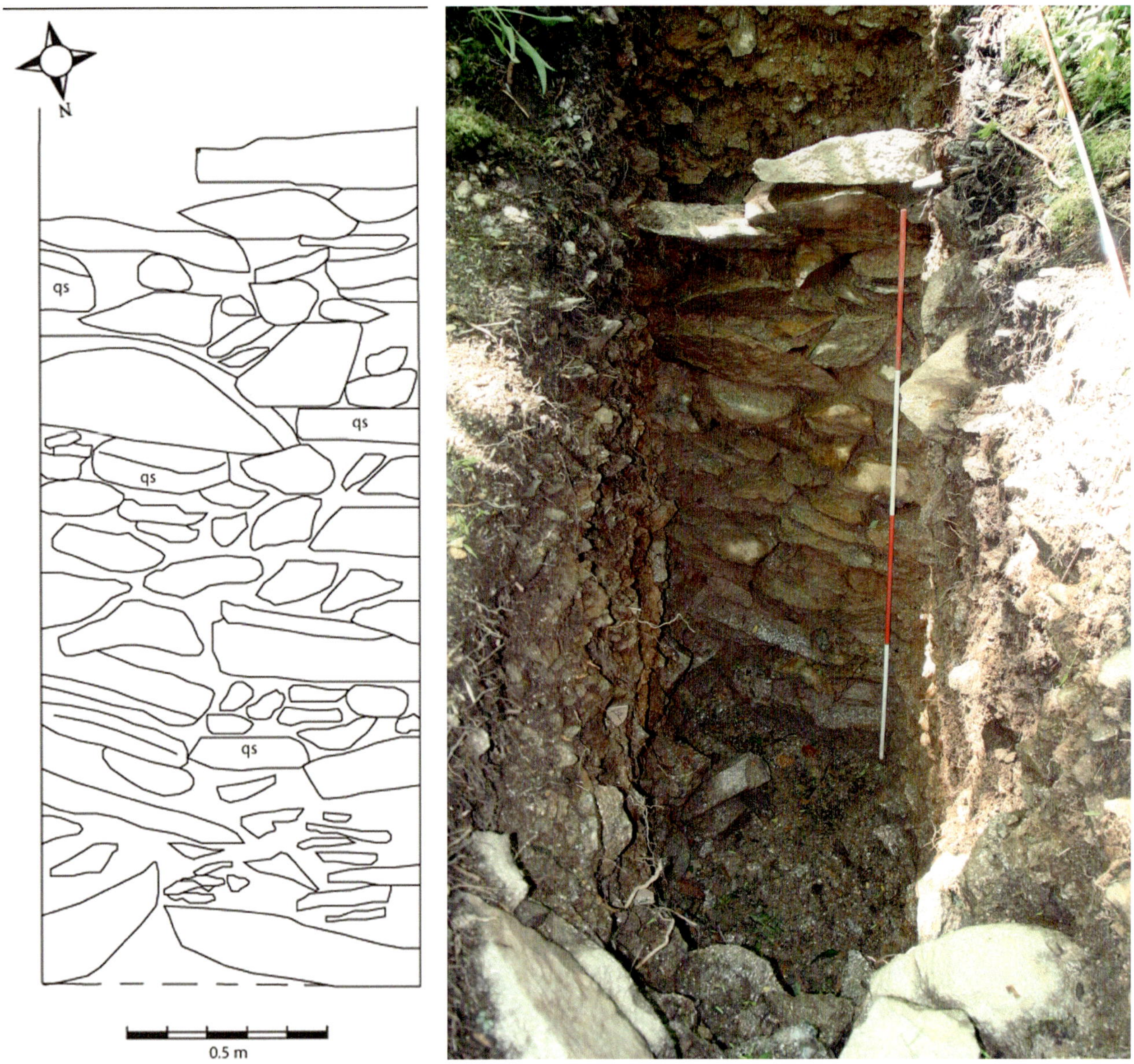

FIGURE 5.15: RØNSET TRENCH 5 WALL (DRAWING: ANJA SÆTRE/ÅSNE HELLEVE, PHOTO: IRENE BAUG, 2008)

higher age compared to other dates of birch, it may give an extraction date that is too early - even though activity here as early as the late Viking period is possible.

Layer 4 is light brown in colour and consists of coarse sand and small flakes, c. 1-30cm in size. Slabs with a diameter of 50-60cm are also present. One quernstone was also found. This layer most likely represents the end of quarry production, which is also supported by the fact that the layer covers the wall supporting the large spoil heap. If production was about to be stopped then the wall would no longer be necessary. One sample, TUa-7930, taken from the bottom of the layer, was dated to 895 ± 35, calibrated to AD1050-1215, which dates the end of production at quarry III widely to the early and high Middle Ages. Layer 5 has a yellow-brown colour and consists of coarse sand and flakes, 10-30cm in size. The layer is interpreted as production waste from extractions of quarry III. One sample, TUa-7929, was taken from the layer, about 10cm behind the wall. The sample consists of birch charcoal and is dated to 950 ± 35, calibrated to AD1025-1160, which concurs with the dating from the deposits in front of the spoil heap. Layer 6 is a red-brown layer with coarse sand and flakes, 10-30cm in size. It also contains some larger slabs, 50-60cm in diameter. One sample consisting of birch charcoal (Tua-7932) was taken from the layer and dated to 1005 ± 35, calibrated to AD 1010-1030, indicating activity around the millennium shift. Layer 7 is similar to layer 6, and may actually be the same layer. Layer 7 is, however, more yellow in colour and was therefore separated from layer 6. Sample Tua-7933 from this layer was dated to 980 ± 35, calibrated to AD 1015-1150, also here an indication of activity sometime in the transition phase from the Viking period to the early Middle Ages (Tab. 5.2; Fig. 5.17).

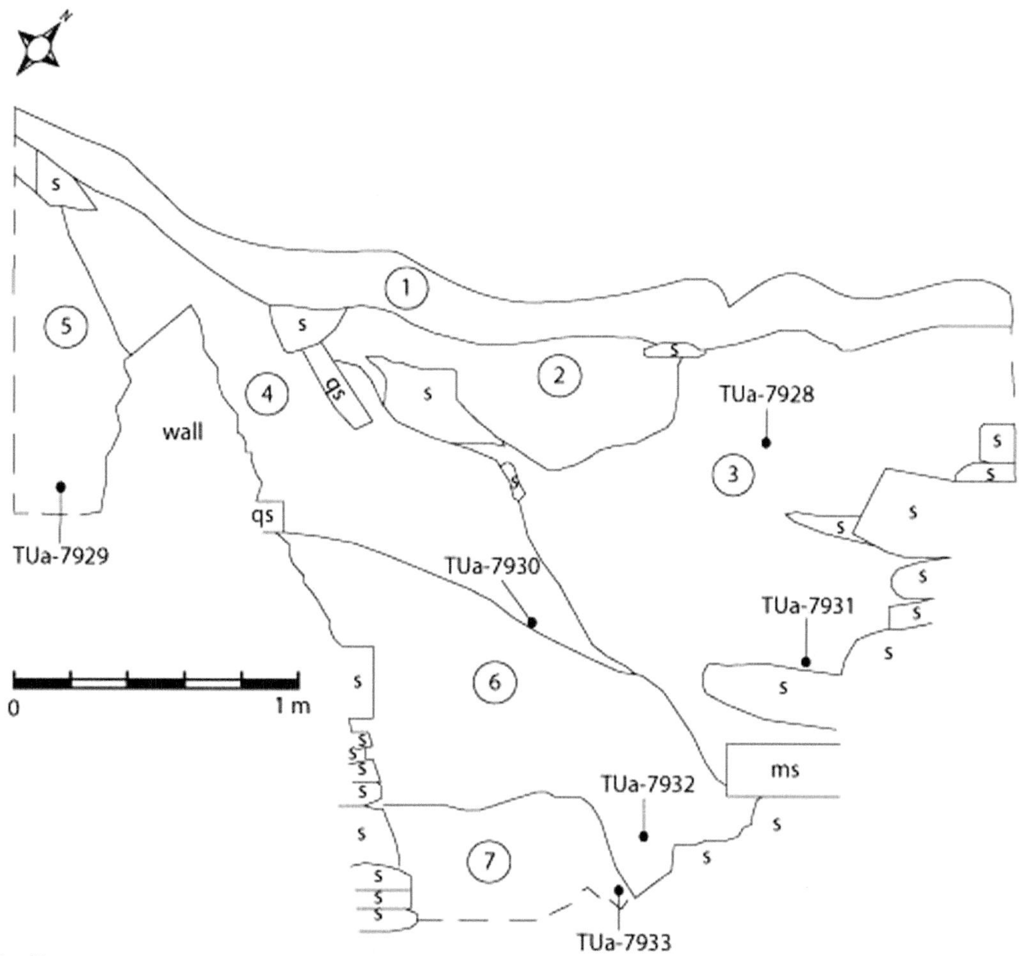

1. Topsoil.
2. Dark-brown layer of humus and flakes. Some small roots. Flakes 2-30cm.
3. Brown layer with humus and small roots. Many flakes (2-3cm) as well as larger slabs, many c. 50cm in diameter. Two millstone fragments were identified.
4. Light-brown layer with flakes from 1-10cm and coarse sand. Quernstones were also present.
5. Yellow-brown layer with flakes c. 10-15cm and coarse sand.
6. Red-brown layer with flakes c. 10-15 and coarse sand.
7. Yellow-brown layer with flakes c. 10-15cm and coarse sand. Most likely the same deposit as layer 6.

Figure 5.16: Section of Rønset trench 5 (Drawing: Anja Sætre/Åsne Helleve, 2008)

Table 5.2: C14-analyses from Rønset trench 5

Lab. Ref.	Context	Age BP	Calibrated Age	Material
Tua-7928	Rønset trench 5, layer 3	1030 ± 35	AD 990-1025	pine charcoal
TUa-7931	Rønset trench 5, layer 3	870 ± 35	AD 1165-1225	birch charcoal
TUa-7930	Rønset trench 5, layer 4	895 ± 35	AD 1050-1215	birch charcoal
TUa-7929	Rønset trench 5, layer 5	950 ± 35	AD 1025-1160	birch charcoal
TUa-7932	Rønset trench 5, layer 6	1000 ± 35	AD 1010-1030	birch charcoal
TUa-7933	Rønset trench 5, layer 7	980 ± 35	AD 1015-1150	birch charcoal

Layers 4-7 may all be extraction remains from quarry III, representing its end phase. Layer 5, behind the wall, was deposited while the wall and quarry were still in use. It does, however, go beyond the height of the wall. Layers 6 and 7 may also have been deposited while the wall was in use. To reach the top of the wall, and so be able to dump spoil behind it, the workers needed to fill in the area in front of it. This area, therefore, seems to have been developed simultaneously with this activity. As layer 4 covers the top of the wall, it must have been deposited after this went out of use, representing the end of extraction in quarry III.

All the six C14-dates from the trench are close in time, and indicate intensive activity from the end of the Viking period into the early Middle Ages (Tab. 5.2 and Fig. 5.17). The quarry seems to have been rather large, which make it difficult to estimate the start phase. Most evidence points

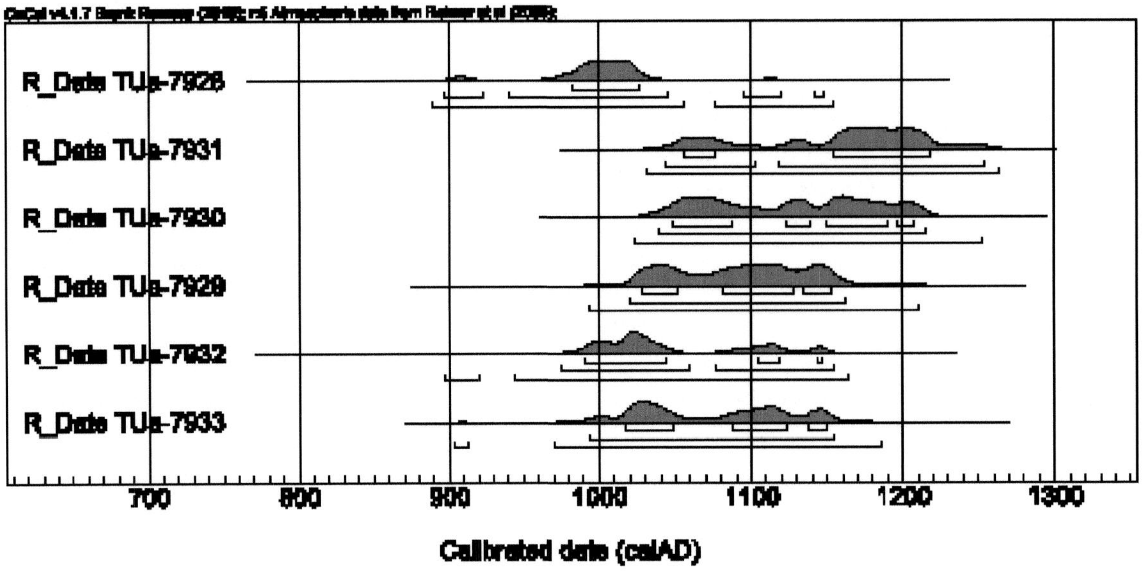

FIGURE 5.17: MULTIPLE PLOT DIAGRAM OF C14-ANALYSES FROM RØNSET TRENCH 5

towards initial activity from the latter part of the Viking period and reached an end in the early Middle Age.

Seven quernstones, most likely hand querns, were found in the trench. The diameters of these quernstones vary between 40-46cm. Two of these stones had diameters that could not be measured as the stones were partly covered in the profiles of the trench. Three of the quernstones have eyes, but the eye is unfinished on one example. The stones varied from 6-13cm in thickness because of fragmentation. Two millstones were also found, one in a fragmentary condition but obviously more than 1m in diameter (Fig. 5.18). The other millstone from the trench profile could not be properly measured. Circular marks on the rock show a diameter of 48-49cm and 80cm for the extracted stones.

5.2.3 The rock shelter at Rønset

About 50m west of this quarry site is a rock shelter consisting of very large blocks of stones, partly on top of each other.[5] The overhang is about 2m high at its highest point and 6.5m long. The western end consists of a large block, and the entrance to the shelter was from the east. At the southern end a slab of stone is placed in an upright position, making the southern wall. In the area between this slab and the stone block to the west there seems to have been a wall built of smaller stones, including quernstones, which has now collapsed.

The floor is intentionally covered with small slabs and also a few fragments of quernstones (Fig. 5.19); this may result from secondary deposition related to a later use of the shelter.

A path, in some places built of stones, is still visible between the quarries and the overhang. This is the same path that continues eastwards towards the ruins of the shed and may be of a younger date. It is possible that the path and the rock shelter may have been connected to animal husbandry. However the location of the rock shelter could also indicate that it was used during quernstone production, perhaps for the storage of tools and other equipment, even though there are no visible traces of such use today. To help provide further information two test pits (50 x 50cm) were dug (Fig. 5.20).

Test pit 1

Test pit 1 was dug 1.4m from the southern end of the rock shelter, 1.4m from the western end and 1.14m from the drop of the eave to the east, and 1.4m below the roof. Immediately west of the test pit some slabs were laid on the floor, but there were fewer stones than on the surface where the pit was made. Here some remains of slag were found on the surface. This area of the rock shelter was therefore considered an interesting site for a test pit. The trench was about 70cm deep, and even though it was not large the pit sections differed in thickness and formation (Figures 5.21 and 5.22).

Interpretation of test pit 1

Test pit 1 consisted of six layers. Layer 1 was a black layer of humus and charcoal, with a few pebbles and some small roots. Some slag was found in the deposit, as it was also on the surface, and identified as slag from smithies (Ole Tveiten pers. comm. 2009). At the base of layer 1, a possible quernstone fragment was identified. One charcoal sample (TRa-785) was dated to 810 ± 35 and calibrated to AD 1215-1275. The shape of layer 1 indicated that before the layer was deposited, a pit had been dug here. Layer 1 most likely represented remains from a forge that was used in the high Middle Ages, and thus a somewhat younger date than indicated for extractions in the investigated Rønset

[5] UTM WGS84, zone 32, east 300781 and north 6790321, 67m.a.s.l.

FIGURE 5.18: FRAGMENTS OF UNFINISHED AND BROKEN QUERNSTONES AND MILLSTONES FROM RØNSET TRENCH 5 (PHOTO: IRENE BAUG, 2008)

FIGURE 5.19: ROCK SHELTER SEEN FROM THE NORTH (PHOTO: IRENE BAUG, 2009)

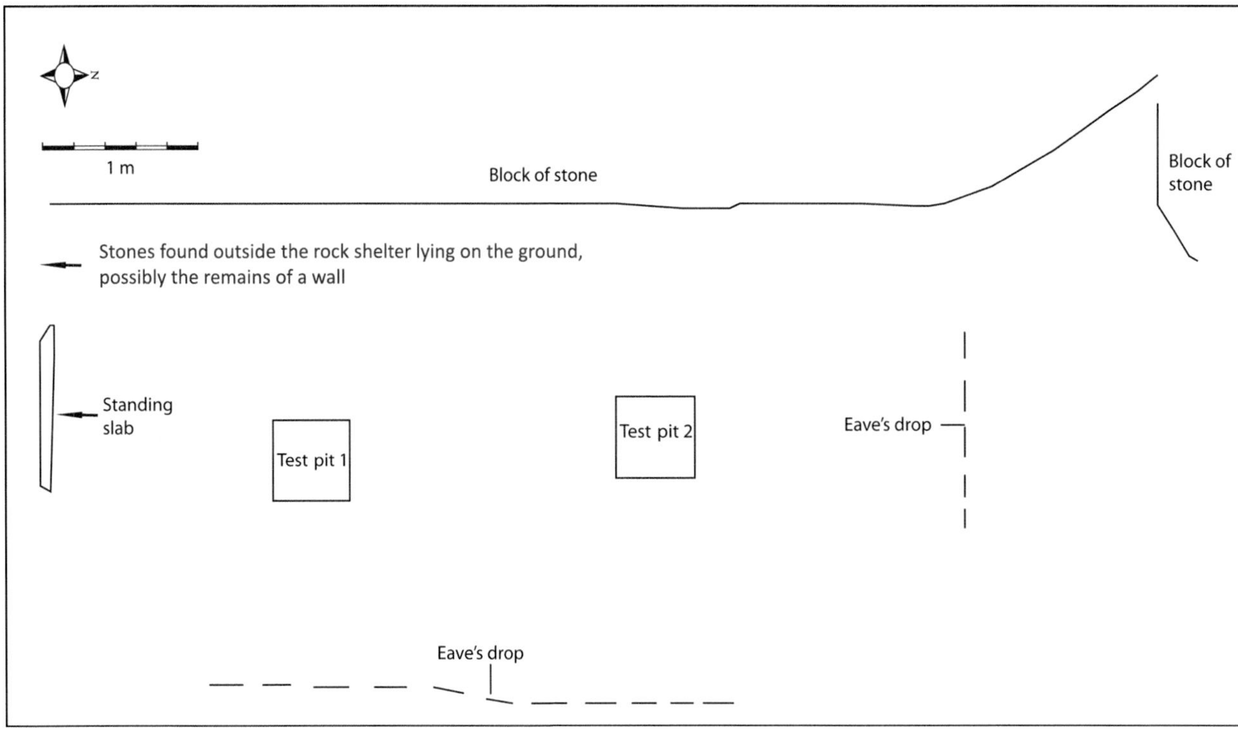

Figure 5.20: Sketch of the rock shelter (Drawing: Irene Baug, 2009)

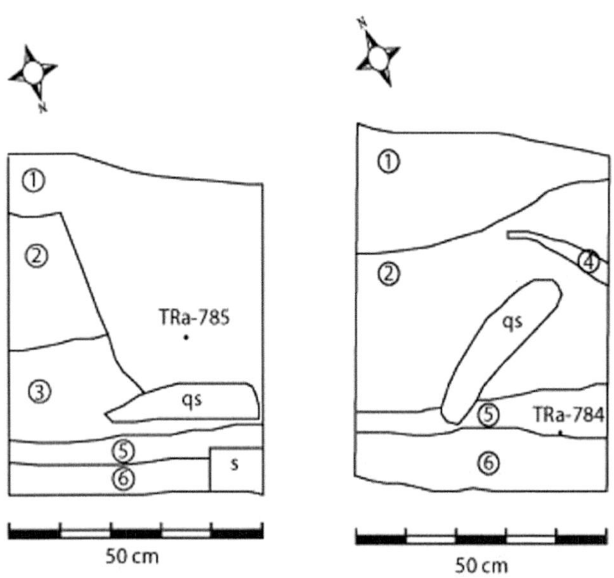

1. Black humus containing charcoal and slag. Some stones were found on top of the deposit.
2. Grey deposit of sand, gravel and small stones, mostly of garnet mica-schist. A few stones are larger than 6-7cm. The deposit is interpreted as production waste from carving of some kind. A possible fragment of an unfinished quernstone was found.
3. Grey deposit similar to layer 2. Most likely layers 2 and 3 are remains from the same activity. This layer contains rather more charcoal than layer 2
4. A pocket of charcoal and humus.
5. Black deposit of humus and some larger stones. Quantities of charcoal and slag were also identified.
6. Brown layer of humus, sand and gravel. The layer also contained stones (2-10cm) of garnet mica-schist. The layer is interpreted as the original subsoil.

Figure 5.21: Sections of test pit 1 (Drawing: Irene Baug, 2009)

quarry (trench 5). The slag and charcoal also identified at the top of the test pit indicate that forging dated to the 13th century is the latest documented activity in the shelter.

Layer 2 was a grey layer with gravel and sand and some stones recognized as garnet mica schist measuring 6-7cm, and the layer was very similar to the debris found in the spoil heaps in the quarries. A slanting fire-cracked stone of garnet mica schist, which may have been used as a *tuyère*, was also found. Layer 3, also grey and of sand and gravel, was similar to the layer above, but had some stones of a red/rust-brown colour. The contents of layers 2 and 3 indicated that they are the remains from some sort of stone carving, possibly quernstones, from loose slabs in the area that were finished inside the rock shelter. Layer 4 was a pocket of charcoal within layer 2, and seemingly the traces of some kind of fireplace used at the same time as the carving. Layer 5 was a black deposit containing charcoal, a few pebbles and some slag, perhaps descending from layers 2 and 3. The contents indicated remains from the same type of activity as represented by layer 1, but from an earlier phase. This is also indicated by

Figure 5.22: Test pit 1: north (left) and south (right) sections (left) (Photo: Irene Baug, 2008)

sample TRa-784, dated to 990 ± 30 and calibrated to AD 1015-1040 (Tab. 5.3). This early date is approximately the same as found at Rønset, trench 5. It is likely therefore that the blacksmithing activities in this period are related to quernstone production. Layer 6 was a brown layer of humus, sand, gravel and stones ranging from 2-10cm, also of the mica-schist type, and interpreted as the original subsoil.

Test pit 2

To investigate the northern part of the rock shelter, test pit 2 was dug 90cm north of test pit 1. Test pit 2 was also 50 x 50cm, but only 25cm deep. The roof of the rock shelter is 1.5m high at this point. The pit was placed c. 60cm east of the western wall in the rock shelter. To the north it was 1.7m from the drop of the eave, and to the east about 1.4 m. As they did in test pit 1, the thickness of deposits varied within the pit, and again two of the sections were documented (Fig. 5.23).

Interpretation of test pit 2

Test pit 2 consisted of three layers. Layer 1, the topsoil, was brown in colour and contained humus and some pebbles, up to 5-6cm in size. Some pockets with more gravel and less humus were also present in the layer. Layer 2 was a black layer of humus, charcoal, some garnet mica schist pebbles (c. 2-4cm), and with a few small roots. Slag, most likely from a smithy, was also identified. The deposit was similar to layer 1 in test pit 1, and most likely the remains of the same kind of activity. Layer 3 was a grey deposit of a little humus but mostly gravel and pebbles of garnet mica schist. Layer 3 contains most likely the remains from the same activity as layers 2 and 3 in test pit 1, possibly some sort of carving. A large slab located at the bottom of the pit prevented further excavation. One sample, TRa-786, was dated to 940 ± 30 and calibrated to AD 1030-1165, and from the same time as the investigated quarry nearby (Tab. 5.3). The degree of finishing of the quernstones found in the quarry indicates, however, that they were almost finished at the quarry site and not transported to other locations for final finishing.

Activity in the rock shelter

Both test pits, therefore, indicate different activities inside the rock shelter. Traces of slag and charcoal indicate smithies, and substantiate that the shelter was used as a workplace for forging and sharpening those tools related to quarry site work with large-scale production close by. The lowest layer of charcoal and slag in test pit 1, layer 5 was synchronous with the extraction at the quarry site, which supports this hypothesis. At a later stage, in the high

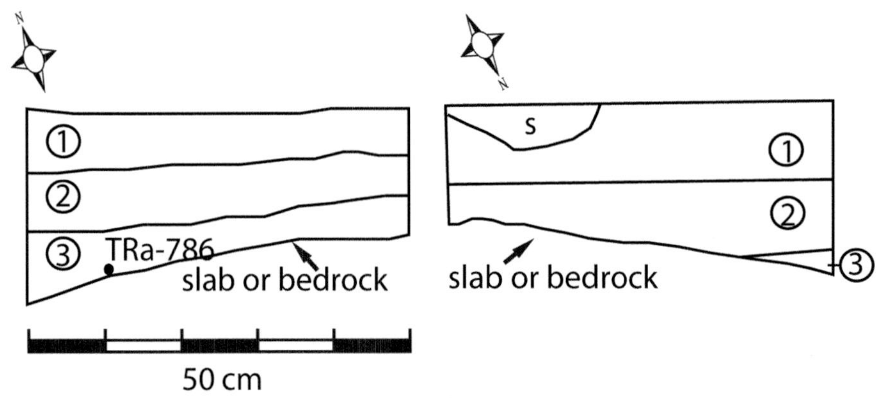

1. BROWN DEPOSIT OF HUMUS AND SMALL STONES, THE LARGEST BEING 5-6CM. SOME AREAS OF GRAVEL WITH NO HUMUS. SOME SMALL ROOTS.
2. BLACK DEPOSIT WITH HUMUS AND CHARCOAL. SOME SMALL STONES (2-4CM), SOME OF GARNET MICA SCHIST. SOME SLAG FRAGMENTS AND SMALL ROOTS.
3. GREY DEPOSIT OF GRAVEL, SMALL AMOUNTS OF HUMUS, AND SMALL STONES OF GARNET MICA-SCHIST.

FIGURE 5.23: TEST PIT 2 (DRAWING: IRENE BAUG, 2009)

Middle Ages, forging activities seem to have restarted. Even though it cannot be directly confirmed by C14-dating, it is possible that this later activity could also have been related to quernstone production (Fig. 5.24).

Correct temperature in the forging process was needed to ensure good quality iron with sharp and solid edges. During forging, the glow of the iron would indicate its temperature. It was perhaps easier to estimate the heat of the iron in a protected and darker place. When the correct temperature was reached, indicated by the colour, the iron was put into water to decrease its surface temperature; after removal from the water the surface of the iron regained some of its heat. This process led to a hardening of the iron so that it could withstand rough work (Lars Stenvik pers. comm. 2010). This rock shelter was surrounded by large blocks of stone that kept it shaded and also protected the occupants from wind and rain, and would serve as a suitable place for a smithy.

Some sort of working, carving or finishing of stone products, also seems to have taken place below the overhang, and the two possible fragments of quernstones indicate that small-scale quern production may have taken place here. It is also possible that loose slabs from the area were used as raw material for quernstones from the late Viking period or the early Middle Ages, which more or less concurs with production at the quarry. Layers 2 and 3 in test pit 1 are only about 40cm thick, indicating shorter and less intensive activity compared to production in the quarries. In the 13th century the rock shelter again seems to have been used for forging.

5.3 Archaeological investigations at Myklebust (no. 79)

Myklebust farm also contains one of the largest concentrations of quarries in Hyllestad. With its 73 identified quarries and five trial extractions, Myklebust represents, together with Rønset, the main area for quernstone and millstone production at Hyllestad. The quarries appear in clusters, mainly at the south-eastern part of the farm area (Fig. 5.25). This is also where the investigated quarry complex is located (Fig. 5.26).

As part of this author's Master's project, one quarry complex was excavated at Myklebust (Myklebust trenches 2 and 3). The quarry, located c. 40m west of the present main road and some 180m north of the present road to Kolgrov, consists of a 23m-long rock face with production marks, partly covered by spoil heaps. It is a deep quarry with high carved walls, with a step-like morphology in some places. The main products here seem to have been millstones (c. 1m in diameter and about 20cm thick), judging by the last production, but some quernstones were also extracted. The production has been dated to the late Viking period, AD 890-1015, and the extraction

Table 5.3: C14-analyses from the rock-shelter and test pits 1 and 2

Lab. Ref.	Context	Age BP	Calibrated Age	Material
TRa-785	Rock-shelter, test pit 1, layer 1	810 ± 35	AD 1215-1275	birch charcoal
TRa-784	Rock-shelter, test pit 1, layer 5	990 ± 30	AD 1015-1040	birch charcoal
TRa-786	Rock-shelter, test pit 2, layer 3	940 ± 30	AD 1030-1165	pine charcoal

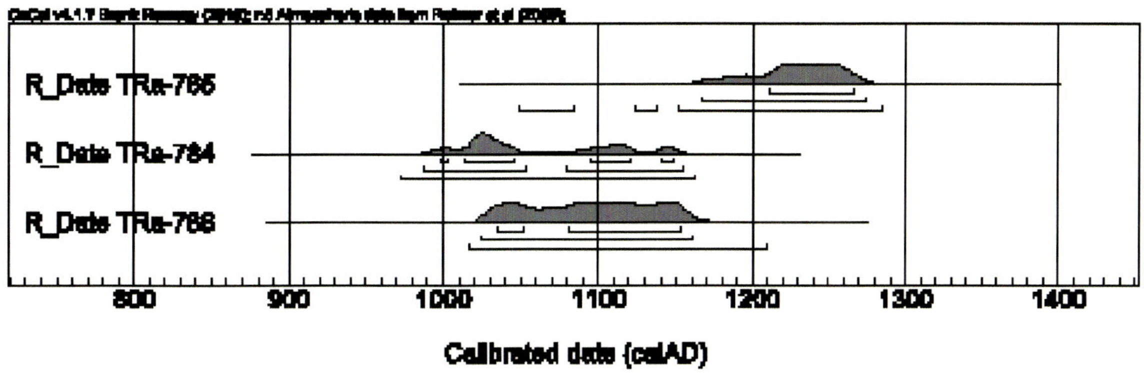

FIGURE 5.24: MULTIPLE PLOT DIAGRAM OF C14-ANALYSES FROM THE ROCK SHELTER, TEST PITS 1 AND 2

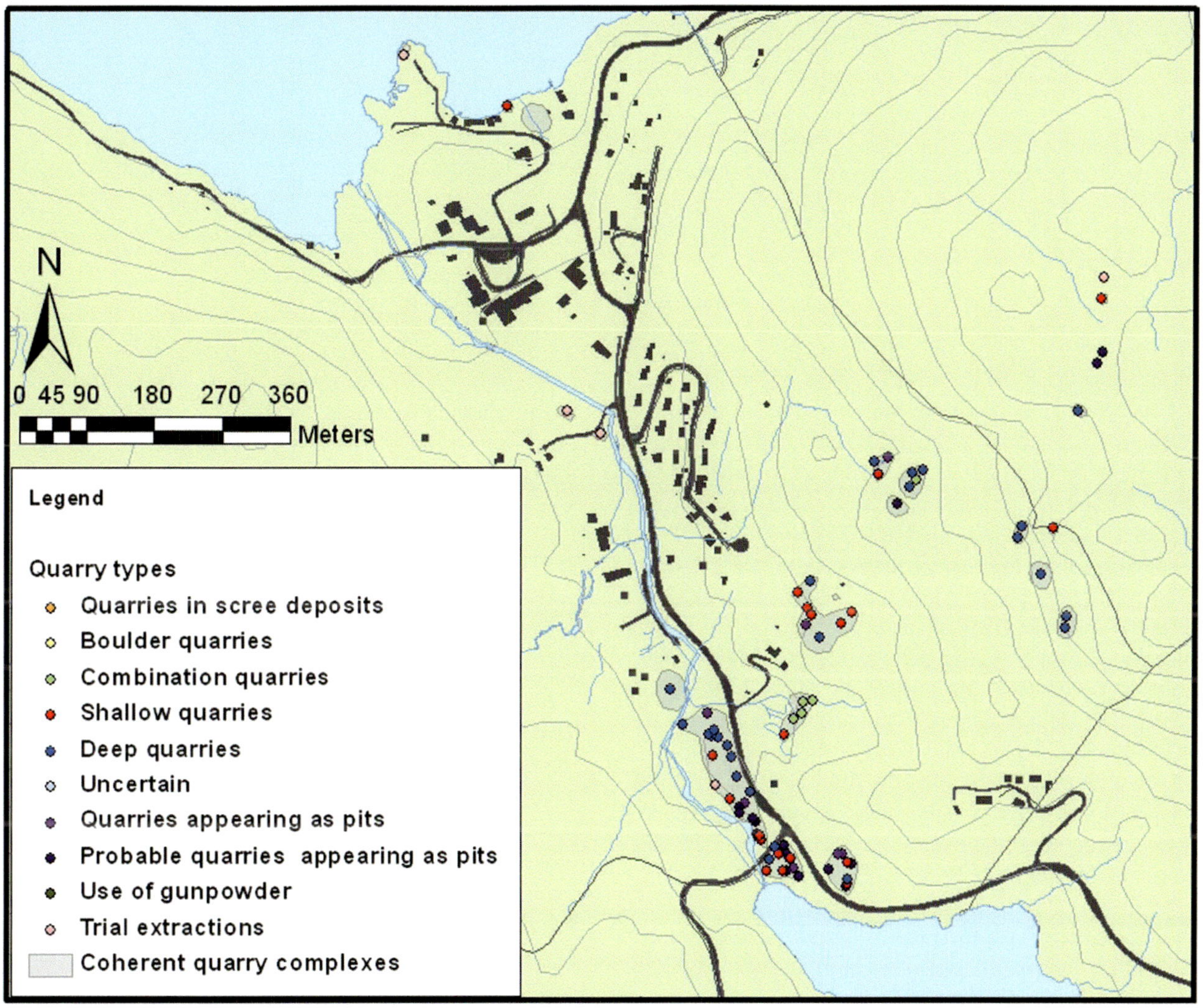

FIGURE 5.25: MAP OF DIFFERENT QUARRY TYPES AT MYKLEBUST (MAP: IRENE BAUG, 2013) (MAP-DATA FOR HYLLESTAD QUARRIES FROM A SURVEY BY TOM HELDAL, NGU)

FIGURE 5.26: THE QUARRY COMPLEX AT MYKLEBUST SHOWING THE LOCATION OF TRENCHES 2-5 (PHOTO: KIM SØDERSTRØM AND JØRGEN MAGNUS©RIKSANTIKVAREN)

of millstones may be this old. The base of the spoil heap, the oldest production phase, was not reached, making it difficult to ascertain when millstone production began in the quarry. One should be careful not to determine the date of millstone production by one single C14-analysis. Another C14-sample is dated to AD 1010-1160, while one from just below the topsoil gives AD 1165-1275 (Baug 2002:53). The C14-dates indicate, then, a somewhat early date for millstones. Millstone extraction may have taken place in the late Viking period or early Middle Ages, making it likely that it represent the initial phase of this type of production. Output of quernstones and millstones in the quarry complex continued into the 15th century (AD 1315-1435), indicated by a C14-date from trench 3. The spoil heaps covered an older worked-out area, meaning that production at the site must be older than the C14-dates from the two trenches (Baug 2002:51-54).

In 2007 an area at a rock shelf northeast of and close to Myklebust (trench 3) was excavated by schoolchildren from the Hyllestad primary school and two trenches (Myklebust trenches 4 and 5) were excavated (Løland 2007). One sample for C14-dating was taken from each of the two trenches so as to study any temporal connection to the spoil heaps from Myklebust trenches 2 and 3.

Trenches 4 and 5 were located on a carved shelf, about 3m up on the carved wall and between 2.7 and 3m below the top of the wall. The spoil on this shelf most likely came from extraction from a higher level, and not from any production on the shelf itself. The spoil seems to have been thrown down over a previously worked-out area of the quarry, covering the shelf as well as the bottom of the quarry, where trench 3 was located. Trenches 3-5 may therefore contain spoil from the same extraction period. Myklebust trench 2 is located slightly further west. This spoil heap also covers an older quarry, but derives from an older production phase than the production revealed in trenches 3-5.

Two samples for C14-dating were taken, one from each of the trenches, and dated to AD 1400-1430 and AD 1280-1295. This may indicate disturbances from clearing and movements in connection with production. It is likely that older spoil heaps were moved in connection with later extractions and thrown into an older part of the quarry. Production in this quarry complex continued at least into the first part of the 15th century.

5.3.1 Myklebust trenches 2-5: some temporal aspects

The site represents a large quarry complex, and, according to the various C14-dates from the trenches, the production phases covered a long period of several hundred years, from the late Viking period to the late Middle Ages. Movements related quarrying seem to have disturbed the investigated spoil heaps, indicated in trenches 4 and 5. It

Figure 5.27: Carved rock wall (4.3m long), with signs of slab production, from the eastern end of trench 6 (Photo: Kim Søderstrøm, Jørgen Magnus © Riksantikvaren)

has therefore been difficult to establish a clear relation between carved rocks and the spoil heap.

With regard to the C14-dates from the earlier excavated trench 2 (Baug 2002), the sequence indicated by the C14-dates concurs with the documented relative stratigraphy, which indicates that this spoil heap has been stable. The production is dated broadly to the Viking Age (AD 810-1015), and continued into the Middle Ages, with AD 1165-1275 as the youngest date (Baug 2002:52-53). This indicates that several production phases are represented in the spoil heap, but it is also possible that some of the spoil comes from quarrying at a higher level in the quarry. This seems to be the most likely situation judging by the other trenches. Nevertheless, this spoil heap represents production phases earlier than documented in trenches 3-5.

5.3.2 Myklebust trenches 6 and 7: production of quernstones and stone crosses

The investigated quarry is located in the outlying field of holding 1 (Nygård) at Myklebust farm, about 95m east of the present road, and 60m south of the farm houses at holding 15.[6] This is a deep quarry that differs from most of the deep quarries in Hyllestad by its vertical carved rock wall. The shape indicates extraction of products other than quernstones and millstones, which made it an interesting new research site (Fig. 5.27). The quarry is located on a slope where production waste was thrown down towards a bog to the west. The secondary rock cleavage is of poor quality in some places, resulting in a large number of destroyed quernstones in comparison with the quarries investigated at Rønset. The site is rather large and two trenches were made to shed light on production there. Extractions were carried out at two different levels in the rock. The first level has a 34m-long carved rock wall stretching in a north-south direction and from c. 60cm to 3.2m in height. Production waste in front of it covers the lower part of the wall and stretches more than 25m westwards from the rock. On the top of the rock wall, a 4m-wide and somewhat slanting area shows traces of carving, followed by another carved rock wall, c. 2.5m high and 11m long (level 2). The area on top of level 1 reveals traces from the extraction of small slabs, partly covered by soil and vegetation, where one measurable slab was rather large, 70cm wide and more than 1m long.

The northern part of the quarry site is now covered with spruce, planted about 50 years ago, replacing the earlier deciduous trees (Gunnar Nygård pers. comm. 2008). To the north runs a brook westwards towards the present road; a few metres west of the carved rock a wall was built some decades ago to make a small watermill, but this was never finished (Gunnar Nygård pers. comm. 2008). Slabs and flakes from the older production, as well as damaged

[6] UTM WGS84, zone 32, east 301175 and north 6786890, 69m.a.s.l.

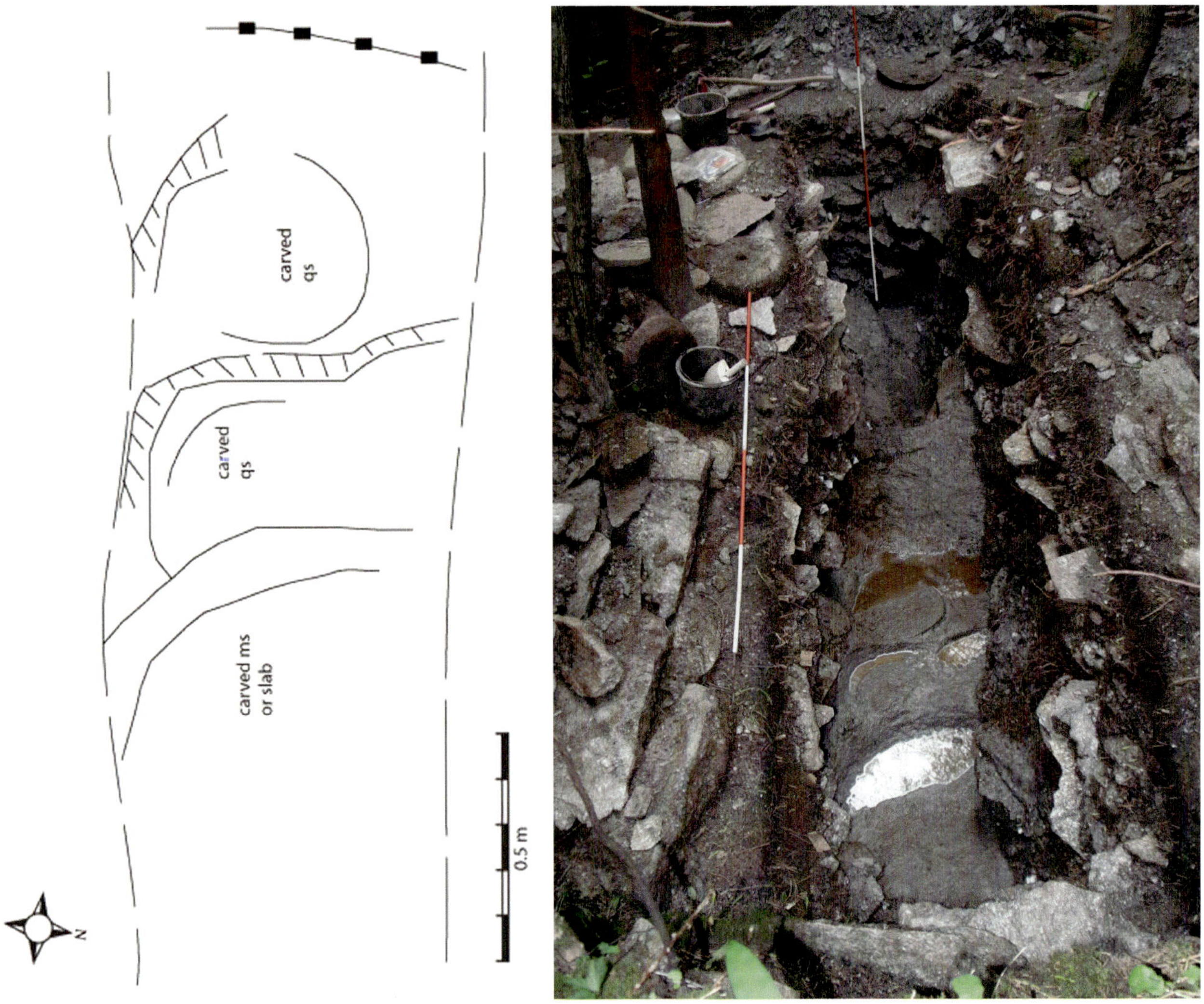

Figure 5.28: Carved base of Myklebust trench 6, from the east (Drawing: Irene Baug/Åsne Helleve; photo: Irene Baug, 2008)

products, mostly quernstones but also a millstone, are still visible in the brook. More unexpected finds were remains from three stone crosses at the northern end of the quarry. One fragment, most likely the base/foot of a cross, was found in the small brook 6m north of trench 6, and the head of another cross was found c. 4.5m south of the trench, while a third cross was found just 1.2m west of trench 6.

This part of the quarry has no clearly defined spoil heap, but about 6m from the rock the topography slants west and a slope shaped by production debris falls away about 25m to the west. This part of the quarry seems to have been more or less undisturbed by human activity since production ended. About 6m from the brook at the northern end of the quarry, close to the rock wall, a further trench (6) was cut to try and establish the chronological relation between the rock and the spoil heap.

At the southern part of the quarry two spoil heaps are still visible, although covered with deciduous trees, mostly rowans and sallow, but also ferns. The southern end of the carved rock wall is disturbed over a 12 x 4m extent, where a bog has spread over the previous extracted area. A recent drainage trench from the bog in a north-western direction separates the two spoil heaps, where spoil as a consequence has been moved from the lower parts of the uppermost heap (heap I) to the lower one (heap II) (Nygård pers. comm. 2008). Heap I is located c. 3.6m from the rock. It is c. 9m wide in an east-west direction and 10.5m wide, north-south; the drainage trench is located at the western side of the heap. Trying to avoid the disturbances at the lower part of the heap, trench 7 was made on top, about 6m west of the carved rock wall. The objective was to solve the chronological relation between the northern (trench 6) and southern parts (trench 7) of the quarry, and possibly reveal different production phases.

Heap II (in the south) measures 9 x 37m east-west and north-south. A second drainage trench dug at the north-western end of this heap had left it too disturbed for any meaningful archaeological investigation.

Myklebust trench 6

A trench, about 5m long and 1.5m wide (east-west), was dug c. 6m from the northern end of the quarry. To reveal

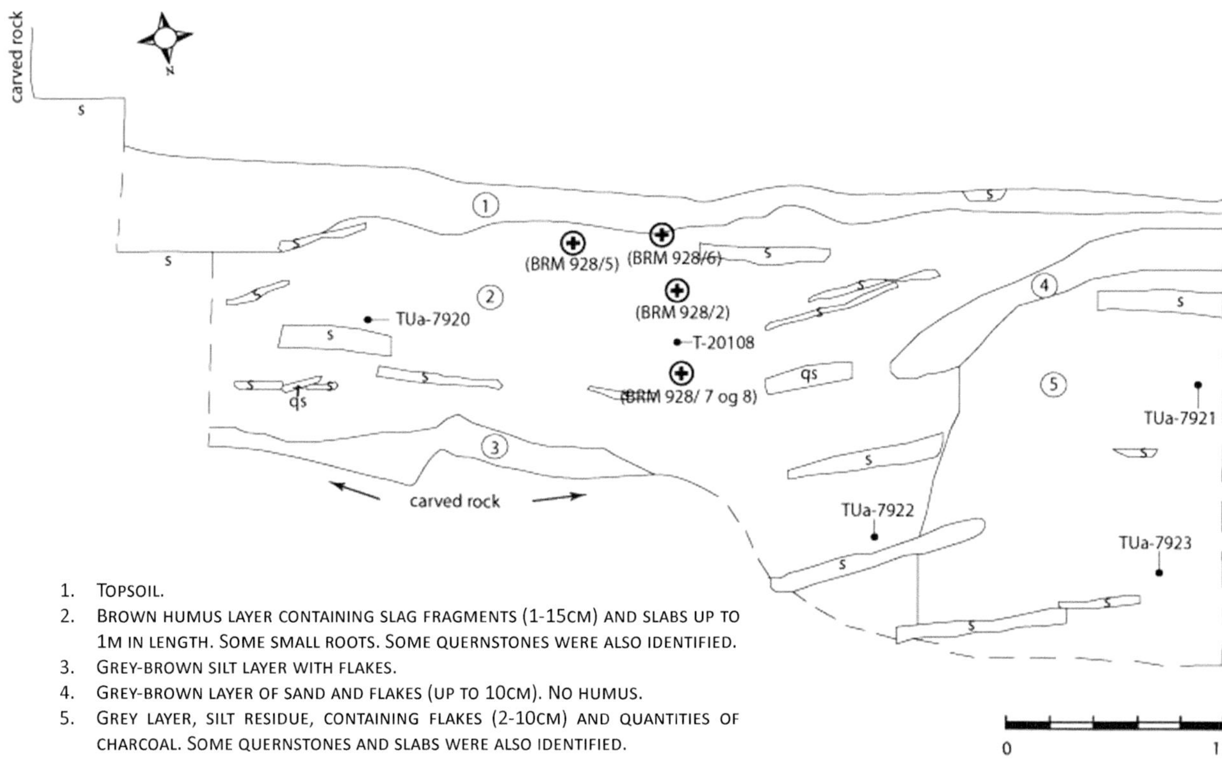

Figure 5.29: Section of Myklebust trench 6 (Drawing: Irene Baug/Åsne Helleve, 2008)

the height of the carved rock wall, the eastern end of the trench was cut in the flat area in front of the rock, as this area was interpreted as a previously worked-out part of the quarry, later covered with production waste. The trench was from 50cm to 2.10m deep. In the eastern part the base was uncovered revealing the surface of the carved rock (Fig. 5.28). In the western part the excavation had to be stopped before the bottom was reached, due to the danger of collapsing profiles.

Extraction from the rock wall at the eastern end of trench 6 must have stopped before the spoil heap was made there. As the rock shows traces of extracted slabs (Fig. 5.27), production of products other than quernstones should be expected, and most likely stone crosses.

Profile interpretation: Myklebust trench 6

Trench 6 consisted of five layers including topsoil (layer 1) (Fig. 5.29). Layer 2 was a brown layer consisting of humus, small roots, and 1-15cm large flakes and slabs about 1m in size. Areas of pure humus, with no stones, were also detected, especially at the eastern end, most likely the remains of rotted trees or plants. The quarry must have been cleared of trees and vegetation before extraction could take place. The layer was interpreted as the result of dumping after production had ended at this part of the quarry, and may derive from extractions at level 2 of the rock wall, or perhaps from the rock wall further south. The size of the quarry may indicate that production waste had been moved within the quarry site to make space for further extraction. It is also possible that layer 2 represented a secondary dumping of spoil. Nevertheless, the layer seems to represent the end of production at this quarry.

In layer 2, several quernstones and the remains of four stone crosses indicate that querns and crosses were extracted (Fig. 5.30). Parts of the tops of three crosses were found. The forth cross found is represented by two rectangular forms, nicely hewn and quite similar pieces, carved on three sides, while the forth side is broken. The two forms are interpreted as damaged cross arms, probably from the same cross.

Three samples for C14-analyses were taken from this layer. The uppermost sample (TUa-7920) from about 45cm below the topsoil is dated to 800 ± 35 and calibrated to AD 1225-1280. The second sample (T-20108) was from the middle of the trench, in an area where the stone crosses were found. This sample of birch charcoal is dated to 870 ± 65 and calibrated to AD 1050-1245. The third sample (TUa-7922) was taken further east in the trench, and deep down in layer 2. It is dated to 845 ± 35 and calibrated to AD 1175-1250. The three dates concur with the relative stratigraphy, and the charcoal was probably deposited more or less simultaneously with the spoil and destroyed products that ended up here. In the multiple plot diagrams for the C14-analyses (Fig. 5.31), sample TUa-7920 shows a peak (1σ) in the middle of the 13th century, making this the most probable date. The other samples from the trench indicate a somewhat earlier date. Sample T-20108 shows its peaks in the middle of the 11th century and in the late 1100s/early 1200s. Sample TUa-7922 indicates a

FIGURE 5.30: TRENCH 6 DURING EXCAVATION SHOWING QUERNSTONES AND CROSS IN SAME DEPOSIT (PHOTO: IRENE BAUG, 2008)

date from the beginning of the 13th century. Overall, the evidence indicates cross extraction from the latter half of the 12th century, or the first half of the 13th century, but possibly as early as the 11th century.

Layer 3 was a grey-brown silt layer, most likely representing crumbling rock from the base of the trench, mixed with humus and flakes from layer 2. Layer 4 is a grey-brown layer of fine sand with flakes c. 10cm in size. There was no humus in the deposit. Layer 5 was a grey deposit consisting of silt and small flakes (1-10cm). A few quernstones were also found, and towards the base, about 1.25m deep, there were several more slabs more than 1m in size. Large amounts of charcoal were also found, indicating dumping from fireplaces. The shift between layers 2 and 5 was vague, due to the presence of large slabs in both layers, and less distinct than indicated in the drawing.

Layer 5 may represent a spoil heap from extractions in the carved rock west of the trench, with a distance of 4m between the two. Broken slabs were deposited here. It is also possible that the layer represents a working site for the finishing of crosses, as these most likely were made of extracted slabs. The finds of nearly finished, but broken, crosses indicate that the finishing was done here or close by. Layer 5 may therefore be the result of the shaping and finishing of stone crosses extracted from quarried slabs - perhaps also from the rock at the eastern end of the trench. None of the crosses, however, were recovered from the deposit itself. One large, broken cross was found 1.2m from the south-eastern end of the trench, just below the topsoil, representing a rejected item. Two samples for C14-dates were taken from the layer, indicating production in the transition phase from the Middle Ages onwards. The uppermost sample, TUa-7921, was dated to 980 ± 35, calibrated to AD 1015-1155, while the second sample (TUa-7923) had a wider dating range to 965 ± 35, calibrated to AD 1060-1215. As the dates may indicate a reverse sequence, it could imply disturbances. The two samples to a large extent concur in time, indicating that they have been made more or less at the same time. Layer 5, then, seems to represent a somewhat older extraction phase than layer 2 (Tab. 5.4 and Fig. 5.31).

Table 5.4: C14-analyses from Myklebust trench 6

Lab. Ref.	Context	Age BP	Calibrated Age	Material
TUa-7920	Myklebust trench 6, layer 2	800 ± 35	AD 1225-1280	birch charcoal
T-20108	Myklebust trench 6, layer 2	870 ± 65	AD 1050-1245	birch charcoal
TUa-7922	Myklebust trench 6, layer 2	845 ± 35	AD 1175-1250	birch charcoal
TUa-7921	Myklebust trench 6, layer 5	980 ± 35	AD 1015-1155	birch and hazel charcoal
TUa-7923	Myklebust trench 6, layer 5	890 ± 35	AD 1060-1215	birch charcoal

In the eastern part of the trench the base rock revealed traces of both quernstone and slab extractions. One unfinished quern (42-43cm in diameter) with an unfinished eye was never loosened from the rock. The stone indicates that the eye in some cases was carved, at least partially, before the stone was removed. The stoneworkers also cut holes underneath the quern in an unsuccessful attempt to loosen it. In the middle of the trench, where the rock slopes downwards, a carved rock wall with production groves was revealed. The garnets in this wall are smaller than in most of the products recovered in the quarry, indicating varying qualities of quarry rock. Because of the danger of collapsing trench profiles work had to finish before the base of the western part of the trench was reached.

Altogether 19 quernstones, or fragments of them, were found in trench 6, located within layers 2 and 5. Four were too fragmentary to indicate the diameter. Of the remaining 15, the diameters vary from 36 to 42cm, and of these seven

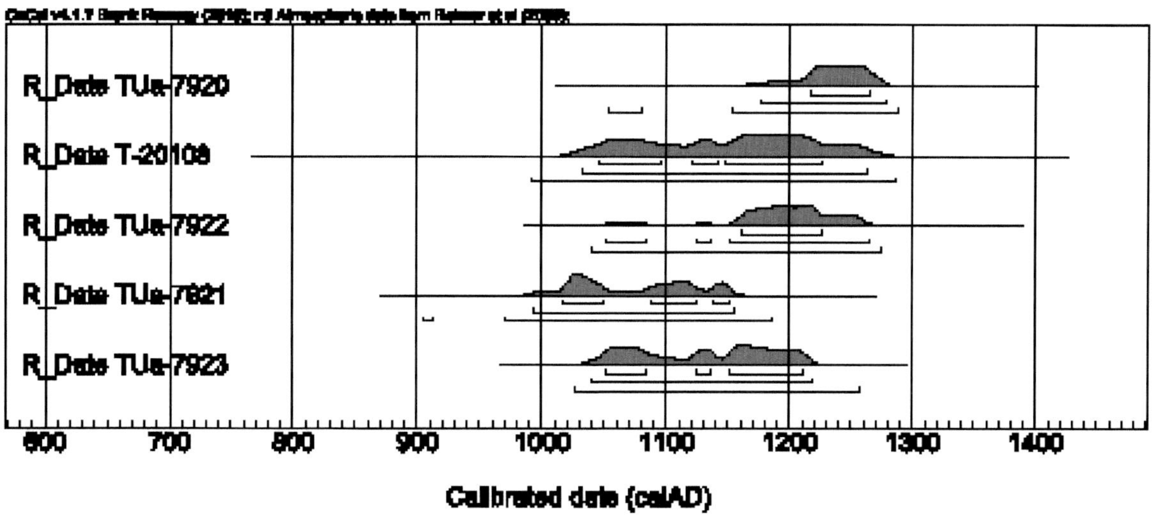

Figure 5.31: Multiple plot diagram of C14-analyses from Myklebust trench 6

have diameters of from 39 to 40cm. The stones were from 4 to 11cm thick, due to various degrees of fragmentation. Nine quernstones had a finished or partly finished eye, indicating that this operation was mainly done on site, but often after the stone was taken from the rock. This is also indicated by quernstones and millstones still attached to the bedrock at several places within the quarry landscape. The holes were made below the stones as a method to loosen them from the rock, but the eye was not necessarily carved at this stage. Different stone cutters may have had different preferences for how to finish stones.

The remains of the crosses vary in shape and size, but a common feature for all of them is the V-shaped cutting below the arms (Fig. 5.32) (Cross type C according to Gabrielsen 2007 [2002]: fig. 3). As all the crosses are damaged and found as fragments, it is generally difficult to estimate the intended height. This is only possible for BRM 928/5, where the whole cross is preserved, measuring no more than 1m in length. The cross is small and would most likely have stood only 50-60cm above the ground, with 40-50cm of the lower section representing the base and which would have been sunk in the ground (Fig. 5.32). The shape of the two crosses BRM 928/5 and 6 is similar, indicating a common size. These are, however, the only two crosses of identical shape; the others seem to be different. The BRM 928/2 cross, also with V-shaped cuttings below the arms, has a somewhat larger crosshead compared to BRM 928/5 and 6. The fourth specimen (BRM 928/3) has a much larger crosshead compared to BRM 928/2, 5 and 6, with the top part measuring c. 45cm from the cross arms, indicating that this cross was meant to be taller. The fifth specimen, BRM 928/1, also V-shaped below the arms, tends towards a semi-circular shape (Cross type A, Gabrielsen 2007: fig. 3). The semi-circular shape is, however, less pronounced, and the V-shaped cutting below the arms represents the most striking feature. This cross is 1.2m tall, with an intact crosshead and a broken beam, measuring 30cm from the top to the arms, and 78cm from arm to arm. The three measurable churchyard crosses still standing at the old church of Hyllestad measure 54, 56 and 65cm from arm to arm, indicating that BRM 928/1 was intended to be somewhat larger and higher than this. The size of BRM 928/1 is similar to some of the smallest freestanding crosses, such as the cross at Gard in Haugesund (82cm between the arms), the cross at Njærheim in Hå (72cm), the cross at Vilnes in Askvoll (76cm), and the cross at Ryggøyra in Gloppen (74cm). These crosses are from 1.17m to 2.95m high above the ground (Birkeli 1973:130, 165, 195, 201), indicating the intended height for BRM 982/1. A similar size should also be expected for BRM 982/3.

The two rectangular fragments, BRM 928/7 and 8, most likely represent two cross arms, probably of the churchyard type. As the top part and beam of the cross are missing, the shape of the cross is somewhat difficult to indicate, but here also V-shaped cuts below the arms are likely. BRM 928/4 is probably a fragment of the base-part of a cross, the sunken part, and the shape and size is quite similar to the base of BRM 928/5 (Fig. 5.32). The crosses from the quarry, therefore, have certain similarities in terms of form (with V-shaped cuts below the arms) but differ in size. Nearly all standing crosses with V-shaped arm cuts found along the western coast of Norway were extracted at Hyllestad, and vary normally in height from 1.5 to 3m above the ground, even though taller crosses are also known (Gabrielsen 2007:146-148, 151). The crosses from Myklebust seem to belong to the latest production phase at the quarry site, most likely extracted from slabs cut from the rock. The carved rock wall east of trench 6 shows evidence of slab production, and this production must be somewhat older than the spoil heap of trench 6. It is therefore possible that cross production at the quarry is somewhat earlier than the C14-dates given from trench 6. The time difference between the start phase of the production and the C14-dates presented may, however, be shorter – perhaps not more than a few years or decades.

Myklebust trench 7

The seventh trench at Myklebust was through the top of the spoil heap, 10.7m south of trench 6 (heap I) and about 6m west of the rock wall (Fig. 5.33). The deposits in trench 7 most likely came from the extractions of the southern part of the quarry. The trench was c. 1.5m wide, 2.6m long and 1.3-2.1m deep. The large quantity of stones in the profiles made them unstable, and digging had to stop due to the danger of collapsing walls, and the bottom of the spoil heap was not reached.

FIGURE 5.32: THE STONE CROSSES (PHOTO: SVEIN SKARE, UNIVERSITY MUSEUM OF BERGEN)

BRM 928/3

BRM 928/4

Figure 5.32: The stone crosses (Photo: Svein Skare, University Museum of Bergen) contd.

FIGURE 5.32: THE STONE CROSSES (PHOTO: SVEIN SKARE, UNIVERSITY MUSEUM OF BERGEN) CONTD.

Profile interpretation: Myklebust trench 7

Trench 7 consisted of nine layers including topsoil (layer 1) (Fig. 5.34). Layers 2-5 were deposited partly on top of layer 6, indicating that these layers represent a younger production phase than layers 6-9. This is also supported by the different C14-dated samples.

Layer 2 was a brown layer of humus and flakes and may represent gradual sinking from the topsoil into the spoil. The layer varied from sand to gravel, with flakes of 15-20cm in size. Layer 3 was grey in colour without humus, consisting of flakes, mostly 2-20cm. One sample for C14-dating from the layer, TUa-7924, is dated to 965 ± 35, calibrated to AD 1025-1155, concurring in time with the spoil heap represented in trench 6. Layer 4 consisted of some humus and grey-orange sand with small flakes, c. 1cm, and some around 20cm. The orange colour most likely reflects iron oxide, a phenomenon that also occurs at the southern part of the carved rock wall. Layer 5 was also grey, similar to layer 3, probably a result of the same extraction as represented in layers 3 and 4. Layer 6 was grey with orange areas, most likely the remains of iron. The grey parts are without humus while the orange areas contain more humus. The deposit varied from fine sand to flakes, 2-10cm in size. One sample for C14-dating from the layer, TUa-7927, is dated to 1030 ± 30, calibrated to AD 995-1025, indicating production at the end phase of

BRM 928/6

BRM 928/7 AND 8

FIGURE 5.32: THE STONE CROSSES (PHOTO: SVEIN SKARE, UNIVERSITY MUSEUM OF BERGEN) CONTD.

Figure 5.33: Excavation of Myklebust trench 7 (Photo: Irene Baug, 2008)

the Viking period. Layer 7 appeared as a long, orange line with humus, fine sand and some small flakes less than 1cm in size, and a few flakes around 20cm. Layer 8 had a red colouration and consisted of humus and small flakes (2-4cm) and a few larger, around 30cm. A piece of slag, most likely from a forge (Ole Tveiten pers. comm. 2009), was identified. One sample for C14-dating, TUa-7925, is dated to 985 ± 30 and calibrated to AD 1020-1040, signifying a somewhat younger dating than the uppermost sample TUa-7927. Layer 9 is a grey-orange deposit comprising fine sand and flakes around 2-4cm in size, and also some slabs 50cm in diameter and small amounts of humus. One sample was taken, TUa-7926, and dated to 1105 ± 35, calibrated to AD 895-990, indicating production in the late Viking period. By expanding the given intervals to a standard deviation of 99.7% (3σ), the dating frames overlap, indicating activity in the 900s and early 1000s (Fig. 5.35). It is more likely perhaps that layer 9 represents a somewhat older phase compared to the other layers. The base of the trench was not reached however and the starting phase of production is not known. Extraction may have begun somewhat earlier then the oldest date represented, that is to say the 9th or 10th centuries (Tab. 5.5). However, as the date frames are rather wide, it is possible that they also include the initial phase of production at the quarry.

Table 5.5: C14-analyses from Myklebust trench 7

Lab. Ref.	Context	Age BP	Calibrated Age	Material
TUa-7924	Myklebust trench 7, layer 3	965 ± 35	AD 1025-1155	birch charcoal
TUa-7927	Myklebust trench 7, layer 6	1030 ± 30	AD 995-1025	birch charcoal
TUa-7925	Myklebust trench 7, layer 8	985 ± 30	AD 1020-1040	birch charcoal
TUa-7926	Myklebust trench 7, layer 9	1105 ± 35	AD 895-990	birch charcoal

The stratigraphy indicates that layers 6-9 represent the oldest production, and layers 1-3 younger activity. This coincides with the dating of sample TUa-7924 from layer 3, calibrated to AD 1025-1155 and the early Middle Ages, and the samples TUa-7927 and TUa-7926 from layers 7 and 9 which indicate production in the latter part of Viking period. The sample TUa-7925 from layer 8 indicates, however, a somewhat younger dating frame. This may reflect some kind of disturbance, perhaps spoil movements not been detected in the relative stratigraphy identified in the section. The time difference is, however, rather small, being based on the rather long time intervals of the C14-analyses.

All layers consisted of slabs about 60-70cm in size and many flakes around 20-40cm, mostly c. 1-10cm. Several quernstones were recovered from the different deposits (Fig. 5.36).

Altogether 33 quernstones were found in trench 7. Six were too fragmentary to be measured, and eight quernstones in the profile of the trench could not be measured. Of the remaining 19 stones the diameters varied from 36-46cm, while half of them (11 stones) had diameters of 40-42cm. Only five quernstones have eyes, fully or partly finished, again an indication that this was mainly done after loosening the stones from the bedrock. The stones varied from 2-11cm in thickness due to fragmentation.

The relation between Myklebust trenches 6 and 7

No stone crosses were found in trench 7. This indicates that slabs for extraction of crosses were moved to a separate workplace, where unsuccessful crosses were also dumped - not in the spoil heaps as with the quernstones. This may explain the concentration of damaged crosses around trench 6, and their absence in trench 7. A large quarry with an intensive and comprehensive production would have needed correctly organized work processes. Shaping and cutting of crosses would hardly have occurred *ad hoc* at the quarry site. This kind of specialist work would call for suitable finishing areas. Quernstones on the other hand probably did not require special work sites as they were usually extracted directly from the bedrock, even though the eye in most cases was carved after extraction. In this quarry some quernstones also seem to have been made

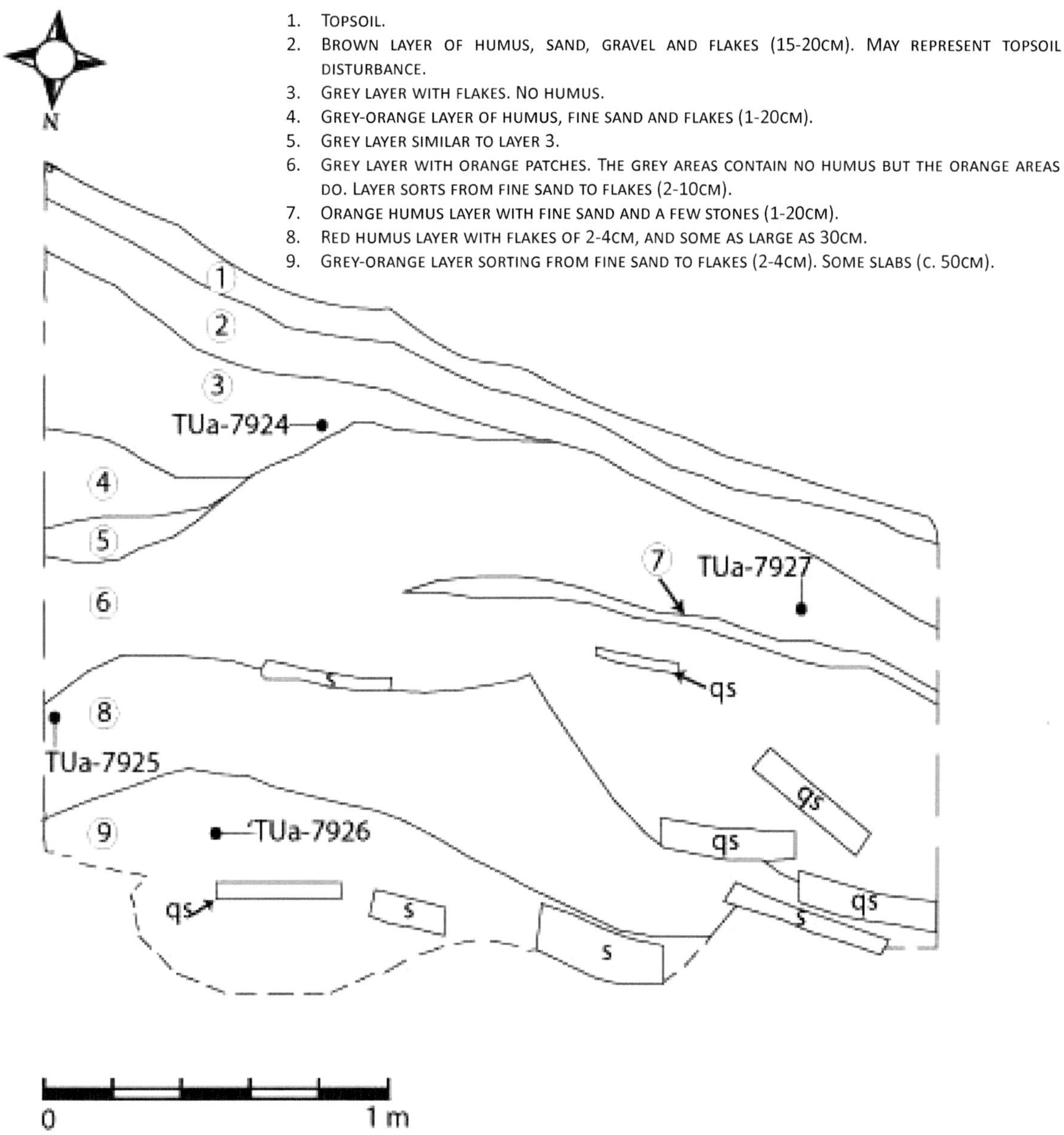

Figure 5.34: Section of Myklebust trench 7 (Drawing: Irene Baug/Åsne Helleve, 2008)

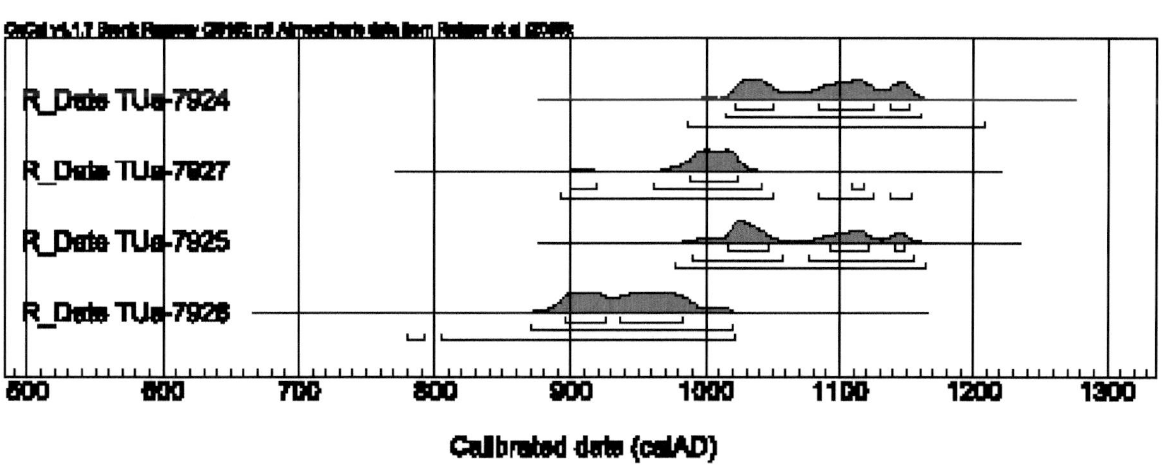

Figure 5.35: Multiple plot diagram of C14 analyses from Myklebust trench 7

FIGURE 5.36: MYKLEBUST TRENCH 7 DURING EXCAVATION (PHOTO: IRENE BAUG, 2008)

from slabs wedged out of the quarry, as seen in trench 6. Some slabs may have been too small and unsuitable for making stone crosses, but large enough for quernstones.

It is also possible that the spoil heap in trench 7 reflects an older production phase before extraction of crosses had begun, which may explain their absence in the trench. The production of quernstones probably started some time before cross production, but continued into the Middle Ages when the range of products was extended to include crosses as well. The shape of the carved rock indicates extraction of slabs also in the area around trench 7. Slabs intended for production of crosses may therefore have been extracted here too, but most likely transported to the work place close to trench 6. The large quantities of charcoal found in trench 6, including some large pieces (1-3cm) as well as ashes, may also reflect dumping from fireplaces. This may support the hypothesis of a special workplace close to the trench. Forges and other fire places were most likely located in one area.

Large amounts of damaged stones for hand querns were found in both trenches, indicating a poor quality of rock. Compared to the minerals in most quarries at Hyllestad, the garnets in this rock seem fewer and more diffuse, which may have reduced the grinding quality. There is also a slight difference in mean quernstone diameter between the two trenches: most stones from trench 6 ranges between 39-40cm and from trench 7 most stones have diameters of 40-42cm, but this may be coincidental. The stones from trench 7 were also somewhat more fragmented than those found in trench 6. Except for one possible extraction mark of a millstone from the bedrock at the base of trench 6, and one millstone found in the brook north of trench 6, no further traces of millstones were noted. This type of production seems to have been of minor, or no importance. The quality of rock could explain this, as different mineralogical qualities were needed for millstones. There may have been different requirements for the large and more costly millstones compared to quernstones, an aspect that will be discussed in chapter 10.

The extant rock walls of the quarry indicate slab extraction, representing the end phase of production. The quarry may have begun as a provider of quernstones in the Viking period. Later, in the middle of the 13th century at the latest, but perhaps as early as the 11th or 12th centuries, the product range changed, including stone crosses alongside quernstone production, a development that is also reflected in one of the still-standing Hyllestad crosses at Eivindvik, with traces of quernstone extraction on its beam (Baug 2002:69). Evidence for products other than quernstones and millstones also seems to exist at the Saltdal quarries, where a gravestone of the garnet-mica-schist type has been identified (Trones 2003:46).

5.4 Archaeological investigations at Sørbø (no. 32)

Only five quarries are known at Sørbø farm, all of them rather small (Fig. 5.37). Four only reveal quernstone production, while one - the investigated quarry - has traces of both millstones and querns. Some of the quarries are located on cultivated land (Heldal and Bloxam 2008:99), and it is likely that more quarries may have been opened there, and that cultivation and expansion of the settlement have destroyed older production sites, something also documented at Rønset.

5.4.1 Sørbø trench 1 - quarry with production of quernstones and millstones

The investigated site at Sørbø is a shallow quarry located in the outlying fields of the farm, c. 40m southeast of the house at holding 5 and about 30m from the fjord.[7] It has not been disturbed by human activity since production ended, making it an interesting research site. Today the whole area is covered with a dense forest of deciduous trees, in a slightly sloping landscape towards the fjord, with a fine beach of sand and pebbles, as well as a smooth, coastal rock slope into the sea. At high tide it would probably be relatively easy to load stones from this rock into boats and ships.

The quarry consists of a c. 8m-long and 2m-high rock wall in an east-west direction. At the eastern end the rock turns

[7] UTM WGS84, zone 32: east 294060 north 6793954, 18m.a.s.l.

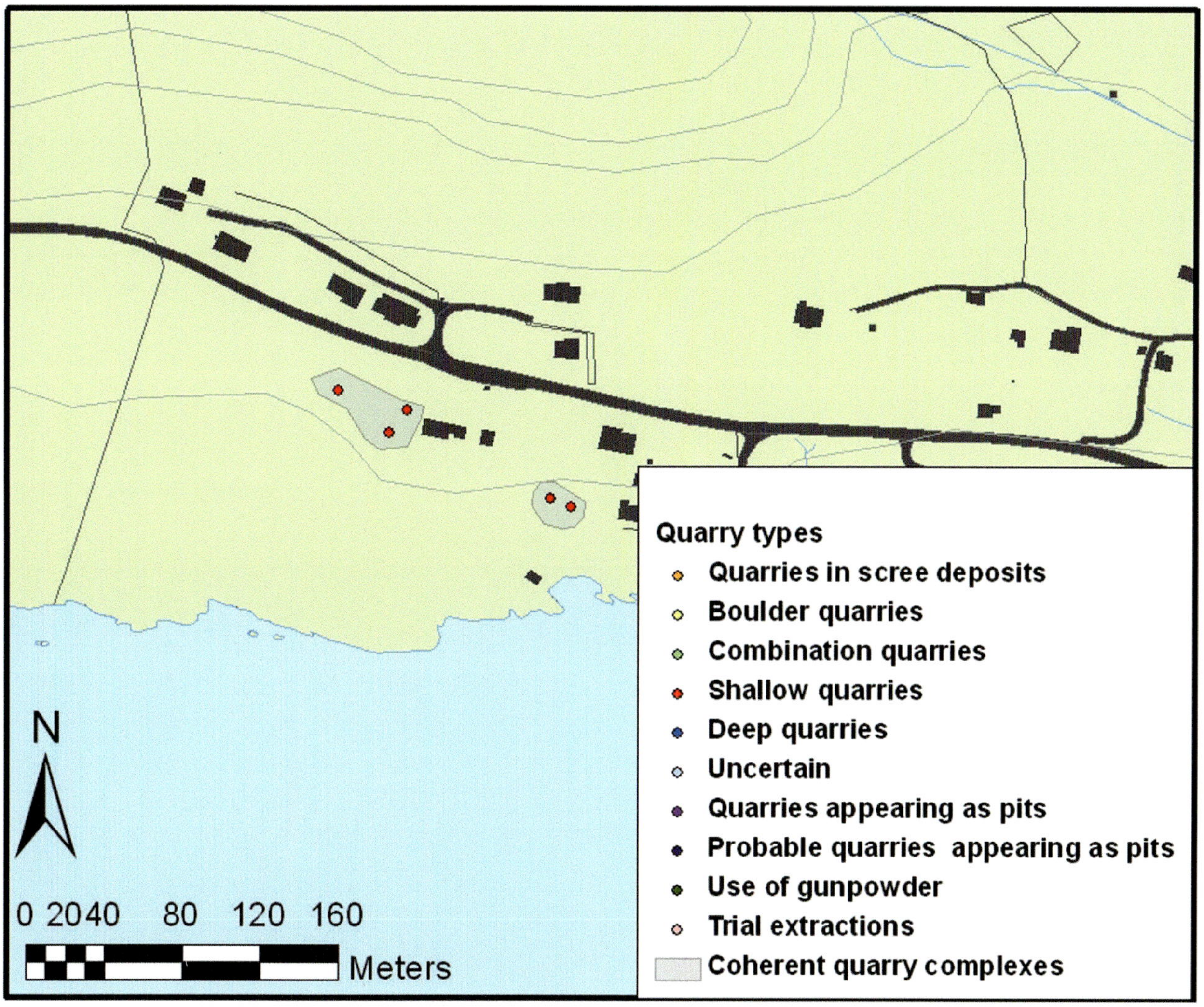

Figure 5.37: Map of quarry types at Sørbø (Map: Irene Baug, 2013) (Map-data for Hyllestad quarries from a survey by Tom Heldal, NGU)

northwards, where a natural split in the rock has resulted in a large slab of stone lying upon the rock. An unfinished millstone, 1.20m in diameter and 20cm thick, was left on the slanting slab. Holes for extractions along the millstone, c. 10-20cm apart, were made in an attempt to loosen it, but the work was never completed. Remains from the extraction of quernstones (38 and 49cm in diameter) were located close by.

There is a natural split in the southern side of the rock, where a large slab of stone, c. 15-25cm thick is located towards the ground, containing grooves from quarrying at the lower part. The working is small compared to most of the quarries found along the Åfjorden, and quarrying traces are also few.

Approximately 6m south of the rock, a natural mound with a flat area between the mound and rock is located, with an enormous block of stone at the eastern end and a smaller rock at the western. The flat area in the middle was considered part of a previously worked-out quarry covered with production waste. A trench was opened here, along the southern side of the rock, in an east-west direction.

In the quarry, as well as in the area immediately surrounding it, remains of unfinished quernstones (diameters of 38-50cm) were found; they had been extracted from small slabs. Quarrying of quernstones from slabs is a rather rare phenomenon at Hyllestad. The many natural splits in the rock may have made slabs of stone more convenient for quernstone production here. It is, however, also possible that slabs were intentionally extracted from the rock to make production easier, but so far no remains from slab extraction have been recorded in this area.

Sørbø trench 1

As the flat area between the production rock and mound was considered to contain production waste, a 4m-long, 1.5m-wide and about 60cm-deep trench, in an east-west direction, was cut parallel to the rock, c. 1.5-1.6m south of it. In the base (east end), the bedrock was uncovered but there were no remains of quarrying. In the southern end of the trench a millstone (1.1m in diameter) was located (Fig. 5.38). Natural stones (30-40cm) were found in the western part of the trench underneath the production waste. As

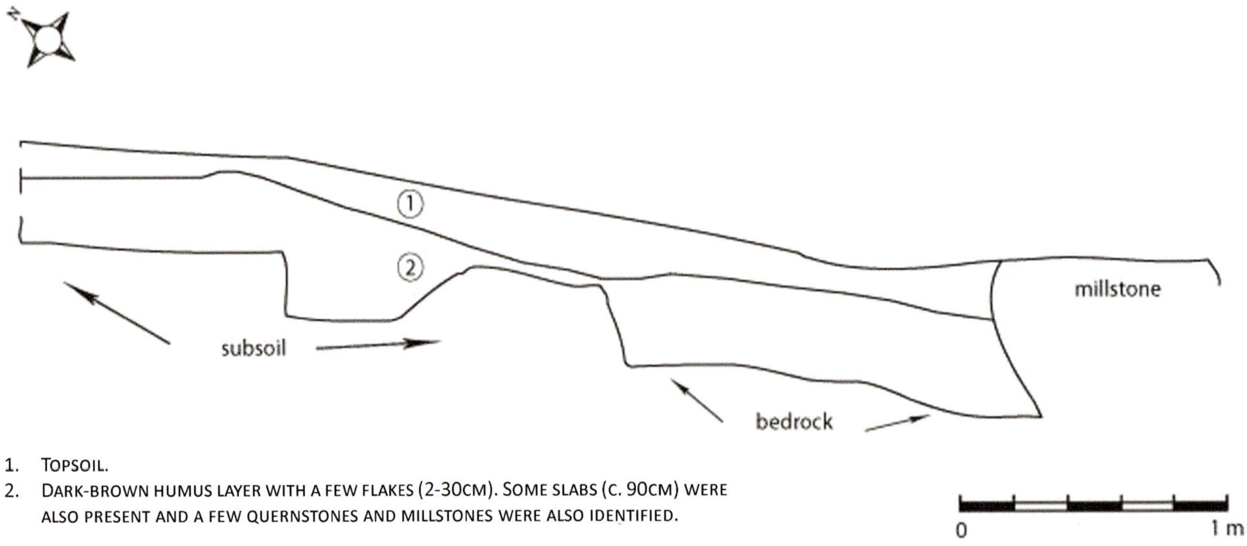

1. Topsoil.
2. Dark-brown humus layer with a few flakes (2-30cm). Some slabs (c. 90cm) were also present and a few quernstones and millstones were also identified.

Figure 5.38: Section of Sørbø trench 1 (Drawing Irene Baug, 2006)

the stones were not of garnet mica-schist they seemed to represent the base layer of the cultural deposits.

Section interpretation: Sørbø trench 1

The trench consisted of only two layers, including topsoil (layer 1) (Fig. 5.38), interpreted as a mixture of spoil and natural deposits, with no remains from quarrying at the base. Most likely quernstone extraction was limited to the rock north of the trench, where spoil seems to have been deposited in front of the rock.

Layer 2 contained dark-brown humus and flakes from production, varying in size, from a couple of cm up to 20-30cm on average. Slabs of 90cm in diameter also occurred, as well as fragments from one millstone and several smaller quernstones. The millstone measured 90cm in diameter and 22cm in thickness (Fig. 5.39).

The trench contained only small amounts of production waste, together with stones of rock types other than garnet-mica schist, with no traces of hewing, which indicate low activity in the quarry. This is also supported by the absence of charcoal. The production in the quarry was probably so small that no forge or fireplaces were needed in its vicinity. It is also possible that the production remains represent a more sporadic extraction, perhaps stretching over a longer period. Altogether, the quarry seems to represent a different form of production than in the central production areas of Rønset and Myklebust. Whether this was a quarry producing quernstones and millstones solely for domestic or local use, or if they were commodities for further export, is difficult to ascertain, as it was impossible to date the activity and then compare the quarry with other datable sites.

The results from Sørbø may perhaps indicate that production sites had to be of a certain size in order make

Figure 5.39: Sørbø trench 1 from the north (Photo: Irene Baug, 2006)

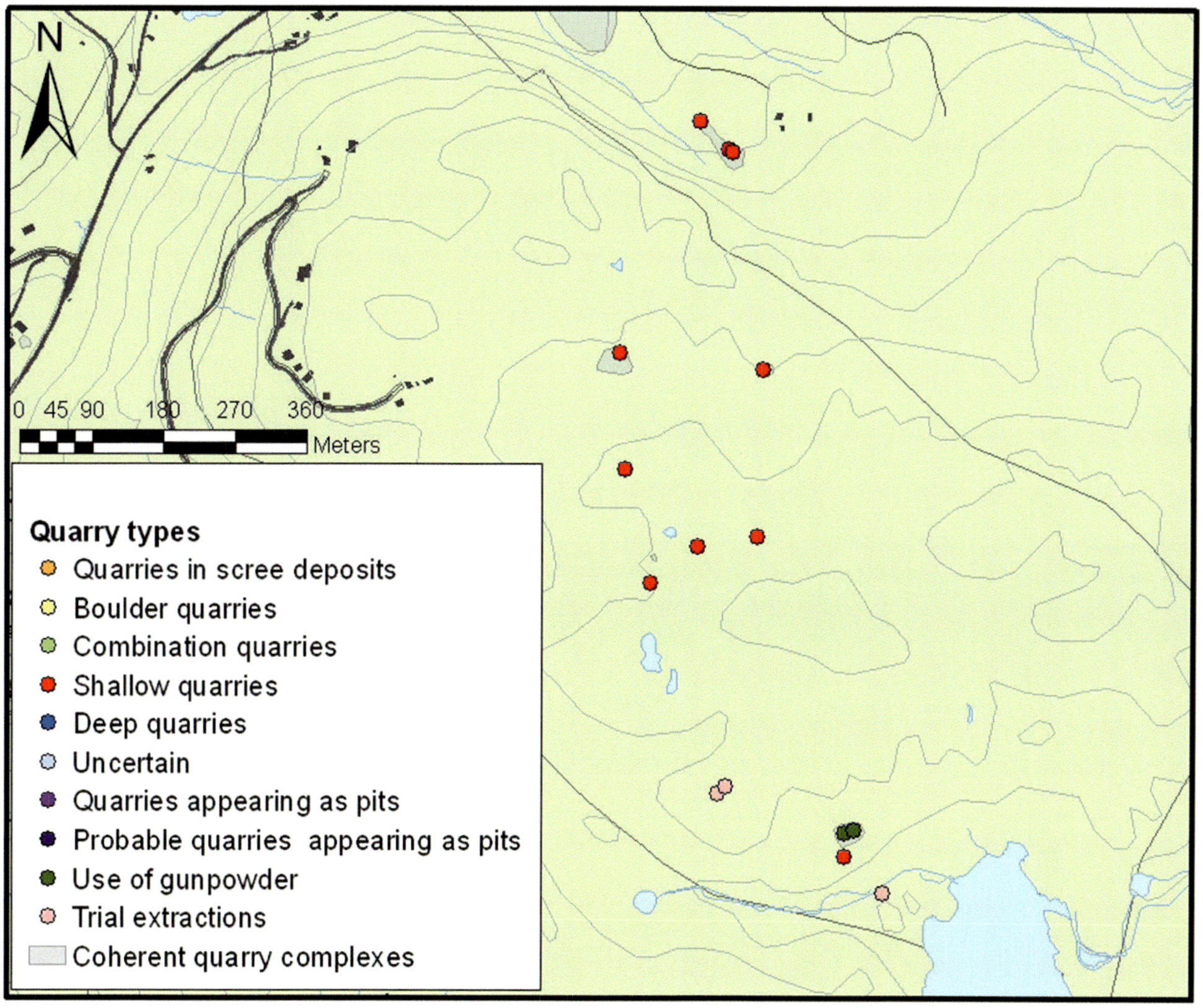

Figure 5.40: Map of the different quarry types at Sæsol (Map: Irene Baug, 2013)
(Map data for the Hyllestad quarries from a survey by Tom Heldal, NGU)

finds of charcoal remains. If this is the case, it may make it difficult to investigate and date the smaller and marginal quarries in Hyllestad, where many have the same size as this one, some even smaller.

5.5 Archaeological investigations at Sæsol (no. 78)

The farm at Sæsol (no. 78), where the quarries also are more dispersed and smaller in size than the quarries of Rønset and Myklebust, is interesting in relation to the issue of marginal and central quarrying. The quarries are scattered in the outlying fields of the farm, about 200m.a.s.l., and altogether 13 quarries were identified (Fig. 5.40). Rocks of sufficient quality for quernstones are more dispersed at Sæsol compared to Rønset and Myklebust (Tom Heldal pers. comm. 2009). This may also explain the more scattered distribution of quarries compared to the central areas. Today the terrain is overgrown and boggy, and several of the quarries, both the rocks and associated spoil heaps, are almost entirely covered with vegetation, making it difficult to get a clear overview of the production landscape. It is likely that more quarries are located at the farm than have been identified so far.

Most of the quarries at Sæsol show production of quernstones for hand querns, but at least two quarries also produced millstones - but to a much smaller degree than hand querns. Additionally, two quarries that used gunpowder are located at the farm, indicating a long production period in the area. Some of the quarries are located in the mountain, about 1km in a straight line from the sea and the nearest harbour. Yet, in the early 20th century, large millstones were transported down the rugged and steep terrain to the fjord (Audun Oppedal pers. comm. 2012).

It is possible that quernstones, and perhaps also millstones, were transported across the mountain. The old road between the communities of Hyllestad, Gaular and Guddal in Sunnfjord went northwards across the heath area at Sæsol. The region is still called Dalemannheia, after the Norwegian expression *dalemann*, referring to people from the inland communities. People and horses went across the heath down to the fjord with commodities for exchange, and some of the quarries are located along this route (Aaberg 1929:11; Audun Oppedal pers. comm. 2008).

FIGURE 5.41: OVERVIEW OF THE AREA AROUND SÆSOL TRENCH 7 (MARKED WITH AN ARROW); TAKEN FROM THE EARLIER INVESTIGATED QUARRY, SÆSOL TRENCH 6 (PHOTO: IRENE BAUG, 2008)

One quarry in this area (Sæsol trench 6) was the subject of earlier archaeological investigation (Baug 2002).[8] This is a shallow quarry with circular marks in the 12-13m long rock where the stones were extracted. Only quernstones were produced. The quarry is located 500m south of the farm houses at holding 2, at an altitude of about 200m.a.s.l. Production at the quarry is dated to 715-890 AD, indicating that the activity dates to the beginning of the Viking period (Baug 2002:55-58).

An archaeological investigation of the agricultural landscape, mainly in the infield of the farm, carried out by Anja Sætre in 2008, revealed evidence of activity, in the form of animal husbandry and pasture, from the late Neolithic and early Bronze Age. There is, however, no sign of arable farming in the early Iron Age, and the farm seems to have been established around the 8th century at the earliest, AD 720-890 and AD 785-975. From this period, there is evidence of a more intensive use of the land and fields than previously, indicating a permanent settlement with farming activities that included harvesting and animal husbandry (Sætre 2008:65-77). An interesting question is how this activity relates to quarrying at the farm.

5.5.1 Sæsol trench 7: quarry with production of quernstones

The archaeologically investigated quarry at Sæsol is a shallow site located some 300m southeast of the farm houses at holding 4, on a plateau c. 190m.a.s.l.[9] Today the area is boggy and covered by deciduous trees (Fig. 5.41). About 145m separate this quarry from the earlier investigated quarry at Sæsol (Baug 2002:55-58), located to the south on a higher plateau, about 210m.a.s.l. It was of interest to look at the relationship between these two quarries, whether they were in use at the same time, or whether they represented different production phases.

The site chosen for investigation appeared to be rather disturbed. Large sections of the rock have been unearthed and most of the spoil heap destroyed. Only a small part of the latter was intact and the trench was opened here. Because of the earlier intrusion it was possible to excavate deeper into the spoil heap than otherwise would have been possible. The edges of the spoil heap can still be determined; they originally extended from the western end of the rock to about 14m to the west, covering an area of about 17m in a north-south direction. As only one quarry was located in this area the entire spoil heap originates from extractions at this quarry.

The carved rock is c. 13.2m long, in an east-west direction and presently 1.5m high; it is completely covered with circular marks from quernstone carving (Fig. 5.42). The earlier disturbances unearthed the base level of the quarry in front of this rock, revealing circular extraction marks. This lower part of the rock has been covered up. Close by, 2.5m south of the production rock, another rock appears,

[8] UTM WGS84, zone 32: east 301762 north 6788114, c. 200m.a.s.l.

[9] UTM WGS84, zone 32: east 301743 north 6788255, c. 190m.a.s.l.

FIGURE 5.42: CARVED ROCK WITH CIRCULAR QUERNSTONE EXTRACTION MARKS (PHOTO: IRENE BAUG, 2008)

but only a few hewing traces are visible. It is partly covered with production waste and vegetation.

Sæsol trench 7

Sæsol trench 7 was dug about 9.7m south of the western end of the carved rock. The trench was excavated along the remaining part of the spoil heap, 3.5m in length in a north-south orientation. The heap was excavated from the outer edge and c. 1.5m into the heap. Only one profile was kept (Fig. 5.43).

Profile interpretation: Sæsol trench 7

Trench 7 consisted of five layers including topsoil (layer 1) (Fig. 5.44). Layer 2 was a dark brown layer of humus and gravel with some flakes (up until 20cm in size) and also some small roots, probably the result of topsoil sinking into the deposit. Layer 3 was a light-brown layer consisting of sand and some larger stones, small flakes and roots. One sample for C14-dating, TUa-7934, was taken from the trench, about 10cm under the topsoil and dated to 915 ± 30, calibrated to AD 1040-1170, the early Middle Ages. Layer 4 was a red-brown layer of humus, flakes, and some small roots. There was also coarse gravel and flakes c. 5-20cm in size. A few larger stones (40-60cm) and some quernstones were also found. One sample for C14-dating was taken, T-20110, dated to 995 ± 55, calibrated to AD 1000-1155, concurring in time with the other sample. Layer 5 was a brown sandy layer observed as a pocket between layers 4 and 6, containing soil and humus as well as flakes, c. 2-15cm in size. Layer 6 was a yellow-brown layer of coarse gravel and with larger flakes (5-10cm); some stones were as large as 40-60cm and there was also more charcoal. Several quernstones were found in the deposit, as well as a fragment of one fairly small millstone, 70cm in diameter and 2-6cm thick. The millstone was perhaps meant for a watermill with a horizontal wheel (Norw. *vasskvern*), as millstones of this diameter are known from more recent mills of this type. Two samples for C14-dates were taken from the layer. The southernmost sample, TUa-7935, was dated to 1030 ± 30, calibrated to AD 995-1025. The sample further north, T-20109, was dated to 915 ± 50, calibrated to AD 1035-1215 (Tab. 5.6). By expanding the given intervals to a standard deviation of 95.4% (2 σ), TUa-7935 is calibrated to AD 898-1117, and T-20109 to AD 1024-1215, which gives an overlap within the span AD 1024-1117. Extraction in the early Middle Ages is plausible (Fig. 5.45). Layer 7 was a black soot layer, covering the natural substratum, bedrock and bog. There are no remains of extraction from this rock, and burnt moss still covers it. Layer 7 consisted of moss and vegetation on the bedrock, the result of the mossy bog being burnt. Whether this represents a natural or a cultural event is difficult to ascertain. No soot layer was observed in the earlier investigated quarry at Sæsol (trench 6) which might indicate a larger forest fire, but here the base layer was not reached at the southern part of the trench (Baug 2002:56). The archaeological investigations of the fossilized agricultural structures at Sæsol by Anja Sætre did not reveal any clear layers of charcoal that could indicate larger fires. Charcoal found in

FIGURE 5.43: EXCAVATION OF SÆSOL TRENCH 7 (PHOTO: IRENE BAUG, 2008)

1. TOPSOIL.
2. DARK-BROWN HUMUS LAYER WITH GRAVEL AND SOME FLAKES UP TO 20CM.
3. LIGHT-BROWN LAYER OF SAND, A FEW FLAKES AND A FEW LARGER STONES.
4. RED-BROWN HUMUS LAYER WITH GRAVEL AND FLAKES (C. 2-20CM), AND SOME QUERNSTONE FRAGMENTS.
5. BROWN SAND LAYER WITH HUMUS AND FLAKES (C. 2-15CM)
6. YELLOW-BROWN LAYER WITH GRAVEL, FLAKES (C. 5-40CM) AND SOME QUERNSTONES.
7. BLACK SOOT LAYER COVERING THE NATURAL SUBSTRATUM: BEDROCK AND BOG.

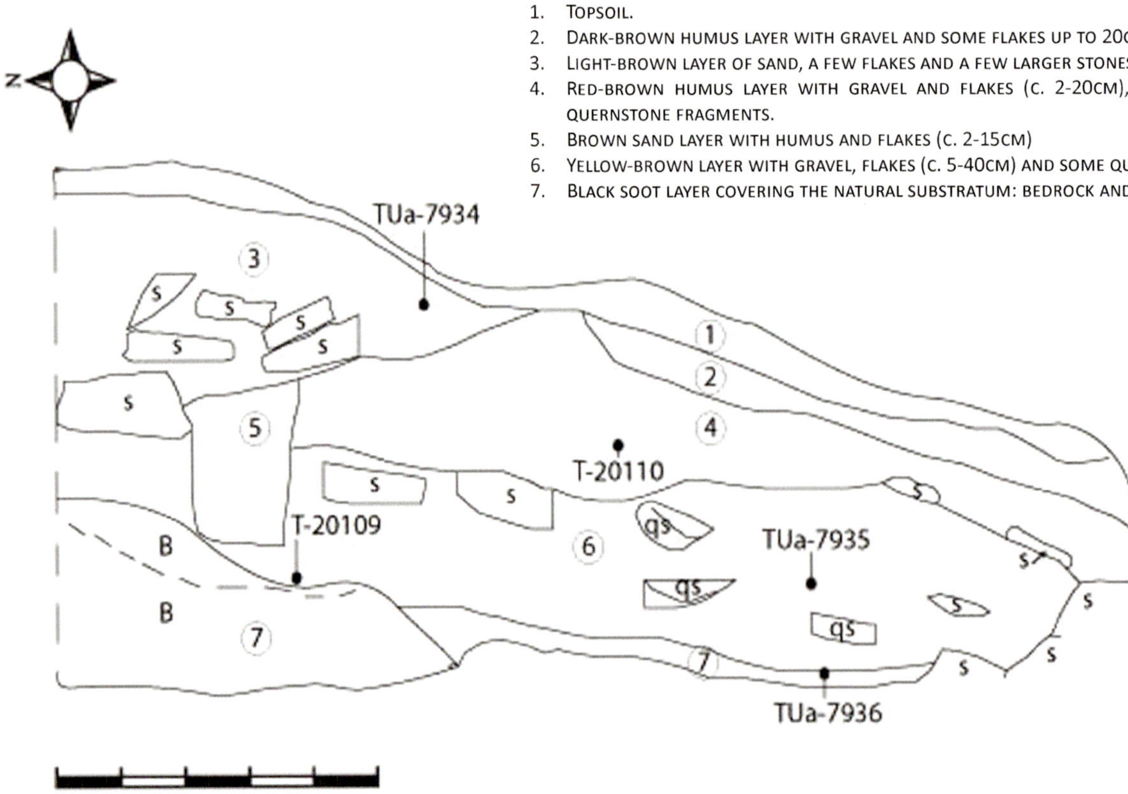

FIGURE 5.44: SECTION OF SÆSOL TRENCH 7 (DRAWING: IRENE BAUG/ÅSNE HELLEVE, 2008)

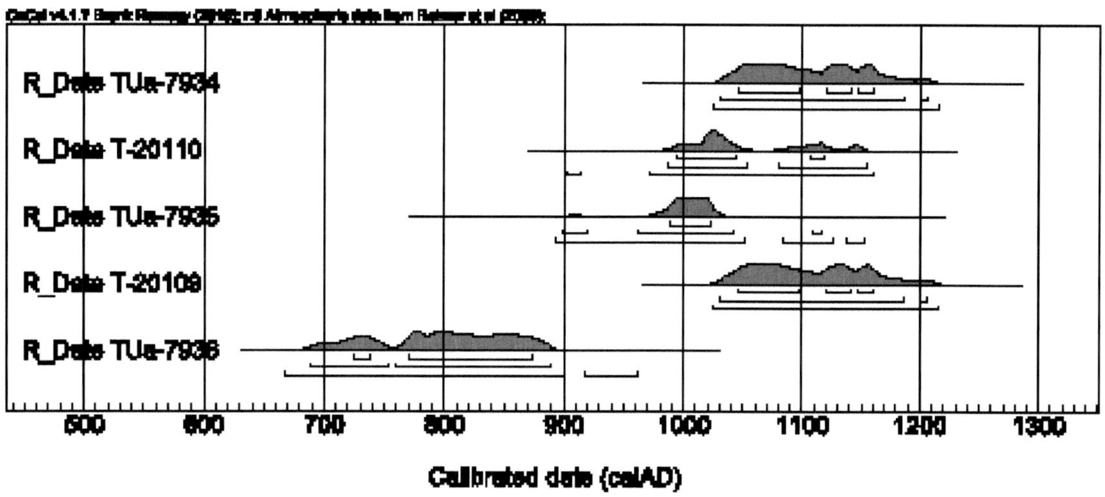

FIGURE 5.45: MULTIPLE PLOT DIAGRAM OF C14 ANALYSES FROM SÆSOL TRENCH 7

these layers was interpreted as land clearing activity, with a following period of cultivation (Sætre 2008:36-77). The possibility that layer 7 represents a controlled burning of vegetation before the start of production should therefore also be considered. One sample for C14-dating was taken from the layer, where only pine charcoal was identified, dated to 1220 ± 35, calibrated to AD 775-880 and the early Viking period, somewhat earlier than production in the quarry. Whether or not layer 7 represents a burning off of the vegetation in connection with the beginning of production is unclear.

Table 5.6: C14-analyses from Sæsol trench 7

Lab. Ref.	Context	Age BP	Calibrated Age	Material
TUa-7934	Sæsol trench 7, layer 3	915 ± 30	AD 1040-1170	pine charcoal
T-20110	Sæsol trench 7, layer 4	995 ± 55	AD 1000-1155	pine charcoal
TUa-7935	Sæsol trench 7, layer 6	1030 ± 30	AD 995-1025	pine charcoal
T-20109	Sæsol trench 7, layer 6	915 ± 50	AD 1035-1215	pine charcoal
TUa-7936	Sæsol trench 7, layer 7	1220 ± 35	AD 775-880	pine charcoal

All five samples for C14-dating collected from the trench were pine charcoal, which, as noted, may be somewhat misleading related to the production. However the samples concur well chronologically and they do not give an unlikely high age of activity compared to dates from other Hyllestad quarries. This makes it feasible that the samples can be used to date the quarry production.

All layers in trench 7 contained large amounts of charcoal, some of it clearly the result of dumping from fireplaces. In the earlier investigated quarry (Sæsol trench 6) only small quantities of charcoal were uncovered (Baug 2002:57-58). This may, of course, be a coincidence, with trench 7 accidentally located near fire sites of some sort. However, the possibility of differences in the work processes around the two quarries should also be taken into account, perhaps because of different time periods of the production. While the dating from the previously investigated quarry, trench 6, may indicate activity in the early Viking period (Baug 2002:57-59), the quarry associated with trench 7 is dated to the early Middle Ages, perhaps with a start in the late Viking period.

Remains of 26 quernstones and one millstone fragment were found in trench 7. The earlier disturbances also uncovered a millstone (c. 70cm) and on this was a circle from the extraction of a quernstone (Fig. 5.46). Ten of the 26 quernstones found in trench 7 were so fragmentary that the diameters could not be determined. The remaining 16 have diameters ranging from 35 to 47cm, with the majority (11) having diameters between 38 and 42cm, indicating a more or less standardized size (however both smaller and larger quernstones are also found). The stone fragments vary in thickness between 2 and 9cm and give little indication of their intended size. One stone with a diameter of 52cm has also been uncovered, most likely intended for a hand quern. Circles on the rock indicate the extraction of stones for hand querns, and no hewing marks from millstones are visible. Altogether, the evidence indicates small-scale millstone production, with quernstones as the main product. The previous disturbances of the quarry uncovered some large slabs, around 80cm long. There is no clear evidence that slabs were extracted. Some of the slabs with traces of extracted quernstones seem to be loose pieces used in the production process.

5.6 Dating the quarries and their products

Of the six investigated sites at Hyllestad it was possible to take charcoal for C14-dating from five of them. The Sørbø quarry is the only site that could not be dated.

The sizes of the investigated quarries and the methods involved make it difficult to date the start of production. Nearly all the sites exhibit large-scale production that may

FIGURE 5.46: MILLSTONE FRAGMENT WITH TRACES OF QUERNSTONE CARVING (PHOTO: IRENE BAUG, 2008)

indicate long periods of activity at each quarry, with the exceptions of Sørbø and Sæsol, which seem to represent more marginal production. Generally, the results from the C14-analyses correspond well with the relative quarry stratigraphy, indicating a primary deposition of the spoil. Only a few discrepancies have been documented, and in most of these cases an expansion of the standard deviation to 95.4% makes the dating framework for absolute chronology fit within the relative stratigraphy. Because of the sizes of the quarries and spoil heaps, some disturbance and movement of older spoil heaps in connection with later activity is to be expected, which may have led to a mix of different production phases in the same spoil heap. Because of the relatively wide dates given by the C14-analyses such processes may be difficult to identify more accurately. At Rønset trench 5, the dating of the uppermost layer seems to indicate a reverse chronology, and the deposit is interpreted as collapses from the spoil heap. At Myklebust trench 4, there is also a certain discrepancy based on C14-analyses and this may also indicate some disturbance. At the other quarries investigated, all the samples show a certain degree of overlap, concurring with the relative stratigraphy. Altogether, this stratigraphy, combined with the C14-analyses, gives a good indication of both the relative chronology and dating of quarry production.

Extractions from three of the quarries, Rønset trench 5, Myklebust trench 7, and Sæsol trench 7 seem to have begun in the late Viking period in the 10th century, at the latest (Tables. 5.8-5.10). C14-analyses from Myklebust trench 7 indicate production from the 10th century, but also here the initial phase of quarrying may be older. The C14 analyses from Myklebust trenches 6 and 7 have date spans from the 11th to the 13th centuries. The quarry is large, and it is highly possible that production continued over centuries, as indicated by the analyses. However a more exact dating is difficult to give. The quarry at Sæsol trench 7 is smaller than the investigated quarries at Rønset and Myklebust. Most evidence points to an activity there sometime during the 11th and 12th centuries, the late Viking period and early Middle Ages. Production at Rønset trench 5 is dated to the 11th to 13th centuries, but the initial phase may be older than the oldest dating. The workplace underneath the rock shelter at Rønset, dated to the 11th and 13th centuries, may have been used in connection with the quarrying at Rønset trench 5. One of the investigated quarries, Rønset trench 4, is clearly dated to the 13th century, which seems to be the end phase of production. How far back the quarrying dates is not known. The size of the quarry indicates a large-scale and probably long period of production. Samples for C14-analyses from Myklebust trenches 4 and 5 indicate extraction sometime in the 13th and 15th centuries, and concur with the earlier investigations at the quarry site. This represents a large-scale and long lasting period of production in the area, where extraction started in the late Viking period, as indicated in the C14-dates from Myklebust trench 2 (AD 890-1015) (Baug 2002:53-54).

For some of the investigated quarries pine charcoal was the only available material for C14-dating (Tables 5.7-5.9), which could give high ages for the quarrying, but this does not seem to be the case. In terms of the spoil heaps, where both pine and birch were used as dating material, they concur in time, i.e. Rønset trench 5 and the rock shelter. Rønset trench 4 is dated to the 13th century by its pine charcoal only, and is somewhat younger than several of the other quarries. This is also the case with sample TRa-787 from Myklebust trench 4, calibrated to AD 1400-1430. At Sæsol only pine charcoal was used to date production at both of the quarries (trenches 6 and 7), making it difficult to evaluate the dates. But compared to the Hyllestad production landscape as a whole the pine dating material does not give an unexpected high age. Sæsol trench 6 is dated to AD 715-890, indicating activity in the Viking period, possibly early. This seems also the case at Otringsneset, where birch and hazel both date production to AD 730-965 (Baug 2002:50, 57).

Table 5.7: C14-analyses from Rønset trenches 4-5 and the rock shelter test pits 1-2

Lab. Ref.	Context	Age BP	Calibrated Age	Material
TUa-6709	Rønset trench 4, layer 2	73.8 ± 0.6% activity	after AD 1960	hazelnut
TUa-6710	Rønset trench 4, layer 2	700 ± 35	AD 1285-1300	pine charcoal
TUa-6711	Rønset trench 4, layer 3	810 ± 35	AD 1220-1280	pine charcoal
TUa-7928	Rønset trench 5, layer 3	1030 ± 35	AD 990-1025	pine charcoal
TUa-7931	Rønset trench 5, layer 3	870 ± 35	AD 1165-1225	birch charcoal
TUa-7930	Rønset trench 5, layer 4	895 ± 35	AD 1050-1215	birch charcoal
TUa-7929	Rønset trench 5, layer 5	950 ± 35	AD 1025-1160	birch charcoal
TUa-7932	Rønset trench 5, layer 6	1000 ± 35	AD 1010-1030	birch charcoal
TUa-7933	Rønset trench 5, layer 7	980 ± 35	AD 1015-1150	birch charcoal
TRa-785	Rock-shelter, test pit 1, layer 1	810 ± 35	AD 1215-1275	birch charcoal
TRa-784	Rock-shelter, test pit 1, layer 5	990 ± 30	AD 1015-1040	birch charcoal
TRa-786	Rock-shelter, test pit 2, layer 3	940 ± 30	AD 1030-1165	pine charcoal

Table 5.8: C14-analyses from Myklebust trenches 4-7

Lab. Ref.	Context	Age BP	Calibrated Age	Material
TUa-7920	Myklebust trench 6, layer 2	800 ± 35	AD 1225-1280	birch charcoal
T-20108	Myklebust trench 6, layer 2	870 ± 65	AD 1050-1245	birch charcoal
TUa-7922	Myklebust trench 6, layer 2	845 ± 35	AD 1175-1250	charcoal of birch
TUa-7921	Myklebust trench 6, layer 5	980 ± 35	AD 1015-1155	birch and hazel charcoal
TUa-7923	Myklebust trench 6, layer 5	890 ± 35	AD 1060-1215	birch charcoal
TUa-7924	Myklebust trench 7, layer 3	965 ± 35	AD 1025-1155	birch charcoal
TUa-7927	Myklebust trench 7, layer 6	1030 ± 30	AD 995-1025	birch charcoal
TUa-7925	Myklebust trench 7, layer 8	985 ± 30	AD 1020-1040	birch charcoal
TUa-7926	Myklebust trench 7, layer 9	1105 ± 35	AD 895-990	birch charcoal

Table 5.9: C14-analyses from Sæsol trench 7

Lab. Ref.	Context	Age BP	Calibrated Age	Material
TUa-7934	Sæsol trench 7, layer 3	915 ± 30	AD 1040-1170	pine charcoal
T-20110	Sæsol trench 7, layer 4	995 ± 55	AD 1000-1155	pine charcoal
TUa-7935	Sæsol trench 7, layer 6	1030 ± 30	AD 995-1025	pine charcoal
T-20109	Sæsol trench 7, layer 6	915 ± 50	AD 1035-1215	pine charcoal
TUa-7936	Sæsol trench 7, layer 7	1220 ± 35	AD 775-880	pine charcoal

Do the different quarry types represent different production periods?

Earlier investigation at Hyllestad (Baug 2002:29-60) indicated that production might go back to the early Viking period. A shallow quarry at Sæsol (trench 6) and a combination site at Rønset, Otringsneset (trenches 1 and

2), which most likely started as a shallow quarry, are also dated to this period. As only one sample for C14-dating was found at Sæsol trench 6 it was difficult to assess this date. The sample consisted of pine charcoal and may show a too early date. However most evidence indicates that the shallow quarry type represents the first form of extraction. Shallow quarries were also in use in the early Middle Ages, as indicated by C14-analyses from Sæsol trench 7. Shallow quarries have not been dated later than the early Middle Ages.

The archaeological investigations indicated that the deep quarry type was also introduced from the late Viking period. This is the case for the quarry complex at Myklebust trench 2, and Myklebust trench 7 (cf. Baug 2002:53-55; tab. 5.9). This quarry type continued into the Middle Ages, and at least for Myklebust trenches 2 and 3 a long lasting period of production is documented. The deep quarries at Rønset (trenches 4 and 5) have been dated to the early and high Middle Ages. However only the last production phase has been investigated and the date of the start of extraction is not known. The sizes of both quarries may indicate lengthy production, and at least for Rønset trench 5 a start phase in the Viking period can be expected.

Even though the production at Sæsol can be documented from the early Middle Ages, no deep quarries here have been identified. The extent of suitable quernstone rock is limited compared to the central production area and the site may also have been less suitable to justify deeper quarries (Tom Heldal pers. comm. 2009).

The shallow quarry type therefore seems to represent the oldest Hyllestad technology, but the deep quarry type began not long afterwards. Archaeological investigations also indicate a certain overlap of the chronology at the different quarry types (Fig. 5.47).

Three samples for C14-dating from Rønset trench 1 and Sæsol trench 6 have been dated to the pre-Roman and Roman Iron Age, giving an unexpectedly high age for the quarrying. One sample from Sæsol trench 6 was even dated to the Neolithic, giving an even more unlikely high age. All these samples are from pine charcoal, most likely from wood that died quite a long time before burning, and therefore possibly misleading in relation to the quarrying (Baug 2002:41-42, 57, 59-60); recent Hyllestad investigations support this assessment. The younger dates from Rønset trench 4 and Otringsneset trench 4, indicating production into the early modern and modern periods, are also considered to be misleading in relation to the quarrying here.

Chronological development of product types

The production of quernstones for hand querns at Hyllestad may go back to the early Viking period (Baug 2002:49-50, 56-58). The new Hyllestad investigations substantiate production of quernstones in the Viking period, at least at Myklebust trench 7, but perhaps also at Sæsol trench 7, however dating here from the late Viking period.

Millstone production has previously been loosely dated to the end of the Viking period or the transition to the Middle Ages at Myklebust, and possibly also at Rønset, where the largest stones have diameters of c. 1m, representing stones for watermills (AD 890-1015 and AD 990-1160) (Fig. 5.48) (Baug 2002:37-38, 52-53). As the conditions for dating single objects from the spoil heaps were not ideal, these C14-analyses may give a somewhat early age for millstone production and should therefore be used with caution. Viking production of millstones at Hyllestad should not, however, be entirely excluded. In the deep quarry at Rønset trench 5, fragments of the stones from the trench, and the circular marks on the carved bedrock,

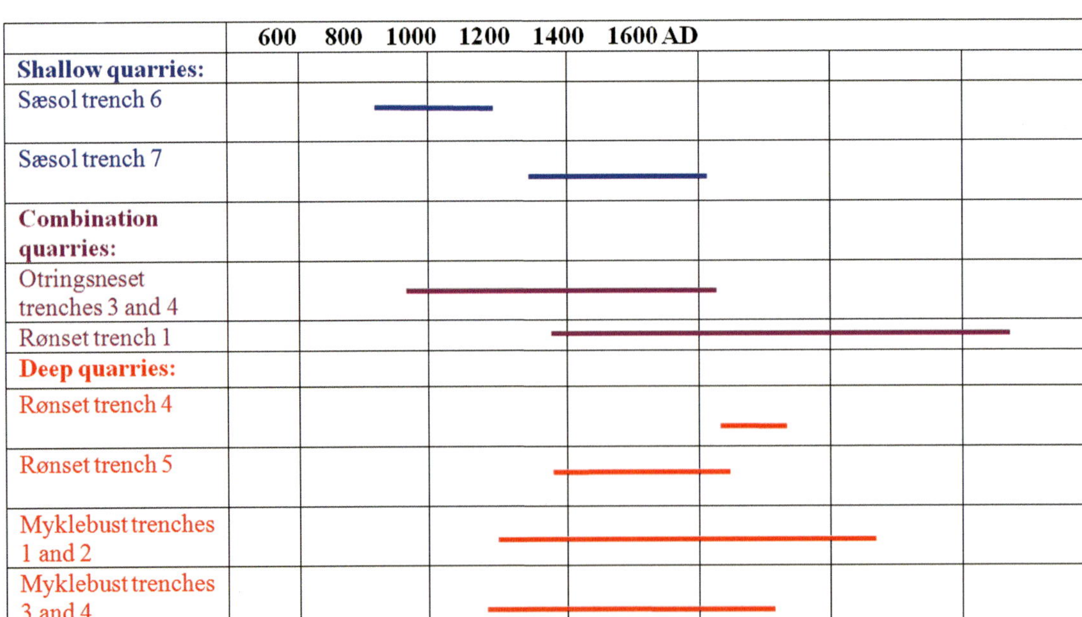

Figure 5.47: Overview of production dates based on archaeological investigations

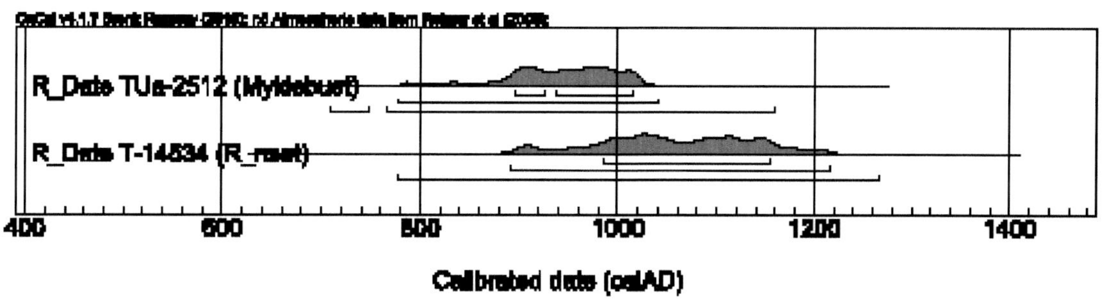

FIGURE 5.48: MULTIPLE PLOT DIAGRAMS OF C14-ANALYSES OF MILLSTONE PRODUCTION AT MYKLEBUST (TUA-2512) AND RØNSET (T-14534)

indicate extraction of stones with diameters from 40cm to more than 1m, and therefore production of both quernstones and millstones in the early Middle Ages. As the initial phase of quarrying has not been fixed, production might go back at least to the 11th century. However it is difficult to assess if this is also the case for millstone extraction. The millstone fragments were found in layer 3, with dates from the 10th to the early 13th centuries, but the oldest C14-dating sample (AD 990-1025) was pine and therefore open to some doubt. Millstone extraction may have begun later than quernstone production, but this could not be ascertained from the excavation evidence.

Deep quarries relate to the extraction of millstones (cf. Heldal and Bloxam 2008:51-53; Grenne *et al.* 2008:57-58). The quarry at Myklebust, represented by trenches 2 and 3, is the only site where millstones seem to have had the major role. Millstones from this quarry measured c. 1m in diameter, and example still attached to the bedrock near trench 3 is 1.10m in diameter and 20cm thick. At two of the other deep quarries, Rønset trench 5 and Myklebust trenches 6 and 7, smaller quernstones are more frequent. Quernstones dominate at Rønset trench 5 (7 examples) while only two fragments of millstones are identified there with a diameter of more than 1m, indicating a larger output of querns than millstones. This is also the case at Myklebust trenches 6 and 7, and even though extractions of stone crosses took place there, the main product seems to have been quernstones, represented by 52 fragments. No millstones were found in the two trenches but one was found close to trench 3. In the third deep quarry investigated, Rønset trench 4, only slab remains were identified, and both quernstones and millstones were absent from the trench.

The production of quernstones began in the Viking period and increased throughout the Middle Ages; the stones were extracted in both shallow and deep quarries. Querns seem in fact to represent the most important product from Hyllestad in both the Viking period and the Middle Ages. Comparisons with stones from other Hyllestad contexts may clarify this issue, and it is further discussed in Chapters 7 and 10.

By the first half of the 13th century at the latest, most likely earlier, the product range had expanded at Hyllestad, including extractions of stone crosses. This is documented at Myklebust trench 6 by finds of six crosses of different sizes, all damaged. Judging by the sizes of four of the finds they probably represent churchyard crosses. Two, BRM 928/1 and 3, are somewhat larger and may have been intended for different use, but still seem to date to the same period as the smaller crosses. The deep quarry at Rønset trench 4 contained neither quern nor millstone, and the activity seems to have been concentrated around slab extraction in the high Middle Ages. It is not known what products resulted from the Rønset slabs.

In the Myklebust quarry, represented by trenches 6 and 7, large quantities of quernstones were extracted, along with slabs and crosses, while at Rønset there are no remains of querns or millstones. This may be the result of different product extraction, where the quarries at Rønset seem to have been more specialized. However, as only small-scale excavations were conducted, production of quernstones and millstones too cannot be fully excluded.

The quernstones found in the different trenches vary in size, and it has not been possible to document any development from the Viking period and into the Middle Ages. Quernstones found at Rønset trench 5 may indicate a somewhat standardized size, but relatively few stones (7) were found. Stones of different sizes were extracted both in the late Viking period and early and high Middle Ages (Tab. 5.10), and differences in size are also supported by visible extraction marks at the quarries. At Otringsneset the extraction marks within the same rock wall vary in size, measuring 27, 40, 60, 70 and 90cm in diameter. Most extraction marks measured here are, however, between 41 and 50cm (10 examples). At Myklebust the measured extraction marks (11) in one of the quarries vary between 41 and 45cm, and at Sæsol between 37 and 46cm (10 marks) (Løland in prep.).

Most of the measurable stones found during excavations are, therefore, quernstones with diameters between 35 and 45cm (49 stones). Other sizes seem to have been produced only rarely at Hyllestad during the period AD 900-1200 (Fig. 5.49).[10]

[10] Stones found in connection with the author's Master's thesis are not included, as the full measurements of most of the stones are lacking. They would however have given a higher number of millstones of c. 1m in diameter.

Table 5.10: Sizes of quernstones and millstones found during excavations at Hyllestad

Quarry site	Dating periods	Product type	Diameter
Myklebust trench 7	900-1150	querns	36-46cm
Sæsol trench 7	1000-1100	quernstones and millstones	35-52cm 70cm
Rønset trench 5	1000-1100	quernstones and millstones	40-46cm < 1 m
Myklebust trench 6	1000-1200	querns	36-42cm

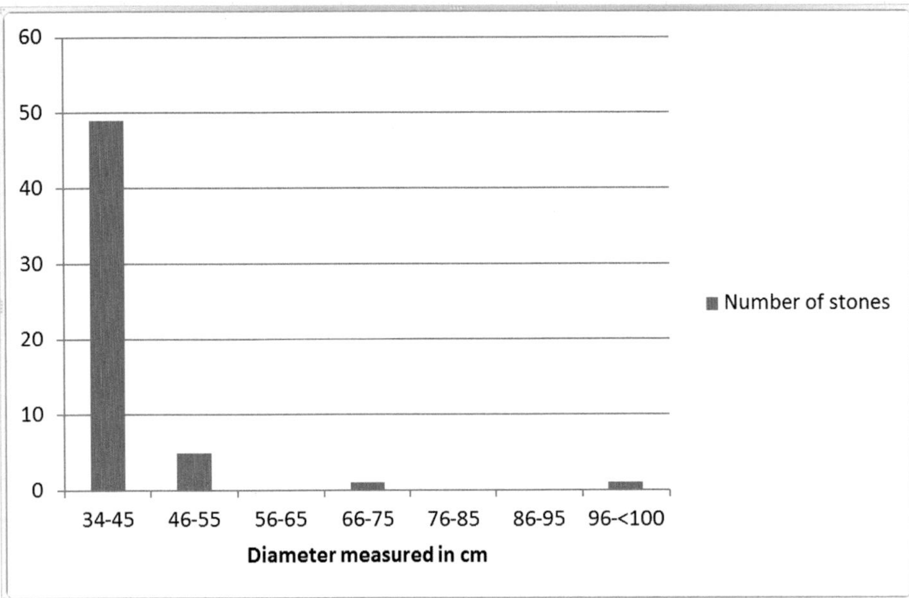

Diameter measured in cm	Number of stones
34-45	49
46-55	5
56-65	0
66-75	1
76-85	0
86-95	0
96-<100	1

Figure 5.49: Diameters of quernstones and millstones (AD 900-1200) found during excavations.

Quarry location

Quarries with different distances to the infields of farms, as well as to harbours and transportation routes, are represented. Looking at the quarry dates, including quarries from earlier investigations, several of the sites seem to have been in use at the same time (cf. Fig. 5.47). The oldest documented quarries are located at Otringsneset, close to the sea and a good harbour, and possibly at Sæsol, about 200m.a.s.l., and about 1km from the nearest harbour. It appears that distances to infields and harbours were not crucial for quarry location.

Quarrying versus farming at Rønset and Sæsol

Agricultural activity at Rønset seems to have begun in the early Iron Age (Foyn 2008:92) and thus long before quarrying started, probably at the beginning of the Viking period. Quarrying in all likelihood had to adapt, therefore, to the agricultural activities already established, which may have affected, and to a certain degree limited, the location of the production sites. Quarrying is also documented within the infield and in conflict with farming activities. Intensive and large-scale quernstone production was carried out with regular farming activities for centuries. Production continued at least until the early 18th century, when stone cutters at Rønset are mentioned in the tax register (Rønset 1977:9), but now at reduced scale compared to earlier periods. At Sæsol a different situation occurred, where the farm may have been established during the Merovingian period, possibly with a permanent settlement from the beginning of the Viking Age (Sætre 2008:76); the latter may be simultaneous with the initial quarrying phase. At Sæsol, quernstone production and farming were carried out simultaneously for centuries. C14-analyses indicate activity sometime between the beginning of the Viking period and the early Middle Ages. However quarrying here went on in a much smaller way than at Rønset, and quern production only took place in the outlying area the farm, therefore not conflicting with agricultural work.

6 Archaeological Investigations of the Quarries of Ølve and Hatlestrand

Archaeological research at selected sites within the larger quarry landscape of Ølve and Hatlestrand has helped our understanding of the quarries in terms of their scale, production techniques, product types and chronologies. In this chapter the results of archaeological researches at the quarries of Ølve/Halestrand are presented and discussed, forming an empirical basis for further analysis.

6.1 Aims and methods

As a starting point the quarry landscape of Ølve/Hatlestrand was surveyed and surface structures, such as quarries, carving traces, spoil heaps, products and remnants of roads and harbours were identified and mapped to aid selection of areas for archaeological investigation.

The presence of the chlorite-rich green schist utilized, *inter alia*, for bakestones was crucial for quarry location in the Ølve/Hatlestrand area. This rock type is found on the southern and eastern sides of Lake Kvitebergsvatnet, and on the western side of the Kvinnheradsfjorden, in the same area where the quarries are located. A similar rock type also exists on the island of Varaldsøy further north in the fjord, also with quarries, but these are not included in this study.

The main goal for the investigation was to identify and date the extraction and products of the different quarries within the wide production area: if possible to date the beginning and end of production. In the same way as for Hyllestad, the considerable size of the production landscape makes it possible only to investigate selected sites. To obtain an acceptable representation through limited archaeological surveys, the same selection criteria used for Hyllestad were also used at Ølve and Hatlestrand. It was important to incorporate different types of quarries as regards products, size, technology and different geological conditions. Assessment and advice from geologists was therefore important. It was also an objective to investigate quarries at different distances from the lake and the fjord, possible harbours and transport routes. Chronological relations and changes in production and products form central issues in the study.

Four quarries and one workplace were selected for investigation, altogether five trenches and one test pit, all dug in spoil heaps (Fig. 6.1). They were excavated in order to analyze stratigraphical relations and to collect material for C14-dating. Large spoil heaps are usually located a few metres in front of the carved rocks and beyond overhangs and caves. Additionally, large amounts of spoil were deposited inside underground quarries, covering the floors and the lowest parts of the carved rock, sometimes

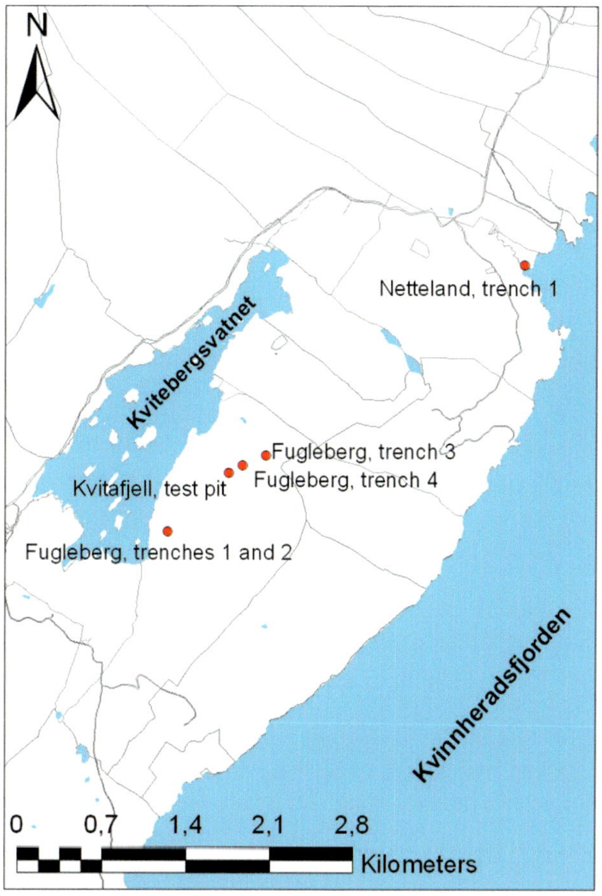

FIGURE 6.1: MAP OF EXCAVATION LOCATIONS (MAP: IRENE BAUG, 2013, NORGE DIGITALT FKB-DATA, 1224 KVINNHERAD COMMUNE, 32_1224EIENDOM_FLATE 21.11.2007)[11]

even blocking the cave entrances. The youngest spoil is therefore most likely to be found in front of and inside the quarries. In an attempt to date both the beginning and end of activities, the trenches were placed close to the caves and overhangs. Cave interiors were not excavated as these remains most likely only originate from the very end of production. The sites are all located in the outlying fields of two historical farms, Fugleberg (no. 17) at Ølve, and Netteland (no. 29) at Hatlestrand. Most of the identified quarries are located in Ølve, which constitutes a central production area with the most quarries. They are located at Fugleberg farm, which was also the focal area of the archaeological survey.

The investigated quarries at Fugleberg are located on the hillside of Fuglebergåsen, c.60-145m.a.s.l., at some distance from the fjord and possible harbours. The farm

[11] The map is based on the single sub-plots in 32_1224eiendom_flate (polygon) to the cadastral units (named farms) by the dissolve-function in ArcGIS by Arne Solli, AHKR, UiB.

does not border on the sea, but had a mooring site at Støo (no. 19) on Kvinnheradsfjorden. In the Middle Ages, Støo seems to have belonged to the farm at Skarvatun (no. 18) (Nysæter 2013:147-148). Stones from the quarries furthest east at Fuglerg may have been brought to Støo for further distribution, but the stones were probably also transported over steep terrain down to Kvitebergsvatnet lake. Here they were transported to the southern end of the lake, c. 350m separates from access to the fjord. To reveal possible differences and similarities in the production landscape, and to obtain a representative impression of products, technology, chronological framework (also over the different distances from the lake), three quarry sites of different scales were chosen for investigation. Furthest north, at Fuglebergåsen, a quarry with traces of bakestones as its main product was chosen, while the two other sites feature fewer carving traces of bakestone production compared to other products. Here traces of other products are also visible: tiles and building stones, possibly also oblong slabs for drying grains, an activity documented in the 17th century (Friis 1632:72). Another issue to look into was whether the extraction of slabs, as well as tiles and building stones, represent a later production than the bakestones, and to what degree it may have obliterated the remains of possibly older extractions.

In addition, a workplace for bakestones at Kvitafjell was chosen as a research site to compare with nearby quarries to the north and south, especially with regard to chronology and product types. It was also of interest to assess technological aspects related to bakestones, a topic that can be better studied at a work place rather than in the actual quarries, where it is difficult to distinguish those flakes from the extraction process and those coming from bakestone finishing.

In the community of Hatlestrand, on the western side of the Kvinnheradsfjorden, there were 17 quarries identified. The production sites are fewer in number and many of them also smaller than the Ølve sites; most are located closer to the fjord. Comparison with the Fugleberg quarries is relevant with regard to chronology, products and quarry types. A quarry at the Netteland farm at Hatlestrand, only 4m.a.s.l., was chosen for investigation.

6.2 The quarry landscape at Fugleberg (no. 17), Ølve

Most of the quarries are located at Fugleberg, representing the main production area. The quarries are located in the mountainous outlying areas of the farm, situated in a row and close together along the Fuglebergåsen hill and stretching in a north-south direction, c. 60-145m.a.s.l. Many of the quarries are so close that they partly overlap each other, and it is not always clear where they start and end. Each quarry site may thus comprise several quarries, probably used at different periods. In the middle of this area is the workplace chosen for investigation, underneath the overhang at Kvitafjell.[12]

[12] Id. no. 45427, Askeladden.

On the western side of Fuglebergåsen, east of the 3km-long lake at Kvitebergsvatn, 43 quarry sites were identified. The forested terrain slants steeply down the hill to the lake. The forest consists mainly of elm, aspen, rowan and holly. The vegetation is dense, with smaller bushes and ferns, making some areas almost inaccessible; hatchets and chain-saws had to be used to access some of the quarries.

In the region's centre is the mountain of Kvitafjell, forming a natural overhanging ridge, stretching north-south at an altitude of 112m.s.l.[13] This location seems to have been used as a workplace for finishing bakestones brought here from the quarries.

Furthest north, at Fuglebergåsen, (118m.a.s.l.), an open quarry was identified, with only a small part of the production rock still visible, and where large amounts of spoil indicate large-scale production. Further south a larger, open quarry is located by a brook, and carving traces are visible on both sides of it, flowing straight through the spoil heap. South of this brook, a small underground quarry was uncovered, c. 2m deep and 5m long, with traces of bakestone production. Further south, c. 113m.a.s.l., a cliff face with three quarries was identified,[14] underneath overhangs. Here there were the remains of extractions from the exterior of the rock, as well as underground quarries. Traces of production from these three quarries indicate manufacture of different products, of which bakestones seem to be most important. In the northernmost and southernmost quarries bakestone extraction seems to have been a secondary manufacture compared with other items, such as tiles, building stones and slabs (larger and thicker than bakestones).

The cliff quarry furthest north includes a 10m-wide, 2m-high and 2m-deep overhang with visible carving traces. A minimum of 40 cubic metres of rock was extracted from the quarry. Rectangular traces can still be seen on the ceiling, some with straight, and others with rounded corners. Between some of the rectangular marks there are traces of carved channels, apparently made for loosening the products from the rock. Two of these production marks measure 48x45cm and 55x65cm respectively, with an 11cm-wide channel between them. As large amounts of spoil were deposited in front of the carved rock, the overhang originally was probably higher. The main spoil heap in front of the overhang was c. 1.5m high.

About 6m southwest is the second quarry in the same cliff, an underground site (9 x 10m) was selected here as a research area. The spoil was located partly inside the quarry and partly in front of the entrance, and originally the quarry was higher. In contrast to most other quarries in the Ølve/Hatlestrand area, traces seem to indicate that bakestones were the main product here.

About 6m southeast of this research area the rock slants eastwards. Here, a third quarry was located on the cliff

[13] Id. no. 45427, Askeladden.

[14] Id. no. 25934, Askeladden.

face with an overhang c. 10m long, 1.5m high and 2m deep. Traces here indicate production of a minimum of 30 cubic metres of extracted products.

Approximately 17m east of this cliff, two small underground quarries were identified next to each other, covering an area of c. 10-11m in length. The opening of the northernmost quarry is c. 80cm high, while the other is c. 1.5m in height, with a 60cm-wide rock pillar separating the two quarries. Traces of bakestone production were found at both sites, but there were also traces of other oblong products with round corners, evidently slab remains. In the southernmost quarry, a nearly finished oblong stone slab still hangs from the roof, probably left there because it was too dangerous to remove it and continue working there. The spoil is located at the entrance to the quarries, indicating that the entrance was originally higher.

Southeast of these quarries, a more or less continuous production area was identified as a quarry complex, covering an area of c. 120m in length. The spoil heaps are located only a few metres west of the quarries, and between the quarries and the heaps is a narrow, flat area. Today it serves as a footpath and was probably also used as a passage between the quarries at an earlier stage for easier transportation. West of the spoil heaps the terrain slopes steeply towards the lake, making product movement difficult.

The underground quarry at Hedlebergshidlaren[15] is located in the middle of this area. It is one of the largest quarries at Fuglebergåsen, c. 50m long, 15m deep and 4m high (Fig. 6.2). A minimum of 3000 cubic metres of stone has been extracted from this quarry. Large blocks have fallen down in front of the quarry, indicating that the overhang was once larger. In the middle of the quarry, by the entrance, a pillar has been left in the rock to prevent the roof collapsing. The pillar is completely covered with concave production traces. Some of these curves have round corners and straight sides, most likely the remains of products other than bakestones, but some of the marks seem to be indicators of bakestone production. Inside the quarry, in an area where the ground slants, there are small terraces of stone and soil on the surface, probably to facilitate production in the quarry.

The production area extends further northeast of Hedlebergshidlaren, to a distance of about 50-60m. Huge collapsed stone blocks now lie in front of the rock. Most of the traces on the carved rock have round corners and straight sides, and indicate the production of tiles or slabs. There are still several smaller overhangs and quarries in the area, many now covered with vegetation and large stone blocks. One of these contains the remains of bakestone production from the lower part of the rock, while the upper sections have traces of tiles or slabs, indicating that different products were extracted over time. Bakestones seem to be the earliest products, with tiles and slabs representing later manufacture.

One of the quarries[16] chosen for investigation is c. 27m south of the Hedlebergshidlaren cave; it is an underground site, c. 11m wide, 7m deep and 2.20m high, with traces of production outside the cave as well. The remaining traces on the carved rock reveal extraction of different products, but with fewer traces of bakestones. The quarry continues north, c. 8m from the underground quarry; in this area only semicircular shaped traces have been identified on the rock, possibly bakestone marks. The area is overgrown, and production waste seems to be covered by vegetation. The rock continues in a northeasterly direction towards the Hedlebergshidlaren, with a 19m-long carved rock wall followed by a 6m-long wall with no traces of carving.

These quarries are all located to the north of the natural overhang of Kvitafjell. Southwest of Kvitafjell is a further bakestone worksite, 9m long and 2m wide. Small fragments of bakestones indicate finishing of these products here. South of these two workplaces, the quarries continue as a nearly integrated production area (underground and open quarries of different sizes) and only the largest and most pronounced quarries are referred to.

Below the southern end of the overhang at Kvitafjell, two large underground quarries were identified in the steep and hilly area of Areskorane, also referred to as Øvre (upper) Areskorshidlarane,[17] c. 25m apart. The southernmost and largest one is a c. 20m-long underground, shallow quarry. It revealed production at two levels in a 6m-high steatite zone, indicating large-scale production. A flat section in the middle of the rock wall separates the two levels. Further south another quarry, Nedre (lower) Areskorhidlaren[18] consists of a c. 13m-deep and 3m-high quarry, with production traces (Figs. 6.3 a and b).

The quarry landscape continues in a south-westerly direction along Fuglebergåsen. Here, Flåbekkshidlaren,[19] a large site, with both surface and underground production, stretches over c. 75m in a north-south direction. At the southern end, production marks are visible in a c. 5m-high zone in the rock, but the underground quarry has now collapsed.

Further to the southwest are the two smaller quarries of Palmehidlaren and Little Palmehidlaren.[20] A forest road made in the early 1990s cuts through the spoil heaps of both quarries, especially the southernmost: Little Palmehidlaren was destroyed by road building.[21]

Southwest of this quarry is another overhang, Bakkhidlaren,[22] representing one of the largest production

[15] Id. no. 6636, Askeladden.

[16] Id. no. 35831, Askeladden.

[17] Id. no. 55162, Askeladden.

[18] Id. no. 35833, Askeladden.

[19] Id. no. 35830, Askeladden.

[20] Id. no. 35832, Askeladden.

[21] http://askeladden.ra.no.

[22] Id. no. 72962, Askeladden.

FIGURE 6.2: HEDLEBERGSHIDLAREN (PHOTO: IRENE BAUG, 2006)

areas of Ølve/Hatlestrand, and a site chosen as a research area. The quarry complex consists of a more or less coherent production zone over a distance of c. 82m in length in a northeast-southwesterly direction. Here, output from underground quarries, overhangs and open quarries has been located, and altogether five underground quarries of different sizes were identified. Inside the largest quarry, a pond reveals that a water source was encountered during manufacture. Product traces here include bakestones, slabs, tiles and building stones.

Three production sites were found south of Bakkhidlaren. In the direction of the Kvitebergsvatn lake, there is a small, open quarry with marks that indicate extraction of not only bakestones, but also slabs or tiles and building stones. Similar marks are observed in a quarry of building stones at Netteland farm in Hatlestrand (see below). The site goes more or less continuously southwards towards Buahidlaren, also called Vasstigbu or Vasstøylbu.[23] The large quarry complex consists of two underground workings separated by a rock wall. The smallest is c. 8m wide and 6m deep, while the largest is 10m wide and 10m deep. There is a pond inside the largest site and an entrance into a northern quarry located at the north wall. The large amounts of spoil make it difficult to enter this quarry. About 40m south of Buahidlaren is the Sauahidlaren quarry,[24] with a 10m-wide and 4m-deep overhang. From the quarries of Buahidlaren and Sauahidlaren the terrain slopes to Ambarevika bay, where large amounts of spoil, still visible through the vegetation, have been dumped. Here, fragments of bakestones were identified.

Further south, some 200m north of holding 1, at Tufta farm (no. 16), is the quarry of Tuftahidlaren.[25] It is 5m wide, 15m deep and 1.5m high, representing a minimum of 112.5 cubic metres of extracted stone. The production waste, however, was most likely removed during the construction of a water pipeline in the 1970s.[26]

6.3 The quarry landscape at Netteland (no. 29), Hatlestrand

The largest concentration of quarries in Hatlestrand is identified at the Netteland farm complex (no. 29). Altogether ten production sites have been identified here, located both in the farm interior and outlying fields. The spoil has been mostly removed from the interior area during later cultivation. Seven of the production sites are open quarries, six quite small. The other four are considerably larger, representing an intensive and varied production.

The large underground quarry of Mannahidlaren[27] is roughly 80m west of the shores of Hellebergsvika, within

[23] Id. no. 25933, Askeladden.

[24] Id. no. 25932, Askeladden.

[25] Id. no. 25935, Askeladden.

[26] Http://askeladden.ra.no.

[27] Id. no. 55167, Askeladden.

Figure 6.3 (A and B): Upper Areskorshidlarane (Photos: Irene Baug, 2006)

the infield of holding 2. In the early 1960s, the spoil heaps in front of the quarries were levelled (Fett 1972:126), and today there are no visible traces of them outside the quarry. Production was carried out in the underground quarries, but the entrances to most of them have collapsed. At the southern end of the site, however, a small opening between large blocks of stones makes it possible to enter one of the other caves. The quarry is c 30-40m deep and wide, and around 1-2m high, which may represent 3,200 cubic metres of extracted stone. Stalactites still hang from the roof in several places. Concave hewing traces are visible on the interior rock, most with rounded corners and straight sides, suggesting the production of items other than bakestones. Large blocks of stone have fallen down from the roof, but at the northern side it is still possible to crawl over the stones and enter a further cave.

At the northern extremity, by Netlandsvågen, is the quarry site of Båtahidlaren ('the boat cave'),[28] only a few metres from the fjord. Here, most of the quarrying was done on the outside rock; the site measures c. 100m in length. At the centre of the site is an underground quarry, c. 5.7m wide and 9.5m deep, with semi-circular traces on the rock, both inside and outside the quarry, many of them seemingly the remains of bakestones. The eastern end of the site, however, appears different from most of the other quarries known so far in the Ølve/Hatlestrand area. The rock is straight and with none of the concave carving traces usually found. The marks on the rock signify the manufacture of quadrangular or rectangular objects, probably building stones. According to documentary evidence from the 19th century, the building stones for Kronborg castle (Denmark) were imported from 'Nættling in Quinherred Præstegjeld', near Hardangerfjord (Buch 2011 [1813]: 14-15), perhaps referring to this quarry. In the 1800s, grave crosses were apparently also extracted here (Knut Johan Nerhus pers. comm. 2006).

6.4 Archaeological investigations at Fugleberg, Bakkhidlaren

Bakkhidlaren[29] is one of the largest quarry complexes in the Ølve/Hatlestrand area, and represents a more or less coherent production area covering a distance of roughly 82m in a northeast-southwesterly direction. Extraction of both open and underground quarries are documented, and five underground quarries of different size (quarries I-V) have been identified. Quarry I is located furthest to the

[28] Id. no. 45431, Askeladden.

[29] Id. no. 72962, Askeladden.

south, and quarry V furthest to northeast. Two enormous spoil heaps were found in front of the rock, suggesting large-scale production.

At the south-western end of the site, quarries I and II are separated by a pillar preventing the roof from collapsing. Behind this pillar (c. 6.5m x 4.2m) a small tunnel connects the two quarries; the feature is entirely covered with production marks (Fig. 6.4).

Quarry I is the smaller of the two, about 13.5m wide, 1.3m high and 6m deep, representing a minimum of 105.3 cubic metres of extracted stone. Originally the quarry was deeper, but its outer area collapsed and large blocks of stones now lie in front of the entrance. Spoil covers the entire quarry floor, reducing its height. Production marks are visible both on the rock walls and the roof inside the quarry, indicating tile extraction. The production marks on the outside, however, point to bakestone extractions.

Quarry II is the largest at the site, 27m deep and 28m wide, with a height varying from 1 to more than 2 m, and a minimum of 1,134 cubic metres masses must have been extracted.[30] Enormous blocks of stones, up to 11m in length, have fallen down, and the only traces from the carving are found on the walls. In the southern end of the quarry, c. 15m inside of it, a 6m wide and about 10-12m long pond appears, most likely stemming from an ore of groundwater that was hit during extraction. Inside the quarry, a small scrape and a wooden wedge (23 x 4cm) was found, evidently representing tools used during the extraction (Figures. 6.5 a and b). The traces inside the quarry indicate the extraction of tiles and slabs, possibly for building stones (Fig. 6.6) as well as bakestones.

At the northern end of quarry II, an 8m wide pillar separates quarry II from quarry III. Carving traces on the pillar also indicate the extraction of building stones. Quarry III has an 11m-wide entrance, but large blocks of stone, over 8m in length, have fallen, blocking the entrance. At the northern end of quarry III, a 3m-wide pillar separates quarry III from IV, which is 3m wide and 8m deep. In the northeastern wall of quarry IV, c. 1m from the entrance, there is a 30cm-high tunnel, which is the only entrance to quarry V. This last underground quarry is c. 6m wide, 6m deep and 1.6m high, representing 57.6 cubic metres of stone extraction. Large amounts of spoil are located inside the quarries, and explain the low opening of the tunnel between quarries IV and V. Most likely the production waste was used to fill up the quarries: to enable extraction from the roof. Outside quarry V the rock slants eastwards, where a c. 10-11m-long and c. 2m-high carved rock wall remains. The spoil is located in front of the rock.

Production traces in the underground quarries of Bakkhidlaren, both inside and outside the sites, indicate the extraction of several products, but with no clear traces of bakestone output in the underground quarries. The carved rock to northeast of quarry V is the only area with clear traces of bakestone extraction. Here several semicircular concave traces are identified, and a remaining fragment, c. 56cm in diameter, is still visible on the rock wall; the size corresponds to bakestones from medieval urban contexts. On the same rock wall there are also traces that may derive from the extraction of building stones. The straight rock wall lacks the characteristic concave traces, and several small pieces of the rock protrude from the wall where the stones were loosened. These indicate that stones c. 16cm thick were extracted here.

Some traces in the rock walls of Bakkhidlaren are 80-90cm in diameter, a size that could fit bakestone removal. However concave traces with round corners and straight sides generally dominate, suggesting the extraction of objects other than bakestones. On the roof of quarry I there are quadrangular or rectangular marks with rounded corners and straight sides. These traces do not, however, always give a clear picture of what objects were extracted here. It is possible that slabs for grain drying, tiles, and building stones were manufactured, items similar to those documented at quarries at Øye in Trøndelag, with their productions of both bakestones and building stones (Lundberg 2007).

The two large spoil heaps at Bakkhidlaren are located at the south-western end of the production site (Fig. 6.7). Here a trial trench was dug in each of the heaps in the hope of revealing possible chronological relations between the heaps and the quarries, and to study possible differences in the products made.[31] The smallest heap (heap A) located furthest southwest, and roughly 12m in front of quarries I and II, is thought to be associated with the outputs of these quarries. The heap, in north-southerly direction, is 17m long and c. 11m wide and 3-3.5m high. Trench 1 was made at the northern side of the heap, c. 13-14m from the pillar between quarries I and II. A 2.5m-wide path runs between the heaps, in an east-west direction; this was probably used for transportation out of the quarry site down to the Kvitebergsvatnet lake. The largest spoil heap (heap B), located just west of quarry III, by a path next to it, is 23m wide, 28m long and 8m high. It is probably extraction residue from quarries II and III. Trench 2 was opened on the eastern side of the heap, 2.8m west of the pillar separating quarries II and III. On the southern side of the heap another path turns towards the top, possibly used in connection with spoil deposition. It is possible that the single large heap originally consisted of two smaller heaps that merged over time. The path, therefore, may represent the top of an older and lower heap that was later covered by more spoil, again indicating a long period of activity.

To the southwest of heap A the terrain is flatter, making it easier to transport the products, possibly by horse down to the lake. By the lake there are remains of an undated stone building and a small stone-built wharf. These structures

[30] An average height of 1.5m is used for calculating cubic volumes,

[31] UTM Zone 33N X:-8655 Y: 6688648.

FIGURE 6.4: PILLAR SEPARATING QUARRIES I AND II (PHOTO: IRENE BAUG, 2006)

may possibly be related to storage and transportation of products from Bakkhidlaren and other quarries in the area.

6.4.1 Fugleberg trench 1

Trench 1 at Fugleberg, 3.10m long and 1m wide, was oriented northeast-southwest, stretching to the south into the heap. It varied from 40cm to 1.6m in depth (Fig. 6.8). Water in the trench made the excavation difficult, and because of the lack of suitable dating material digging stopped before reaching the base layers.

Profile interpretation: Fugleberg trench 1

Trench 1 consisted of five layers, including topsoil (layer 1) (Fig. 6.9). The deposits contained large amounts of slabs and smaller flakes, and a few bakestones. The content in the trench is interpreted as extraction débris from the extractions in quarry I, and possibly also quarry II.

Layer 2, located at the northern end of the trench, in the area with the possible path, contained red-brown humus with fragments of slabs 20-30cm in size, and also some larger slabs and very small flakes. The slabs were lying horizontally, making it easy to walk between the heaps. The straight edge that separated layer 2 from layer 3 may be the result of earlier digging in the spoil heap, perhaps when improving the path between the heaps. Layer 3 was a red-brown sand layer, with a large amount of slabs, 30-50cm in diameter, as well as smaller flakes, c. 2cm in size. Layer 3 was rather similar to layer 2, but dryer and with less humus. Layer 4, of brown sand and silt, at the northern end of the trench, differed in colour from the underlying layer 5, probably caused by drainage from the topsoil into layer 4, however the two layers are most likely the results of similar activity. Layer 5 was a light, grey-brown sand and silt layer, in different shades of yellow. It contained slabs, up to 50cm in diameter, as well as smaller flakes (2-10cm) and the remains of bakestones with different finishes.

The content in the trench matched the production marks on the rock and pointed to only small-scale bakestone extraction. The slabs may derive from tile production of the two quarries furthest south, perhaps also slabs used for grain drying. No building stones were identified, but they cannot be ruled out. As no datable material was found, the time of production is not known.

6.4.2 Fugleberg trench 2

Trench 2, located on the eastern side of the largest heap (B), was c. 3.10m long, 1.5m wide, and between 0.7-1.4m deep; it is oriented east-west (Fig. 6.10). Because of absence of material for C14-dating it was decided to halt the excavation before the base level was reached.

Figure 6.5: (A) wedge (23 x 4cm) and (B) scrape found inside the quarry (Photos: Atle Ove Martinussen, 2012)

FIGURE 6.6: CARVING TRACES FROM THE EXTRACTION OF TILES AND SLABS, POSSIBLY FOR BUILDING STONES INSIDE QUARRY II (PHOTO: ATLE OVE MARTINUSSEN, 2012)

Profile interpretation: Fuglebert trench 2

Trench 2 consisted of three layers, including the topsoil (layer 1) (Fig. 6.11). Layer 2 was a thin, yellow-brown layer beneath the topsoil, containing sand but no humus, only production waste in the form of small flakes (1-10cm), many with cutting marks. Layer 3 differed from layer 2 in colour, being a grey layer of silt and coarse sand, and contained stones of different sizes, large parts of slabs, with both rounded and sharpened corners (10-60cm in diameter and 1-6cm thick), many with cutting marks. Smaller flakes (1-20cm) were the most dominant feature in the layer, but there were only a few bakestone fragments.

6.4.3 The relationship between Fugleberg trenches 1 and 2

Both Bakkhidlaren trenches contained production waste, reflecting quarry activity. The trenches show a different composition of the spoil. Large slabs were the most pronounced element in trench 1. Similar material, but less, was also found in trench 2, but consisted of smaller flakes.

The limited sizes of the excavated trenches made it difficult to assess the contents, and to what degree they represent different activities. Heap B seems, however, to a greater extent to be the result of product finishing, but the extraction of different products should also perhaps be taken into account. Because of the lack of dating material it was not possible to date the activity and chronology of these two large spoil heaps.

6.5 Archaeological investigations at Fuglebergåsen: Fugleberg trench 3

The second site explored is located furthest north on Fuglebergåsen hill, at a cliff face where three small quarries were identified.[32] The central quarry was selected for investigation. It is an underground quarry, 1.6-2.5m high, 9m wide and 10m deep, representing a minimum of 180 cubic metres of extracted stone (Fig. 6.12).[33] The spoil heap is located in front of its entrance, but large amounts of waste also seem to have been dumped inside the quarry. The carved rock walls inside the quarry are partly covered by spoil, and the working must therefore have originally been higher. The topography in the area slopes slightly, and west of the quarry there is a small boggy area at a lower level. The spoil heap, c. 2.5-3m high at its highest point, partly covers the bog area.

Production was mainly done inside the quarry. Most of the extraction marks have concave curves of a semicircular shape, indicating bakestones (Fig. 6.12). The production marks also indicate other manufactured items. The east wall of the quarry has rather straight sides and lacks the

[32] Id. no. 25934, Askeladden; UTM Zone 33N X: -7779 Y: 6689187.
[33] 2m has been used as the average height for underground quarries for calculating volumes.

FIGURE 6.7: THE TWO LARGE BAKKHIDLAREN SPOIL HEAPS FROM THE SOUTH, WITH HEAP A IN THE FOREGROUND (PHOTO: ATLE OVE MARTINUSSEN, 2012)

semi-circular shape typical of bakestone production. On the south-western wall there is an area with marks in the shape of right angles and straight lines of unclear product origin. To the southwest of the quarry there is a small area with cutting traces on the rock exterior: three semi-circular curves indicate bakestone extraction. Below these curves is a small ledge, beneath which the production marks are different, with rounded corners and straight sides, evidently the remains from production other than bakestones. The marks overall, both inside and outside the cave, indicate that bakestones were most likely the original and main product, which also makes it easier to date the bakestone production here.

Quarrying most likely began by hewing circular formations into the rock wall the size of the bakestone to be extracted. The stones were then cut out more or less horizontally, following the cleavage in the bedrock, until the base of the green schist band was reached. This repeated process gradually resulted in overhangs and caves. Production from the roof was probably conducted in the latest phase of quarrying, when it was no longer possible to extract downwards. Spoil left inside the quarry made it easier to reach the roof, covering the oldest parts of the site with débris. Rock cleavage at the quarry site is more or less horizontal, some places slightly slanting, and bakestones and other product types were therefore produced horizontally. Because of the schistose rock, the bakestones were probably made one by one, being a more stable and easier process than carving larger columns for splitting into several blanks. The production process would leave concave traces, in the form of curves of different shapes and sizes, reflecting the objects manufactured. Bakestones commonly left circular marks, usually with diameters of between 55 and 65cm (Fig. 6.13).

Of the two spoil heaps, the main site (I) is located in front of the underground quarry, the result of production within the quarry. Heap I borders heap II, in front of the south-west end of the quarry opening. Heap II, however, is most likely the result of extraction both from the rock exterior and from the southernmost quarry, an overhang in the cliff. To investigate the chronological relationship between the two spoil heaps, and possible differences in the objects extracted, trench 3 was laid out towards the rock at the southern end of the quarry, covering parts of heap II, and to the north reaching heap I. This also made it possible to investigate the height of the rock at the southern end of the quarry (Fig. 6.14).

6.5.1 Fugleberg trench 3

Trench 3 was 4.3m long, 1.5m wide and 0.4-2.2m deep. The base level of the heap was revealed except for a small area (c. 50cm) in the southern part of the trench, where water made it difficult to continue digging (Fig. 6.15 a and b). At the southern end of the trench another rock face with production marks was uncovered (Fig. 6.16). Although the

FIGURE 6.8: FUGLEBERG TRENCH 1 (PHOTO: IRENE BAUG, 2006)

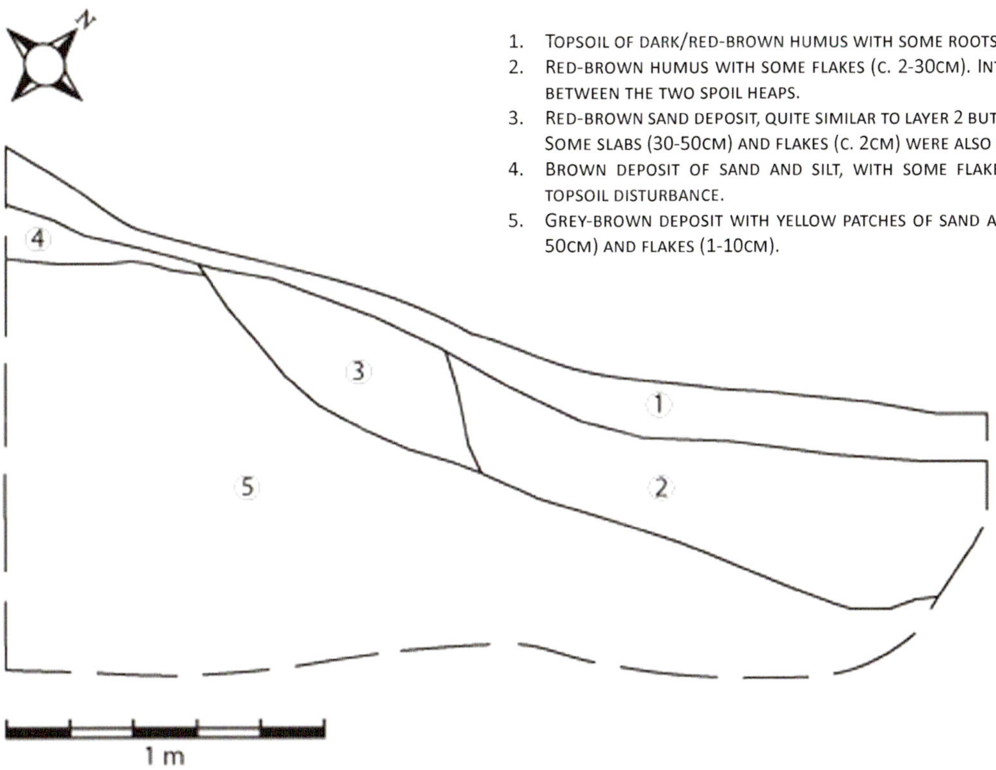

1. TOPSOIL OF DARK/RED-BROWN HUMUS WITH SOME ROOTS.
2. RED-BROWN HUMUS WITH SOME FLAKES (C. 2-30CM). INTERPRETED AS A POSSIBLE PATH BETWEEN THE TWO SPOIL HEAPS.
3. RED-BROWN SAND DEPOSIT, QUITE SIMILAR TO LAYER 2 BUT DRYER AND WITH LESS HUMUS. SOME SLABS (30-50CM) AND FLAKES (C. 2CM) WERE ALSO FOUND.
4. BROWN DEPOSIT OF SAND AND SILT, WITH SOME FLAKES (1-10CM). INTERPRETED AS TOPSOIL DISTURBANCE.
5. GREY-BROWN DEPOSIT WITH YELLOW PATCHES OF SAND AND SILT, CONTAINING SLABS (C. 50CM) AND FLAKES (1-10CM).

FIGURE 6.9: SECTION OF FUGLEBERG TRENCH 1 (DRAWING: IRENE BAUG, 2006)

traces are not completely circular, bakestone extraction could well have caused them. The base level was reached at the northern end of the trench.

Profile interpretation: Fugleberg trench 3

Trench 3 comprised of seven layers, including topsoil (layer 1) (Fig. 6.17). Layers 3-5 were interpreted as the remains from the most recent heap in the quarry (II), whereas layer 6 was regarded as having remains the oldest production remains, represented by heap I. Layer 2 is thought to be a possible secondary production site, covering the old heap, and perhaps contemporaneous with heap II. Altogether eight samples for C14-dating were taken from this trench (Tab. 6.1).

Layer 2 contained large amounts of flakes, as well as roots and humus, grey-brown to yellow in colour. A bakestone fragment, 45cm in diameter, was uncovered, together with some slabs c. 40cm in diameter. Flakes (from a few mm to 2-4cm) were also found. The deposit is considered to be the result of a secondary production phase, as it also covered layer 5 relating to heap II. A sample of charcoal for C14-dating (TUa-6697) from the bottom of layer 2 was dated to 720 ± 35, calibrated to AD 1280-1295, and indicating production in the high Middle Ages. Layer 3, a brown humus layer, contained large amounts of flakes of different sizes, from c. 0.2-10cm.

At the southern end of the trench, layer 4 is thought to be the remains of the most recent spoil heap in the quarry, heap II. It was a red-brown layer with humus and flakes

FIGURE 6.10: FUGLEBERG TRENCH 2 DURING EXCAVATION, SHOWING PRODUCTION WASTE: FLAKES, SLABS AND BAKESTONE FRAGMENTS (PHOTO: IRENE BAUG, 2006)

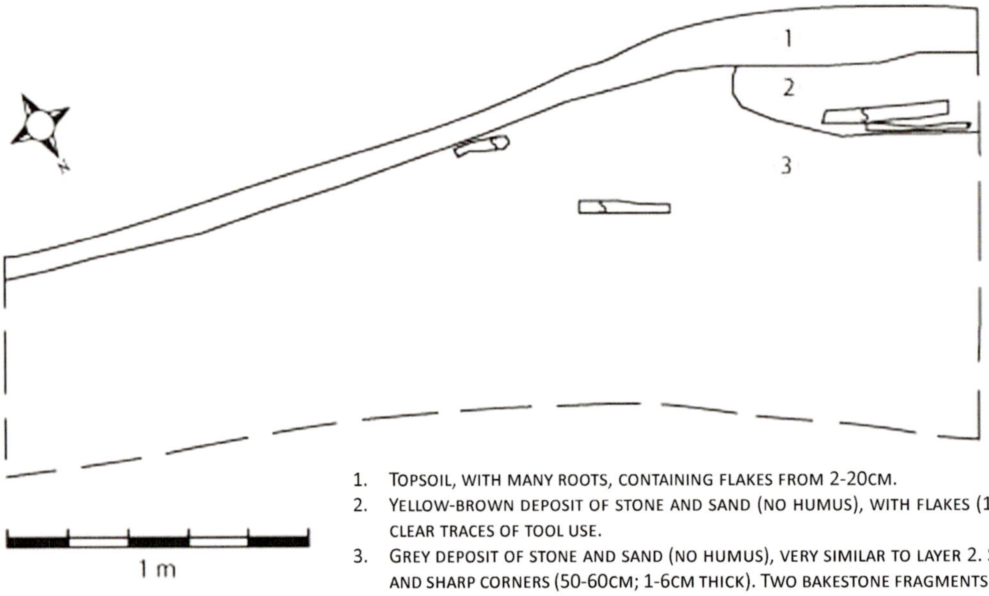

1. TOPSOIL, WITH MANY ROOTS, CONTAINING FLAKES FROM 2-20CM.
2. YELLOW-BROWN DEPOSIT OF STONE AND SAND (NO HUMUS), WITH FLAKES (1-10CM), SOME SHOWING CLEAR TRACES OF TOOL USE.
3. GREY DEPOSIT OF STONE AND SAND (NO HUMUS), VERY SIMILAR TO LAYER 2. SLABS WITH BOTH ROUND AND SHARP CORNERS (50-60CM; 1-6CM THICK). TWO BAKESTONE FRAGMENTS IDENTIFIED TOWARDS THE BASE OF THE TRENCH. SOME SMALL ROOTS ALSO PRESENT.

FIGURE 6.11: SECTION OF FUGLEBERG TRENCH 2 (DRAWING IRENE BAUG/MARIE ØDEGAARD, 2006)

FIGURE 6.12: THE UNDERGROUND QUARRY AT FUGLEBERG TRENCH 3 VIEWED FROM INSIDE. PRODUCTION MARKS OF BAKESTONE EXTRACTION ARE VISIBLE ON THE WALL TO THE RIGHT (PHOTO: ATLE OVE MARTINUSSEN, 2012)

from a few mm to c. 20-30cm in size. Some of the slabs were quite large: c. 45-55cm in diameter and 8-20cm thick. Fragments of bakestones were also identified, varying in form and size, including two drop-shaped examples. The layer was rather humid, probably caused by trickling water from the rock; the layer is located near the edge of the roof overhang at the southern end of the quarry. Furthest south in the layer, below the topsoil, large amounts of charcoal in between several flat stones/slabs were found. This material may have come from fires in more recent periods, not necessarily from the time of production. Layer 4 seemed to be somewhat disturbed at the southern end, while the rest of the deposit appeared untouched by human activity after production had terminated. To clarify these aspects it was decided to extract charcoal samples from different places in the deposits and three samples for C14-dating were taken from the layer. One sample (T-19070) from below the topsoil most likely represents final workings at the quarry and was dated to 370 ± 45, calibrated to AD 1455-1635. This indicates activity in the late Middle Ages or the early modern period. A second sample, T-19071, taken at the most southerly point in the quarry was dated to 190 ± 45, younger than AD 1660. This sample does not necessarily date the time of production, but may reflect further activity after production ended, or possibly sporadic extraction in the early modern period. The third sample, TUa-6699, taken further down in the layer above the carved rock at the southern end of the trench, was dated to 800 ± 35 and calibrated to AD1225-1280. This indicates production in the high Middle Ages, and concurs in time with the sample from layer 2. The C14-samples from layer 4 seem, therefore, to suggest the deposition of spoil over a long period. The bases of layers 4 and 2 represent production dated to the high Middle Ages, whereas the uppermost sample from layer 4 is younger.

Layer 5, in the middle of the trench beneath layers 2 and 3, bordered on layer 4 to the south and layer 6 to the north. Layer 5 consisted of brown-grey production waste, mostly small flakes, less than 1cm in thickness and up to 20-30cm in diameter, and many slanting slabs. Layers 3 and 5 differed from layers 2 and 6, where the stones lay horizontally. Layer 5 was interpreted as part of the latest heap, heap II. The shape of the deposit where it joined layer 6 indicates that parts of the oldest heap (heap I) have been removed, perhaps during a later production phase, in an attempt to uncover good quality rock for production. If this was the case, production from the carved rock at the southern end of the trench therefore represents a later production phase compared to extractions from the underground quarry. This also fits in with the carving traces on this rock, which suggest production other than bakestones (cf. Fig. 6.16).

FIGURE 6.13: TRACES OF BAKESTONE EXTRACTION FROM THE UNDERGROUND QUARRY AT FUGLEBERG TRENCH 3 (PHOTOS: ATLE OVE MARTINUSSEN, 2012)

The area where layers 2, 3, 4 and 5 joined seemed to be quite disturbed and the divisions between them were unclear; consequently the different layers may be rather imprecise. In this area, heap I seems to border heap II. Layers 3, 4 and 5, and possibly also layer 2, may therefore contain material from the youngest heap (heap II).

Layer 6 belonged to what was interpreted as the oldest spoil heap in the quarry (heap I), representing the first production, and older than layers 2 and 4. Layer 6 was located below layer 2 at the northern end of the trench, and bordering layer 5 to the south, which seems to have been dug into layer 6. The latter consisted of grey flakes with no humus. The material varied from silt to flakes, a few mm to 2-4cm in size. Some slabs were found, c. 50-60cm in diameter and generally 1-10cm thick. Several fragments of bakestones with different degrees of finishing, both with and without furrows, were also uncovered. The content of the layer was rather similar to layer 2, but differed in colour, probably caused by elements sieving down from the topsoil. Two samples were taken from the deposits for C14-dating. Sample TUa-6696 from just under layer 2 is dated to 960 ± 35, calibrated to AD 1025-1155. Sample TUa-6698 from above the subsoil may represent a starting phase for extractions in the underground quarry, dated to 905 ± 35, calibrated to AD 1045-1190. Both samples indicate production in the early Middle Ages (Tab. 6.1 and Fig. 6.18), and intensive but probably short-lived activity in this first phase.

Table 6.1: C14-analyses from Fugleberg trench 3

Lab. Ref.	Context	Age BP	Calibrated Age	Material
TUa-6697	Fugleberg trench 3, layer 2	720 ± 35	AD 1280-1295	pine charcoal
TUa-6699	Fugleberg trench 3, layer 4	800 ± 35	AD 1225-1280	birch charcoal
T-19070	Fugleberg trench 3, layer 4	370 ± 45	AD 1445-1635	birch and pine charcoal
T-19071	Fugleberg trench 3, layer 4	190 ± 45	Younger than AD 1660	birch and pine charcoal
TUa-6696	Fugleberg trench 3, layer 6	960 ± 35	AD 1025-1155	pine charcoal
TUa-6698	Fugleberg trench 3, layer 6	905 ± 35	AD1045-1190	pine charcoal
TUa-6700	Fugleberg trench 3, layer 7	2605 ± 35	BC 805-790	charcoal from deciduous trees
TUa-6701	Fugleberg trench 3, layer 7	4440 ± 45	BC 3260-2930	charcoal from sallow, willow/ aspen

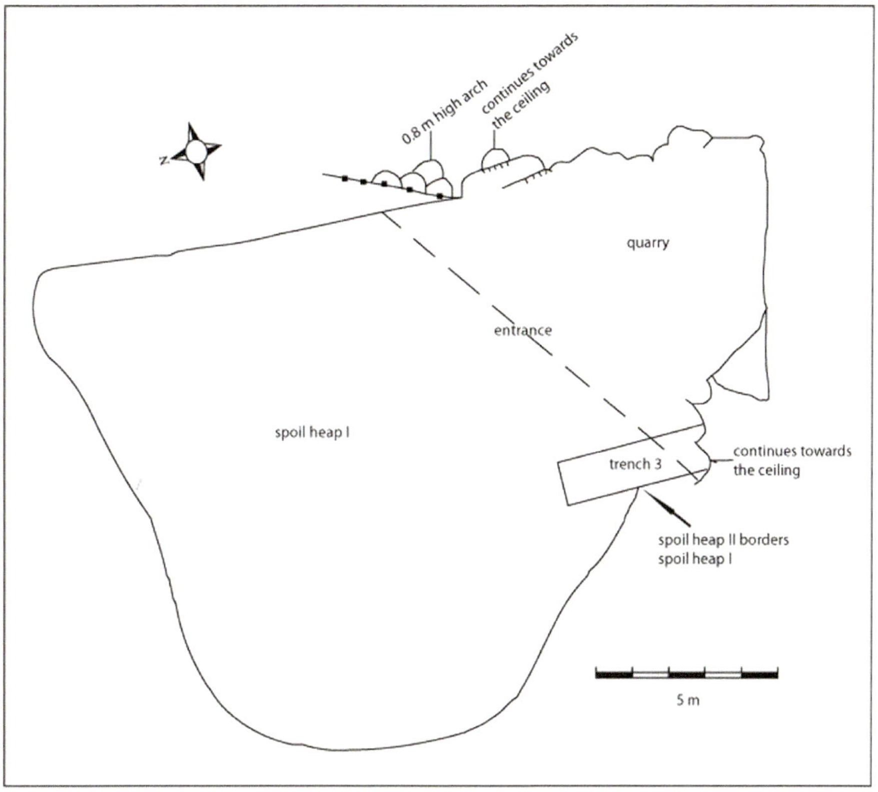

FIGURE 6.14: THE INVESTIGATED QUARRY AT FUGLEBERG TRENCH 3 (DRAWING IRENE BAUG/ØYSTEIN J. JANSEN, 2006)

FIGURE 6.15 A AND B: FUGLEBERG TRENCH 3 AT DIFFERENT STAGES OF EXCAVATION. LEFT (A): UNFINISHED BAKESTONE. RIGHT (B): NORTHERN END OF THE EXCAVATED TRENCH (PHOTOS: IRENE BAUG, 2006)

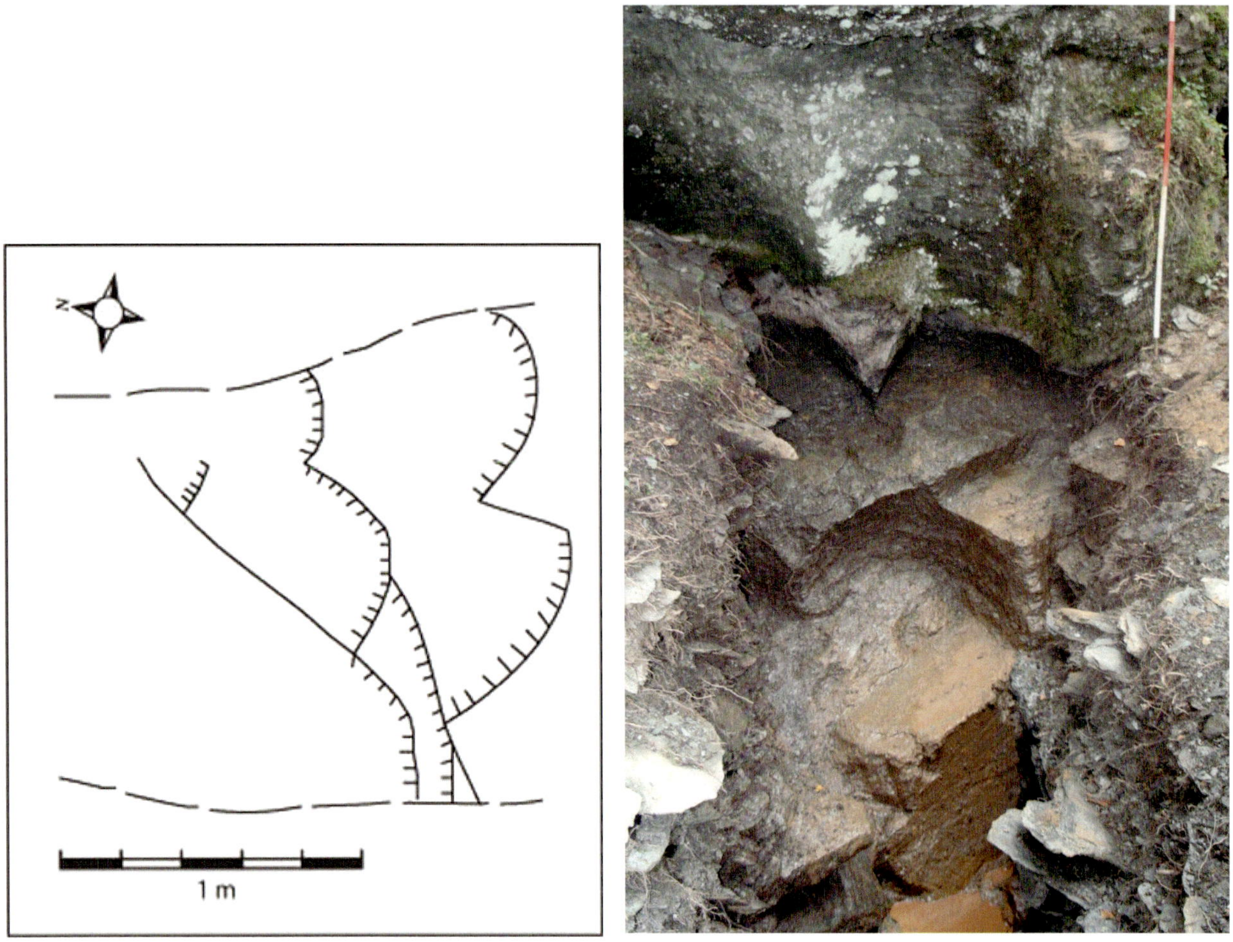

FIGURE 6.16: CARVED ROCK AT THE BASE OF THE SOUTHERN PART OF FUGLEBERG TRENCH 3, SEEN FROM THE NORTH (DRAWING: IRENE BAUG/HALLDIS HOBÆK, 2006, PHOTO: IRENE BAUG)

At the southern end of layer 6 a 60cm-deep, 20cm-wide ditch was uncovered. It was filled with production waste from layer 6. The ditch had straight sides and at the base of the southern profile several flat slabs had been placed in an upright position. Above the slabs, c. 20-30cm from the base, several flat slabs covered the ditch (Fig. 6.19). The structure is interpreted as a possible ditch for draining water out of the production area. Even though the area in front of the quarry seems to have been sloping, it may have been necessary to drain the site. After the production waste was deposited the drainage ditch seems to have lost its function. If this was the case, the ditch may have functioned only in the first phase in the quarry.

Layer 7 represented the natural subsoil. It was a red-orange layer containing large amounts of humus, and also a fine-grained mass, almost silt. Some flakes were also revealed, most likely from the overlying layer 6. Elements from the production phases seem to have been pressed into the soft subsoil as the spoil heap grew, and the largest flakes are about 10-15cm in length. It was therefore decided to dig into the subsoil to try and shed light on the beginnings of production activity here. At the deepest level, about 50cm of subsoil was removed. Excavation stopped when the amounts of charcoal and spoil in the subsoil gradually decreased and disappeared. Two samples for C14-dating from the layer were taken, one from each side of the ditch; TUa-6700, dated to 2605 ± 35, calibrated to BC 805-790, the late Bronze Age, and TUa-6701, dated to 4440 ± 45, calibrated to BC 3260-2930, the Neolithic, and obviously much older than the production in the quarry. Even though small flakes from the extraction were pressed into the subsoil, the charcoal in the layer cannot come from the time of production.

Compared with the other investigated quarries, more fragments of bakestones were identified, which supports the hypothesis they were the main products extracted here. The size of the measurable bakestones found ranged measured 27cm, 28cm, 34cm, 36cm, 45cm, and 55cm respectively. They also varied in thickness: unfinished blanks were c. 3-4cm thick, while the more finished bakestones with furrows on each side were about 1cm thick. Some of the production marks on the rock are however c. 60-70cm in diameter, indicating that there was no standard size for the bakestones produced in the quarry. Finds from the different layers of the trench also show that the plates were made in different shapes - rounded, oval and drop-shape (Fig. 6.20).

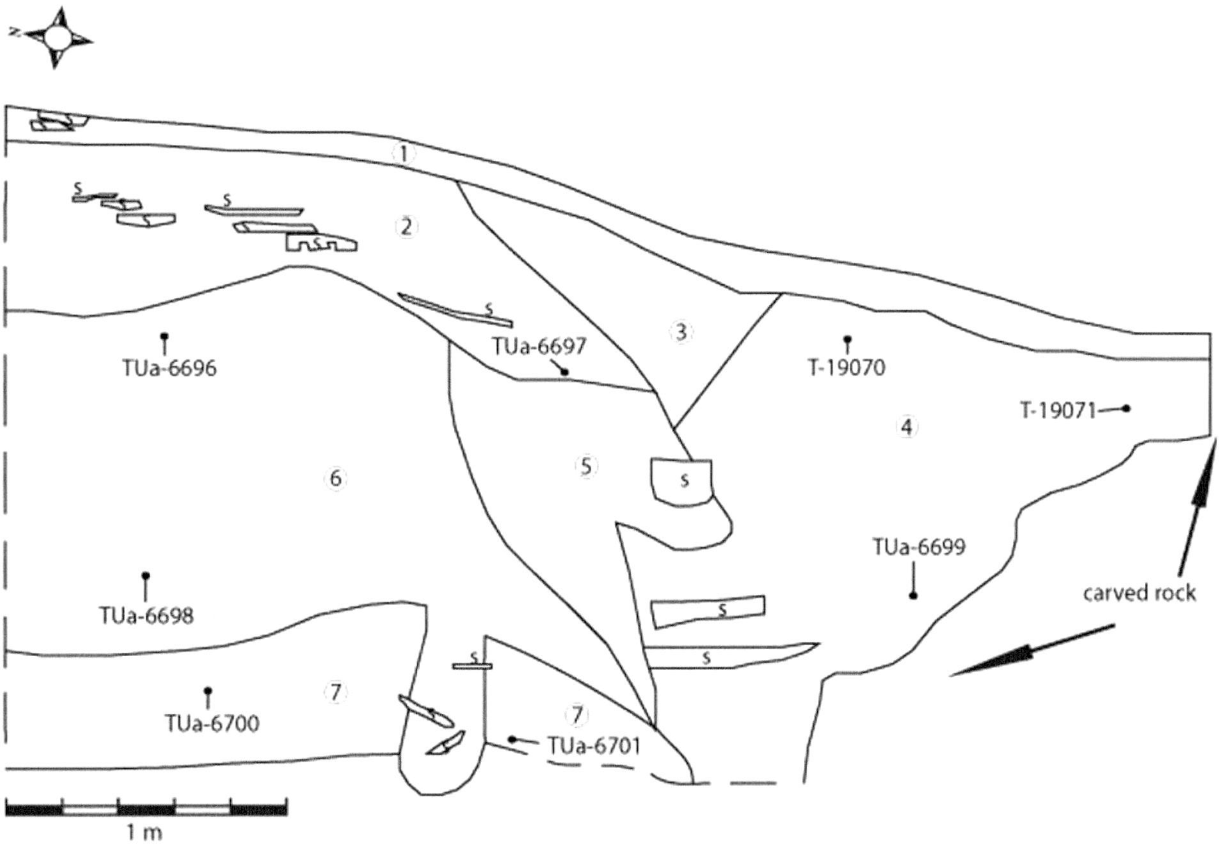

1. Dark-brown topsoil of humus and a few flakes (2-20cm). Some small roots also present.
2. Light-brown/grey deposit of gravel, flakes and some humus, probably representing topsoil disturbance. A bakestone fragment (45cm in diameter) was found. The flakes varied in size from 2-4cm, but some larger stones (c. 45cm) were also present, lying horizontally.
3. Brown deposit of humus and flakes (0.1-10cm). The stones slanted downwards towards layer 4.
4. Dark/red-brown layer of humus, flakes (0.2-30cm) and some larger slabs (45-55cm; 8-20cm thick). The layer was wet, probably the result of water running down from the carved rock to the south. The layer was located at the edge of the overhang.
5. Grey-brown layer with flakes (the largest 20-30cm and <1cm thick). As with level 3, many of the stones were slanting.
6. Grey deposit (no humus) grading from silt to slabs. The flakes range from 0.2-4cm. Some slabs were also present (50-60cm; 1 10cm thick). All stones lying horizontally.
7. Red-orange layer of humus and silt, most likely representing the original subsoil. A few flakes were present (10-15cm) likely to have dropped in from the layer above.

Figure 6.17: Section of Fugleberg trench 3 (Drawing Irene Baug/ Halldis Hobæk, 2006)

The material in the trench indicated that other product types were also made, since some of the slabs found had straight sides and rounded corners. The production marks on the quarry roof are oval and of different sizes, 60x55cm, 55x70cm and 66x56cm, possibly the results of bakestone extraction, but also tiles/slabs. The carved rock at the bottom of trench 3 also suggests extraction of products other than bakestones from the high Middle Ages onwards (cf. Fig. 6.16), e.g. slabs, tiles or building stones. This extraction is more recent than the bakestone production.

6.6 Archaeological investigations at Fuglebergåsen, Fugleberg trench 4

Further south at Fuglebergåsen, c. 200m north of Kvitafjell, is the last investigated quarry at Fugleberg,[34] with remains of manufacture at both an underground quarry and on the exterior of the rock. The underground quarry is c. 11m wide, 7m deep and 2.2m high, with traces of different production marks, both on the walls and roof. A minimum of 170 cubic metres have been extracted from the underground quarry. South of the quarry is a rock wall, without traces of carving, most likely the result of a natural fissure, and probably uncovered during quarry activity. The carved rock wall outside the underground quarry is c. 2.5m in high, but large amounts of spoil along the rock face and inside the quarry cover older extraction traces. To the northeast of the quarry, production marks on the rock exterior are visible over a stretch of c. 33m.

The first eight metres of the rock north of the quarry feature production marks in the form of the semicircular curves typical of bakestone production. The concave curves are large, with diameters of c. 90cm-1.2m, and

[34] Id. no. 35831, Askeladden; UTM Zone 33N X: -7980 Y: 6689127.

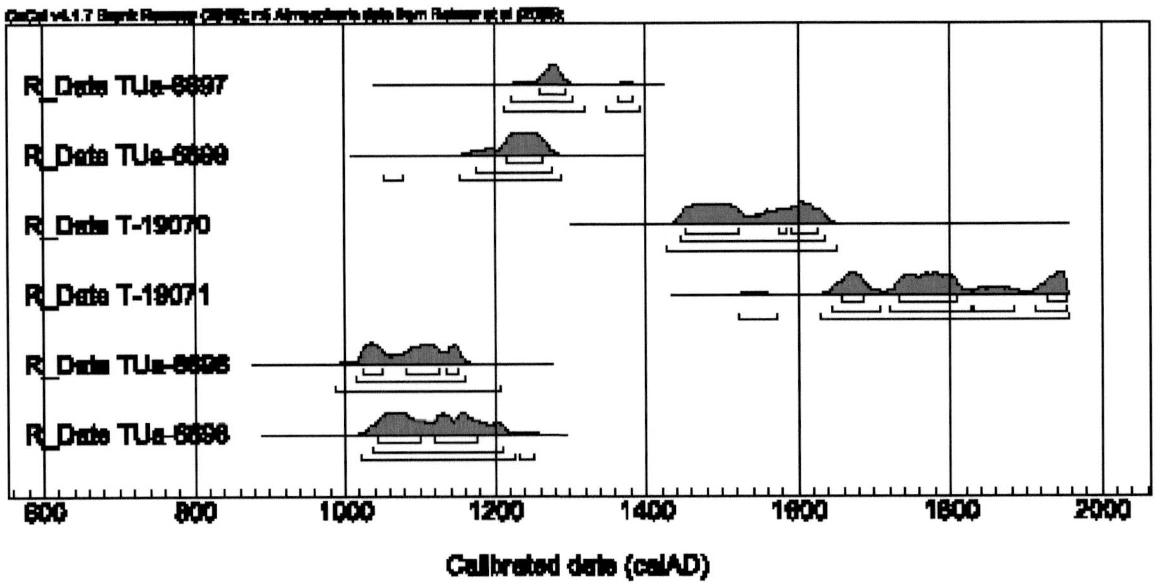

FIGURE 6.18: MULTIPLE PLOT DIAGRAM OF C14-ANALYSES FROM FUGLEBERG TRENCH 3[35]

may derive from the production of large bakestones, or possibly slabs for grain drying or other use. Further to the northeast, at a distance of 6m, there are only sporadic traces of extraction. The dense vegetation in the area may, however, be covering some of the traces. At this point the rock slants slightly eastwards towards another carved rock wall, 19m long. Slanting 6m further to the east the production traces disappear, probably because of a natural crack in the rock. This wall forms part of the entrance to the large Hedlebergshidlaren cave,[36] one of the largest Fuglebergåsen quarries.

The spoil close to the investigated quarry is deposited near the rock, forming a 9m-wide heap that stretches c. 20m in an east-west direction along the production rock face. Close to the rock, a path runs over the spoil and continues north-east towards Hedlebergshidlaren. This path, as already mentioned, may have functioned as a transport route. The area west of the spoil heaps is steep and less suitable for transportation.

The production marks in the quarry vary in shape and size. Some of the concave curves are semi-circular, while others have straight sides with rounded corners, evidently signifying extraction of a variety of products. On the roof there are also production marks running in straight lines, with round corners, probably resulting from the extraction of tiles or slabs. One of the marks on the roof measures 1.3x1.28m, demonstrating a large but unidentified product. These traces, both on the walls and roof, indicate that there was no standard size for the objects made. Bakestone traces are fewer, and totally absent inside the underground quarry. The rock type at this quarry is very hard and does not easily split into flakes (Jansen pers. comm. 2006), making it less suitable for bakestone production. The visible marks on the rock give no clear answers as to what products were produced and only major excavations could identify them.

6.6.1 Fugleberg trench 4

Trench 4 was cut east of the quarry entrance, with the rock surface at the southern end and the production heap at the north-western end. The objectives were to investigate the extent of the rock and try to reach the bottom of the quarry. On the carved rock wall at the southern end of the trench there were semicircular curves with diameters of c. 90cm - 1.2m. These traces may be related to large bakestone extractions, or possibly slabs for grain drying. According to Friis (1632:72) the sizes of these marks were 2 x 1.5 *alen*, i.e. about 120x 94cm, which concur with the traces found.

Trench 4 was 4m long, c. 1.5m wide, and 1-1.4m deep. The base level of the heap was reached and a carved rock appeared (Fig. 6.21). This clearly demonstrates that the spoil heap covers an older worked-out part of the quarry. The rock at the bottom of the trench continues into the heap to the northeast.

The production marks at the base show that stones with straight sides and rounded corners were extracted (Fig. 6.21). Channels were hewn between the different objects, most likely to make it easier to use different tools when loosening them from the rock. The channels are c. 10-11cm wide at the top level and c. 7-8cm at the base, a usual phenomenon with the production of building stones. The

[35] TUa-6700 and TUa-6701 are not included as they show a much older date than the production.

[36] Id. no. 66361, Askeladden.

FIGURE 6.19: THE DITCH AND SLAB COVERING (PHOTO: IRENE BAUG, 2006)

unidentified objects seem to have been larger and different from bakestones, possibly building stones, but apparently of no uniform dimension.

Profile interpretation: Fugleberg trench 4

Trench 4 revealed five layers, including topsoil (layer 1) (Fig. 22). The deposits in the trench contained spoil, most likely from production in the underground quarry. Layer 2, light brown in colour, contained sand, stone and humus; there were assorted flakes (20-30cm) and some larger slabs. Layer 3 was of dark-brown soil and sand with large amounts of flakes and slabs, rather similar to layer 2, but with more humus, and perhaps caused by the overlaying path. Layer 4 consisted of grey sand and smaller flakes. Layer 5 was of grey stone with flakes and slabs, packed with stones of c. 10cm in size, as well as pulverized stones. There was no humus in the layer, only production remains, including unfinished and fragmented bakestones, one just above the rock at the base of the trench.

All the layers contained large slabs, many over 1m in diameter, but also a few unfinished and fragmented bakestones, varying in size from 43-65cm in diameter, measuring 43cm, 45cm, 55cm to 65cm respectively, and with no indications of a uniform size. Fragments of bakestones were few compared to the larger slabs, which suggest that they may not have been the main products at this site. This is supported by the production marks found at the quarry. The different layers may therefore represent different production periods and the carved rock at the base of the trench seems to indicate the production of building stones. Unfortunately no organic material suitable for C14-dating was found, and the date of production is still uncertain.

6.7 Kvitafjell

The last location chosen for investigation at Fugleberg was Kvitafjell,[37] a natural overhang stretching in a north-south direction (Fig. 6.23). The rock type is green schist without elements of steatite. The site represents therefore a workplace where bakestones were finished - a similar but larger workplace than those identified a few metres south-east of Kvitafjell and at Tufta farm (no. 16). At these latter sites as well the workplaces were situated beneath natural overhangs.

Kvitafjell first achieved scholarly interest when Weber and Naterstad surveyed the quarries in 1983 (Weber 1984:153). A trench at the southern end of Kvitafjell was excavated, uncovering a fireplace, a *tuyére*, and 20kg of iron slag. Charcoal was collected from the deposits but unfortunately it does not seem to have been retained and dated (Top. ark. Bergen Museum).

Spoil in large amounts was located underneath the overhang, extending 2m from the rock wall and following the length of the overhang for c. 30 m. The end of the roof is c. 2-2.5m from the rock wall, and it is dry and light underneath the overhang. Beyond the roof the area slopes gradually for c. 2m before the ground falls away steeply. Small flakes of steatite and fragments of bakestones are still visible; the whole slope is probably covered with spoil. A test pit was made here in the production waste, 1m from the rock wall, 13m north of the southern end of the overhang and 6.3m north of a small cliff next to the rock wall. The cliff ends at the south profile of the trench dug by Weber (Fig. 6.24).

6.7.1 Test pit at Kvitafjell

The test pit measured 1x1m and was 1.5m deep; it contained only spoil. At the base of the pit a rock was revealed, showing that the overhang was completely cleared of sand, stones and vegetation before work started. The depth of the production waste indicates intense activity. Four C14-samples were taken for analyses from the test pit.

Profile interpretation

Four different layers were uncovered (Fig. 6.25). Layer 1, brownish, fine-grained, nearly silty, also contained small flakes (<1cm) and a few flakes of 10cm. Layer 2 was grey in colour but otherwise similar to layer 1. One

[37] Id. no. 45427, Askeladden; UTM Zone 33N X:-8094 Y: 6689078.

A: UNFINISHED OVAL BAKESTONE

B: FRAGMENTS OF UNFINISHED BAKESTONES

C: UNFINISHED DROP-SHAPED BAKESTONE

D: FRAGMENT OF AN UNFINISHED DROP-SHAPED BAKESTONE

FIGURE 6.20 A-D: FRAGMENTS OF BROKEN AND UNFINISHED BAKESTONES FROM FUGLEBERG TRENCH 3 (PHOTOS: IRENE BAUG, 2006)

sample of charcoal (TUa-6702) was taken 8cm from the surface, dated to 875 ± 35, calibrated to AD 1160-1220, probably representing the last phase of activity. Layer 3 was similar to 1 and 2, but contained many very small roots. A nearly complete bakestone was identified at the lower edge of the layer. The base layer (4) was the thickest, with large quantities of chlorite-rich green schist. It was greyer in colour than layer 2 and contained silt and flakes (c. 0.5-30cm). More bakestones fragments were found here than at the other sites investigated, but mostly small - only a few cm in size (Fig. 6.27). Some nearly complete bakestones of various sizes were, however, also found (42cm, 50cm, 52cm and 55cm in diameter and c. 1cm thick). Three samples for C14-dating were taken from the deposit. Sample TUa-6703, taken below layer 3, was dated to 845 ± 35, calibrated to AD 1170-1245. The same date was found for sample TUa-6705, taken c. 35cm above the pit floor. Between these two samples, c. 60cm above the floor, a further sample (TUa-6704) was taken and dated to 940 ± 35, calibrated to AD 1030-1165, a somewhat earlier date than the other two samples from the same layer. All three dates concur in time, and there is nothing to indicate that the charcoal C14-samples give dates that are too early. By expanding the given intervals to a standard deviation of 99.7% (3σ) the dates overlap, indicating activity around AD 1050-1250 (Tab. 6.2 and Fig. 6.26). The depth of the spoil heap (1.5m), makes it feasible that a long period of activity took place here. However, because of the wide range it is difficult to suggest a more precise date for the workplace.

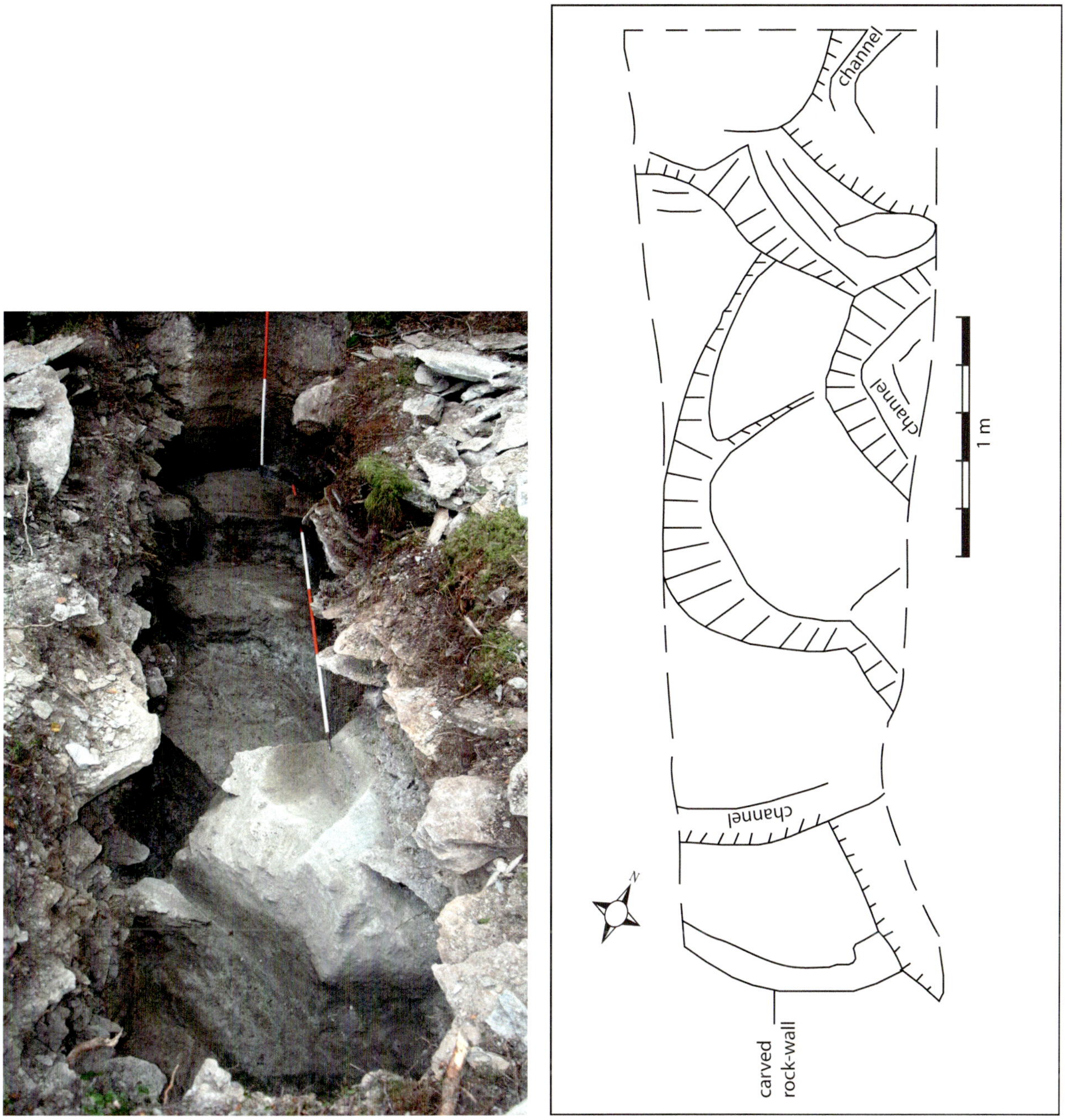

FIGURE 6.21: THE CARVED ROCK AT THE BASE OF FUGLEBERG TRENCH 4, SEEN FROM THE NORTH (PHOTO AND DRAWING: IRENE BAUG, 2006)

Table 6.2: C14-analyses from the Kvitafjell test pit

Lab. Ref.	Context	Age BP	Calibrated Age	Material
TUa-6702	Kvitafjell layer 2	875 ± 35	AD 1160-1220	pine charcoal
TUa-6703	Kvitafjell layer 4	845 ± 35	AD 1170-1245	birch charcoal
TUa-6704	Kvitafjell layer 4	950 ± 35	AD1030-1165	charcoal from pine
TUa-6705	Kvitafjell layer 4	840 ± 35	AD 1170-1245	pine charcoal

At the western end of the test pit an unfinished bakestone was found in an upright position, seemingly intentionally placed directly on the rock at the base of the pit, probably representing an unfinished product ready to be thinned down; there were no furrows on the bakestone and the process had not yet begun. It is possible that the bakestone was accidentally covered with spoil and forgotten, but it may also be a broken object that was dumped. The bakestone that continued into the west profile of the pit was c. 55cm in diameter and 3cm thick (Fig. 6.28).

Layers 1-3 seemed to contain dust from the rock wall of the overhang. Trickling water crystallizes out to the exterior of the rock and shatters, producing a very fine-

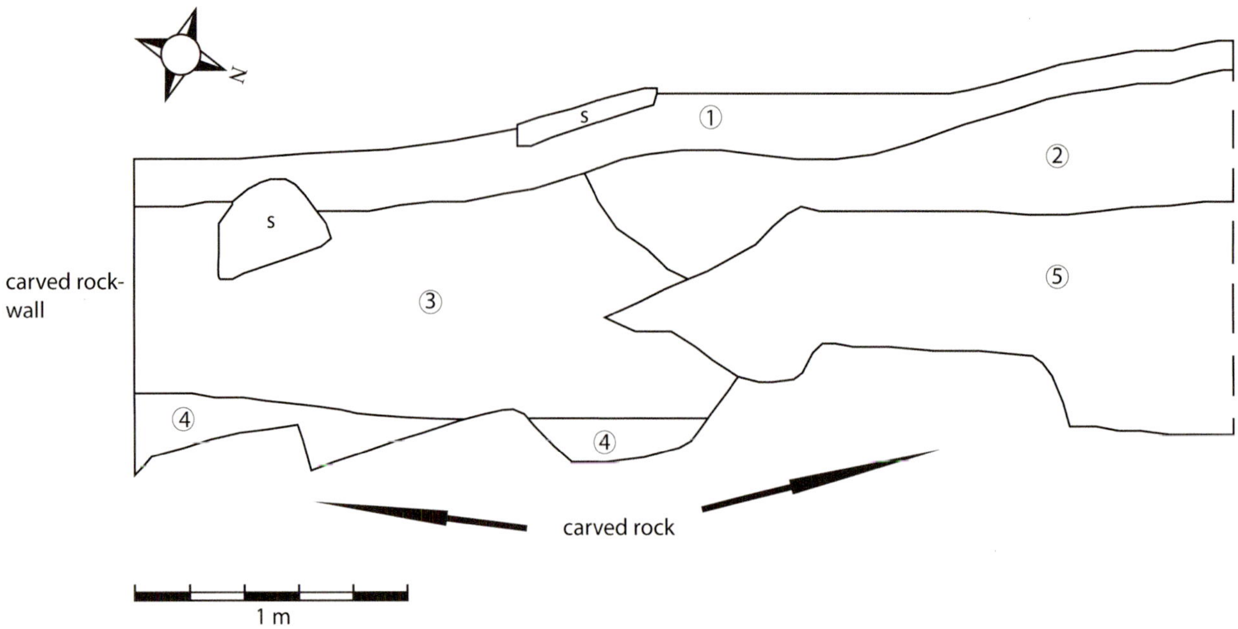

1. RED-BROWN HUMUS TOPSOIL WITH SOME FLAKES (<10CM) AND ROOTS.
2. LIGHT-BROWN DEPOSIT OF SAND AND STONES. SOME SMALL AND LARGER FLAKES (20-30CM) ALSO PRESENT.
3. DARK-BROWN DEPOSIT OF HUMUS, SAND, SIMILAR TO LAYER 2 BUT WITH MORE HUMUS AND CONTAINING MANY FLAKES.
4. GREY DEPOSIT WITH SAND AND SMALL STONES, PROBABLY ALSO CONTAINING CRUMBLED FRAGMENTS FROM THE BASE LAYER.
5. GREY DEPOSIT (NO HUMUS) WITH FLAKES (0.2-10CM).

FIGURE 6.22: SECTION OF FUGLEBERG TRENCH 4 (DRAWING IRENE BAUG, 2006)

grained dust. This process is known as *tafoni* or cavernous weathering (Jansen pers. comm. 2006). The contents of layers 1-3 indicate that dust from the rock wall became mixed with the production waste, a process that to a lesser degree affected layer 4. This indicates that layers 2 and 3 were exposed for some period before they were covered with a new layer (layer 1). Whether bakestones arrived here from one or more quarries is difficult to ascertain.

Bakestones with different degrees of finishing were found, making it possible to study the finishing processes (Fig. 6.29). After extraction from the quarry bedrock as nearly rounded, or oval, thin slabs of various size, the bakestones were transported to the Kvitafjell worksite for final finishing, i.e. thinning and acquiring the characteristic furrows known from bakestones that have been found in different use contexts (cf. Weber 1983; Tengesdal 2010). To thin the bakestones, coarse grooves were carved into the stone and the spaces between the furrows then seem to have been 'peeled' off. This technique is also indicated by the unfinished bakestones found at Fugleberg (trench 3) (cf. Fig. 6.20). Secondary working produced thinner grooves closer together, and obviously made by a finer tool than used for the initial grooving process. This form of grooving is found on finished bakestones with parallel marks, in different directions and patterns. Some bakestones have these grooves on one side only, while others are scored on both sides. It has been suggested that the grooves were required for cooking processes (frying or baking). In medieval Bergen food remains on bakestones

FIGURE 6.23: THE KVITAFJELL WORKPLACE SEEN FROM THE NORTH (PHOTO: IRENE BAUG, 2006)

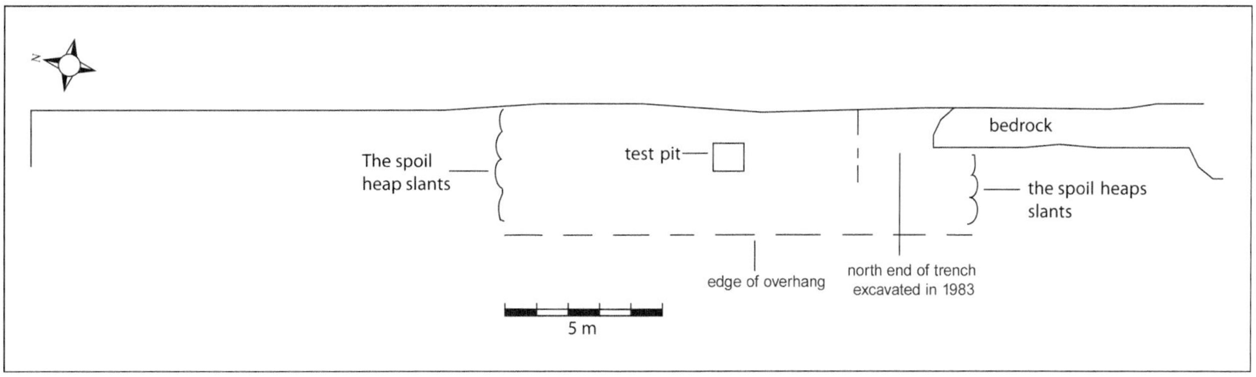

FIGURE 6.24: SKETCH OF THE KVITAFJELL OVERHANG (DRAWING: IRENE BAUG/HALLDIS HOBÆK, 2006)

are only found on the grooved sides (Tengesdal 2010:70), which supports this assumption.

6.7.2 The relationship between Kvitafjell and the Fugleberg quarry site, trench 3

Activity at the investigated quarry at Fugleberg trench 3, and the Kvitafjell workplace seems to have been simultaneous, at least for a period between the 11th and early 13th centuries. However, as already demonstrated, at least two production phases are documented at Fugleberg trench 3. Here, the first period dated to the early Middle Ages, or the beginning of the high Middle Ages, concurs with the activity at Kvitafjell. The latter site was abandoned as a workplace when the later extraction in the high Middle Ages and late Middle Ages/early modern period was carried out - sometime between the 13th and early 17th centuries. But bakestones were also produced in this later extraction phase, possibly along with other objects. Even though the quarry and Kvitafjell site to a certain degree overlap in time, it is uncertain whether unfinished bakestones from the quarry were transported to Kvitafjell for further processing. Nearly finished bakestones found in the quarry signify that at least some of the products were completed at the quarry site. Nevertheless, fewer bakestone remains have been identified at the quarries compared to the worksite. For a period therefore Kvitafjell may possibly have operated as a finishing site for several of the Fuglebergåsen quarries.

6.8 Archaeological investigations at Netteland: Båtahidlaren

The Båtahidlaren quarry, located near Netlandsvågen, 14m north of the main fjord, was chosen as another potentially interesting research area related to bakestone production.[38] It is an underground quarry, 5.7m wide, 9.5m deep and c. 2m high at its entrance, narrowing below the ceiling further inside (Fig. 6.30). Approximately 108 cubic metres of rock have been extracted. Semicircular production marks on the rock wall are identified inside the quarry, some the remains of bakestone extraction. Above, the underground quarry, c. 2.25m in from the entrance, there is another carved rock wall, 1.4-1.8m high, with traces of production at two levels. At the entrance to the quarry some indistinct letters and numbers have been carved into the rock, possibly '1847' on the western side and '1837' on eastern side of the entrance, and additionally a cross. The latter probably represents a property mark, since the quarry is also located on the border between holdings 1 and 2. The numbers probably refers to years, either of production or other events.

East and west of the quarry, the production site continues over roughly 100m, reflecting intensive extraction. The underground quarry is centrally located within this larger area. Substantial amounts of stone have been removed from the rock, and many semi-circular marks indicate the extraction of bakestones. At the eastern end of the quarry site the rock changes. At a distance of 12.3m the rock appears rather straight, without the concave traces typical of bakestone production, but there are traces of straight-sided and right-angled products, evidently building stones (Fig. 6.31). Three small buildings are located within the quarry site, built on the production waste that covers the whole area down towards the sea.

At Båtahidlaren, therefore, bakestone production seems to have been of secondary importance, with building stones and possibly tiles and slabs predominating. This was also substantiated by the contents of the excavated trench. Holes drilled into the rock wall are also seen at Båtahidlaren, close to the underground quarry, a phenomenon observed at two other quarries, the so-called 'Ingridhidlaren' at Tufta farm (Ølve) and at Helleberget (Hjortland farm, Hatlestrand). These holes may represent a process whereby roof tiles were wedged from the rock.

6.8.1 Netteland trench 1

Close to the underground quarry there is an area with more visible traces of bakestones than found further east. These traces appear on the carved walls and as surface remains. A search trench was established here north of the entrance to the underground quarry, along a rock with production marks stretching in a north-south direction, with the underground quarry to the south (Fig. 6.32). The visible rock, c. 2m

[38] Id. no. 45431, Askeladden; UTM Zone 33N X: -5486 Y: 6690529.

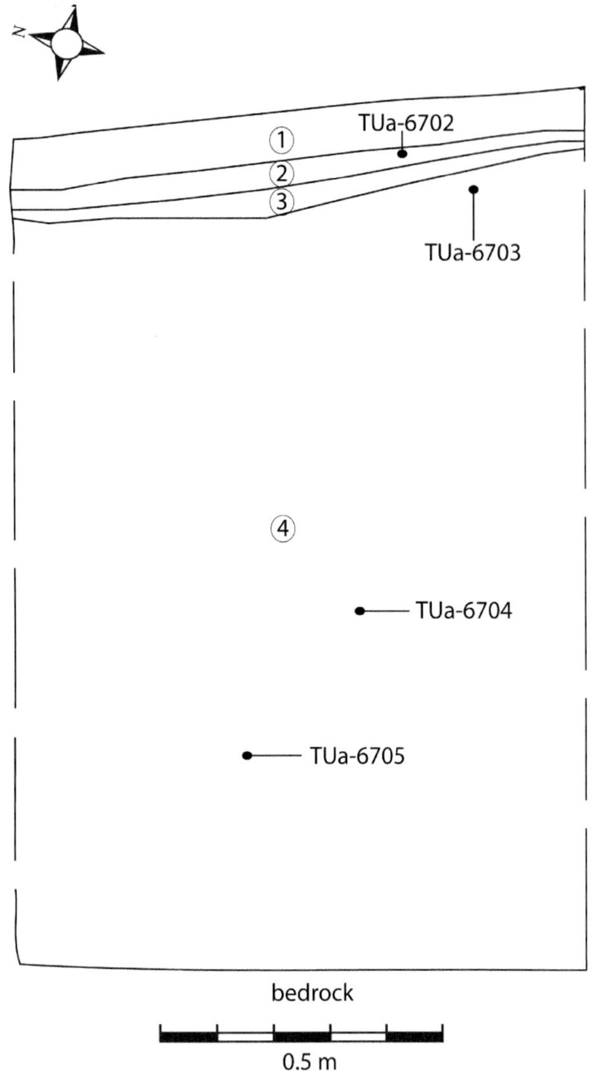

1. Brown layer grading from silt to flakes (0.2-1cm), containing a few flakes (c. 10cm).
2. Grey layer similar to layer 1 but differing in colour.
3. Layer with flakes the same size as layers 1-2 but with many small roots. A bakestone was found at the base of the deposit.
4. Blue-grey deposit of silt and flakes (0.5-3cm) and larger flakes (10-30cm). Several bakestone fragments with varying finishes.

Figure 6.25: Section of the Kvitafjell test pit (Drawing Irene Baug/ Halldis Hobæk, 2006)

east of the trench, was partly covered with spoil and at the southern end only 32cm was visible above the production waste; 1.3m of rock is still visible at the northern end. The area in front of the quarry therefore seems to represent a previously worked-out part of the quarry now covered with spoil from later activities. The material found may come from the underground extractions or from the rock face just above. This makes it difficult to date the beginning of production here.

The trench dug was 3.8m long, 1.5m wide, and 1.4-1.9m deep. The combination of large, heavy stones and slabs and the depth of the trench made it difficult to continue excavating. The base level of the spoil heap was not reached, nor could its depth be estimated. No charcoal or organic material was found at the lowermost level.

Profile interpretation: Netteland trench 1

Netteland trench 1 consisted of six layers, including topsoil (layer 1) (Fig. 6.33). Three samples for C14-datings were taken from two of the trench layers (Tab. 6.3). The topsoil of humus, small flakes (2-4cm) and some roots, also contained slag and a few fragments of modern glass, together with fragments of bakestones. The bakestones show different degrees of finishing, with a few c. 4cm thick. The fragmentary state of the bakestones made it difficult to measure their diameters. Layer 2, grey-brown in colour, contained humus, gravel, flakes, some larger stones and roots. Some large flakes were found, 30-40cm long and 8-10cm thick, along with smaller flakes (2-6cm). A few fragments of bakestones were identified. One sample for C14-dating (TUa-6706) was taken from below the topsoil and dated to 750 ± 35, calibrated to AD 1265-1290. A whetstone and iron nail were found in the southern part of the deposit (Fig. 6.33).

Layer 3 was grey in colour and consisted of flakes of different size, from a few mm to c. 10cm. The deposit varies from stones to fine grained sand, almost silt. The layer also contained larger slabs.

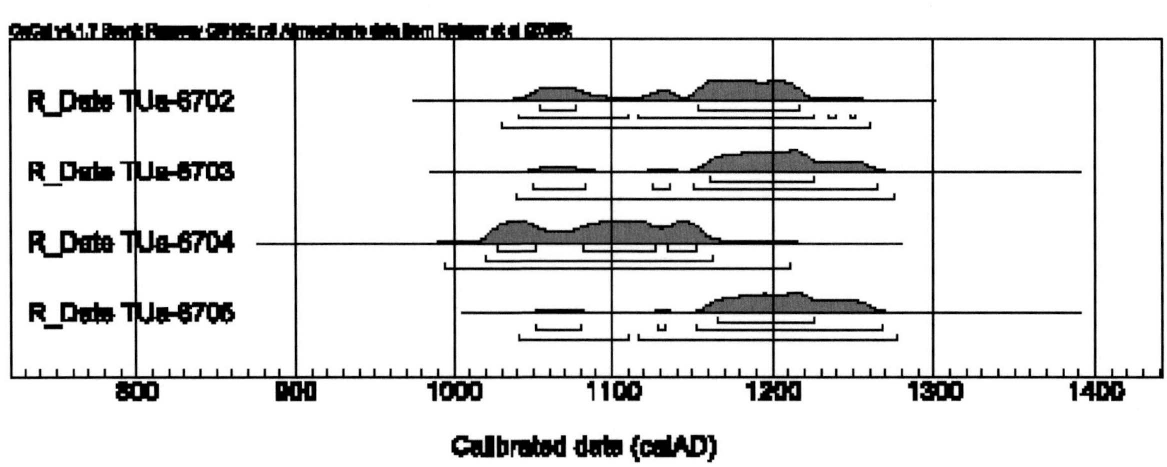

Figure 6.26: Multiple plot diagram of C14-analyses from the Kvitafjell test pit

A: Flakes probably from bakestone thinning

B: Bakestone fragments

Figure 6.27 a and b: Flakes and bakestone fragments from layer 4 (Photos: Irene Baug, 2006)

Figure 6.28: Profile of the Kvitafjell test pit, with an unfinished bakestone at the base (Photo: Irene Baug, 2006)

A: Bakestone fragment showing initial grooving

B: Bakestone fragment showing final grooving

Figure 6.29 a and b: Bakestone fragments showing different grooving patterns. Both fragments are less than 20cm in diameter (Photos: Irene Baug, 2006)

FIGURE 6.30: THE UNDERGROUND QUARRY, BÅTAHIDLAREN (PHOTO: ATLE OVE MARTINUSSEN, 2012)

Layer 4, also grey in colour, contained tilting slabs of different size, and smaller flakes, similar to layer 3, but mostly larger (c. 30cm long and <1cm thick). Slabs, c.1m in size, slanted downwards in the trench in a chaotic manner. The tilted stones/slabs differed markedly from the rest of the trench, where the stones lay horizontally. The same phenomenon was observed at Fugleberg trench 3, interpreted as two different spoil heaps and two different production phases. The same explanation might apply here, with layer 4 possibly the result of production waste deposited on an existing spoil heap.

Layer 5, a yellow-grey mass, contained fairly large stones, some of the slabs more than 50cm in length, as well as smaller flakes, only a few mm in size, and very fine-grained sand, nearly silt. Two samples for C14-dating were taken: TUa-6707 from the uppermost level of the deposit was dated to 595 ± 35, calibrated to AD 1310-1405; TUa-6708 was taken from further down in the deposit and dated to 140 ± 35, calibrated to AD 1680-1940, indicating activity well into the modern period (Tab. 6.3 and Fig. 6.34), unsurprisingly, as building stones were supposedly extracted from here in the 19th century (Buch 2011: 14-15), and the latest stones at Netteland removed in the 20th century (Knut Johan Nerhus pers. comm. 2006). A worked stone, perhaps the remains of a whetstone, was found in the southern end of the deposit.

Layer 6 was a grey layer with large amounts of slabs, many of them c. 60-70cm, and some more than 1m long, 50cm wide and 15cm thick. Contrary to layer 4, the stones in this deposit were lying horizontally. No bakestones were found. Layer 7 had a grey colour and consisted of sand, gravel and flakes (6-20cm).

Three C14-samples showed a reverse chronology, with the oldest dating from the uppermost layer. This indicates that the spoil has been disturbed by later activities, probably caused by material from later extractions at the quarry site. The disturbances in the spoil heap made it difficult to date the bakestone production more exactly, or the date extractions began. Work seems to have been carried out during both the high and late Middle Ages, but also later. The oldest production, however, was not discovered.

The northern end of the trench (layers 2, 3, 4 and 6) contained more large slabs than the southern end (layers 5 and 7). The contents in the layers indicate different types of extractions and product types, probably also various production phases. Due to the lack of datable material in most layers, it was difficult to date the phases more accurately. Except for layer 6, fragments of bakestones were found in all layers, but in rather small amounts. Bakestone production therefore seems to have been on a small scale. Some of the slabs found have rather straight sides and are probably the extraction remains of building stones, tiles, and slabs for grain drying.

Figure 6.31: Straight rock wall, probably the result of building stone extraction (Photo: Irene Baug, 2006)

Table 6.3: C14-anayses from Netteland trench 1

Lab. Ref.	Context	Age BP	Calibrated Age	Material
TUa-6706	2	750 ± 35	AD 1265-1290	oak charcoal
TUa-6707	3	595 ± 35	AD 1310-1405	charcoal from birch, cherry/ rowan
TUa-6708	3	140 ± 35	AD 1680-1940	hazel nut

Figure 6.32: Netteland trench 1. The entrance to the underground quarry can be seen in the upper left corner (Photo: Irene Baug, 2006)

6.9 Dating of quarries and products

Ultimately it proved difficult to obtain samples for C14-dating from some of the Ølve and Hatlestrand trenches, and two of the investigated sites could not be dated. Three of the six sites produced enough material for C14-dating, and altogether 15 samples were dated (Tables. 6.4 and 6.5). The lack of charcoal may also reflect the organisation of the activities undertaken. A workplace with forges or other fire places, perhaps also for cooling, does not seem to have been a necessity at all quarry sites, or they may have been located beyond the investigated quarry sites. Based on the available material it was difficult to understand why charcoal was lacking at some sites and found at others. Large amounts of charcoal were only found at a quarry where bakestones seem to have been the most significant extractions. Tables 6.4 and 6.5 give an overview of all the C14-dates.

Table 6.4: C14-analyses from Fugleberg trench 3 and the Kvitafjell test pit

Lab. Ref.	Context	Age BP	Calibrated Age	Material
TUa-6697	Fugleberg trench 3, layer 2	720 ± 35	AD 1280-1295	pine charcoal
TUa-6699	Fugleberg trench 3, layer 4	800 ± 35	AD 1225-1280	birch charcoal
T-19070	Fugleberg trench 3, layer 4	370 ± 45	AD 1445-1635	birch and pine charcoal
T-19071	Fugleberg trench 3, layer 4	190 ± 45	Younger than AD 1660	birch and pine charcoal
TUa-6696	Fugleberg trench 3, layer 6	960 ± 35	AD 1025-1155	pine charcoal
TUa-6698	Fugleberg trench 3, layer 6	905 ± 35	AD 1045-1190	pine charcoal
TUa-6700	Fugleberg trench 3, layer 7	2605 ± 35	BC 805-790	charcoal from deciduous trees
TUa-6701	Fugleberg trench 3, layer 7	4440 ± 45	BC 3260-2930	charcoal from sallow, willow/aspen
TUa-6702	Kvitafjell layer 2	875 ± 35	AD 1160-1220	pine charcoal
TUa-6703	Kvitafjell layer 4	845 ± 35	AD 1170-1245	birch charcoal
TUa-6704	Kvitafjell layer 4	950 ± 35	AD 1030-1165	pine charcoal
TUa-6705	Kvitafjell layer 4	840 ± 35	AD 1170-1245	pine charcoal

Table 6.5: C14-analyses from Netteland trench 1, Båtahidlaren

Lab. Ref.	Context	Age BP	Calibrated Age	Material
TUa-6706	2	750 ± 35	AD 1265-1290	oak charcoal
TUa-6707	3	595 ± 35	AD 1310-1405	charcoal from birch, cherry/rowan
TUa-6708	3	140 ± 35	AD 1680-1940	hazel nut

Dating the activities

Because of the few datable quarries it was difficult to reach conclusions for the production landscape as a whole in terms of chronological development. The small-scale investigations of this study and the few datable sites can only indicate certain trends. The results however suggest that bakestone production extends back to the early Middle Ages, to the 11th or early 12th centuries (AD 1025-1155 and AD 1045-1190). It is, however, only at the bakestone quarry of Fuglebergåsen trench 3 that the initial production phase was able to be dated. A second production phase at this quarry is documented from the high Middle Ages (AD 1225-1280 and AD 1280-1295), continuing until the early modern period (AD 1445-1635). Large-scale and intensive bakestone production in the early Middle Ages is also documented at the Kvitafjell workplace site, with a 1.5m thick assemblage of spoil from bakestone finishing processes. An estimate of the number of bakestones finished at Kvitafjell is difficult to calculate. The workplace indicates, however, an organized, almost 'mass production' industry in the early Middle Ages. There are no indications of other stone objects being produced at this site, and only bakestones seem to have been transported from quarries to Kvitafjell for finishing.

Dated samples from Netteland trench 1, point to activity in the high Middle Ages (Tab. 6.5). The scarcity bakestones found, along with the presence of larger slabs, suggests the extraction also of other products at this period. Apart from the bakestone fragments, no other identifiable objects were found except for traces of building stones, and perhaps tiles and slabs for grain drying. Both building stones and tiles were extracted from the Middle Ages onwards (see e.g. Helland 1893:23-24), but so far building stones from Ølve/Hatlestrand quarries have only been identified in one medieval building: the church of Onarheim at Tysnes, Hordaland, where geochemical analyses link the stones to Ølve/Hatlestrand (Jansen 2013:79). The church is mentioned in written sources from 1327 (DN X, 25), but was most likely built in the latter half of the 1100s (Hoff and Lidén 2000:267-268). This is further discussed in Chapters 7 and 10. It is not known when the production of slabs for drying grain began, but this form of production may also date to the Middle Ages. The reverse stratigraphy suggested by the C14-analyses of Netteland trench 1 indicates disturbances and spoil removed from the site, probably in connection with later extraction.

As the initial production phase has only been revealed at one of the investigated quarries (Fugleberg trench 3), it is difficult to give a clear chronology for the start of stone production in the Ølve/Hatlestrand area. Activity in the transition phase to the Middle Ages, from the middle of the 11th century, should be considered. How the C14-analyses from the quarries and the workplace correspond with the bakestones found in other contexts will be discussed in Chapters 7 and 10.

Product types

Remains on the carved rock in the quarries indicate large-scale production of products other than bakestones, most probably slabs, tiles and/or building stones. At most quarries this production seems to have been carried out on a larger scale compared to bakestone production, or possibly, represents later extractions that may have destroyed the remains of the older bakestone workings.

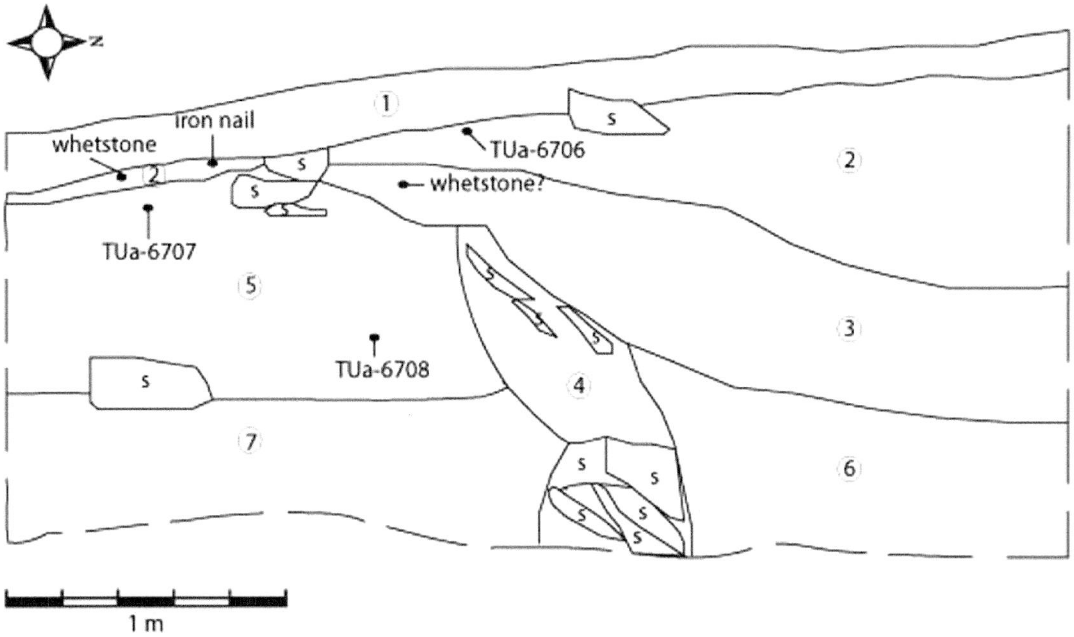

1. Topsoil of humus and a few flakes (2-4cm) and small roots.
2. Grey-brown deposit of humus, gravel, flakes (2-6cm) and some larger stones (30-40cm; 8-10cm thick). Also a few roots.
3. Grey deposit grading from silt to flakes (0.2-10cm). Some slabs also present.
4. Grey deposit containing titled slabs, the largest c. 1m in length, and many flakes (up to 30cm but <1cm thick).
5. Yellow-grey deposit of sand, gravel, flakes and slabs (>50cm).
6. Grey deposit of slabs of varying sizes (60-70cm, with the largest 100x50cm and c. 15cm thick). Contrary to layer 4 the stones were placed horizontally. No bakestone fragments were found.
7. Grey deposit grading from sand and gravel to flakes (6-20cm), but some larger stones were also present.

Figure 6.33: Section of Netteland trench 1 (Drawing Irene Baug/Camilla Jacobsen, 2006)

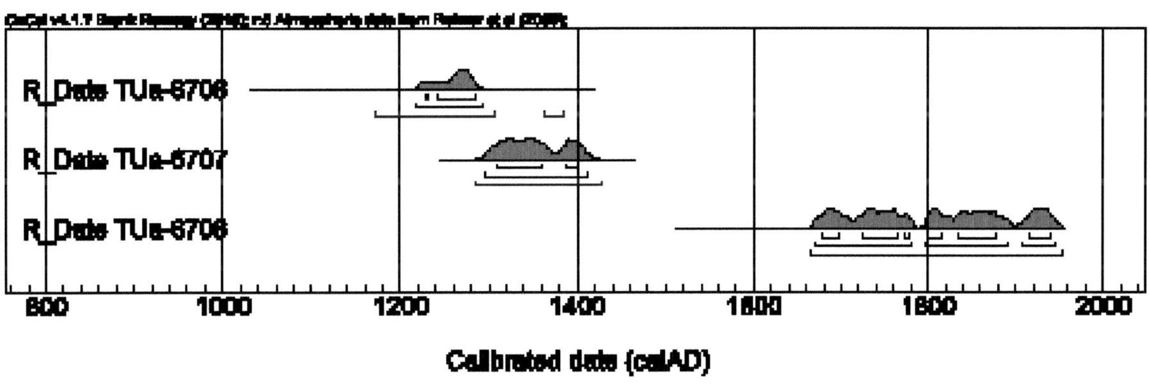

Figure 6.34: Multiple plot diagram of C14-analyses from Netteland trench 1

This is also reflected in the investigation trenches, where bakestone remains occur less frequently. The exception is the investigated quarry site of Fugleberg trench 3, where the remains of bakestone production are more abundant. The investigation also showed a larger volume of bakestone fragments in the spoil heap of this quarry compared to the other spoil heaps.

In the other trenches, larger and less definable slabs dominate, and the production of bakestones was obviously carried out simultaneously with extractions of other products, however these seem less important than the larger products. That being said, no clearly defined object types were identified in the spoil heaps.

The lack of bakestone fragments in the investigated spoil heaps may, however, not reflect an absence of bakestone production at the quarries. The possibility that bakestones were taken to other workplaces for finishing should also be taken into account, and is documented by the workplaces at Kvitafjell, further south at Fuglebergåsen, and at Tufta farm (no. 16). C14-analyses from Kvitafjell signify that sometime in the period between the early and high Middle Ages bakestones were extracted from the quarries and moved to special worksites for final finishing.

Size of the investigated quarries

The dimensions of the underground quarries make it possible to give a rough estimate of the volumes extracted from each quarry. Because of the large amounts of spoil inside the quarries, only minimum estimates can be given. All investigated quarries show evidence of surface recovery, where the carved rock, to a large degree, was covered by spoil, making it difficult to identify the full extent of the carved walls. Only an estimate of the underground quarries has been made, and the amount of rock extracted from each quarry must have been larger than the estimates given in the table below (Tab. 6.6). At some quarries production was on a larger scale than others. The size of the underground quarries varied, often within the same farm complex.

Table 6.6: Sizes of quarries investigated

Underground quarries	**Cubic metres**
Fugleberg:	
Fugleberg trenches 1 and 2; Bakkhidlaren quarries I, II and V	1297
Fugleberg trench 3 (3 underground quarries)	250
Fugleberg trench 4	170
Netteland:	
Netteland trench 1, Båtahidlaren	108

6.10 Brief comparison of Ølve/Hatlestrand and Hyllestad

Investigations of the quarry landscapes of Ølve/Hatlestrand and Hyllestad show both similarities and differences. Both areas represent long-abandoned proto-industrial landscapes with large-scale production developed over hundreds of years, and in both areas quarrying has had a significant impact upon the landscape. Both quarry areas were used for different purposes, with the extraction of several different product types. Nevertheless, they differ significantly with regard to product types, quarry types, and period of activity.

Production in Hyllestad may have started in the early Viking period, and is therefore the result of a different process than quarrying at Ølve/Hatlestrand, which seems to have started first in the early Middle Ages. By the transition to the Middle Ages the stone industry in Norway developed further, with large-scale quarrying and a wider range of products.

The introduction of Christianity and development of new institutions, represented by the king, Church and monasteries, led to an expansion of product types for Hyllestad and Ølve/Hatlestrand. This influenced the production of building stones, stone crosses, and large-scale production of millstones. Both quarry areas should therefore be seen in connection with the societal development in the Viking period and Middle Ages, where, to a great extent, external factors and social conditions influenced the development of the production landscapes, both production and product types, and new conditions for quarrying. The production was also closely tied to trade and exchange, and the distribution and contact networks are investigated and discussed in Chapters 7 and 10. The interaction between natural conditions, geological resources, the development and need for exchange, and the development of ship technology resulted in new conditions for production and exploitation of resources.

7 Spatial and Chronological Distribution of Products

Researches at Hyllestad and Ølve/Hatlestrand signify large-scale and long-lasting exploitation that presupposed access to markets where these products had competitive advantages and qualities compared to local products. The stones extracted represent a scarce, imperishable material, and at the same time a raw material whose provenance can be identified. These characteristics make the material suitable for tracing trade routes and networks which might be difficult to reveal through other sources. Archaeological finds of quernstones and millstones, as well as of bakestones outside the production areas, indicate their distribution as commodities, on a regional and international level.

Studies of the distribution of quernstones and millstones from Hyllestad have been carried out from the 1990s in relation to Bergen and some Scandinavian towns (Elfwendahl and Kresten 1993; Schön 1995; Carelli and Kresten 1997; Hvoslef 2009). Since then, more quarries of the garnet mica schist type in Norway have been identified,[39] suggesting several quarry sites for possible provenances for the distributed material. A more thorough investigation based on both archaeological and geological studies of the finds within different areas is still needed, both in Norway and abroad. Due to the time and cost limitations of the present project, investigations had to be carried out within selected areas. Important trading sites from the Viking period and the Middle Ages have been studied, concentrating on areas with close contacts to western Norway during this period. Wherever possible, sites with reliable information regarding context and dates of finds were preferred.

Together with geologists from the University Museum in Bergen and NGU within the Millstone project, I have studied the quernstones and millstones of the medieval towns of Bergen, Stavanger, Tønsberg, Oslo, and the small urban centre of Borgund on the west coast of Norway, as well as rural districts in western and eastern Norway. Outside of Norway, investigations of quernstones have been carried out in the North Atlantic region: Iceland, the Faroe Islands and Shetland, and also at important Viking and medieval commercial nodes at selected sites in Denmark, and in Hedeby and Schleswig in present-day Germany.

The spatial and chronological distribution of quernstones and millstones from the different sites demonstrate the location from the date of deposition and indicate the extent of the distribution. The stones are found both in *in situ*, meaning the cultural layers from their original and functional contexts, and also in secondary contexts, re-deposited as waste, or used as fill, or in constructions. The distribution of stones from Hyllestad gains the main focus, and the following presentation of quernstones and millstones from different sites is structured according to distances from Hyllestad. The provenance of stones other than those from Hyllestad will only be mentioned incidentally, as these analyses are to be published as part of the Millstone project (Hauken and Andersen, in press; Baug *et al.* in prep. (a and b); Grenne *et al.* in prep. (a and b); Jansen *et al.* in prep.; Meyer *et al.* in prep.).

7.1 Quernstones and millstones from medieval towns in Norway

Quernstones and millstones found within five medieval towns in Norway have been studied: Bergen, Stavanger and Borgund from western Norway, and Oslo and Tønsberg from eastern Norway.

7.1.1. Bergen

The medieval town of Bergen developed around the bay of Vågen, where the Bryggen area on the eastern side of the bay stands out as a main habitation area, characterized by long-distance trade from the early 12th century onwards. From the middle of the 12th century onwards Bergen was known as the largest commercial centre in the country, a major hub offering a wide variety of products (Helle 1982:167; 2006:83-86). Finds of quernstones and millstones in Bergen may therefore be regarded as commodities linked to trading activities. Bergen as a centre for the quernstone industry is also mentioned in the Urban Code of Bergen from 1276, stating that quernstones should be sold from Stranden, on the western side of the Vågen bay (Bl VI:8).

The medieval town of Bergen can roughly be divided into five different socio-economic zones: (1) the royal and ecclesiastical centre at Holmen/Bergenhus; (2) the main commercial and habitation area of Bryggen; (3) Øvregaten/Øvrestretet with different crafts and workshops; (4) Vågsbotn, also populated by craftsmen and; (5) Strandsiden, represented by a more scattered settlement as well as ecclesiastical and monastic institutions (Helle 1982:228-259). Quernstones and millstones from 16 different excavation sites within three of these zones have been retrieved and included in the analysis (Fig. 7.1).[40] The majority of the quernstones found at Bergen are recorded in the Bryggen area, and rarely at other sites within the medieval town area. This material has previously been

[39] See, e.g. www.millstone.no.

[40] In this project the sites are numbered from 1-16, adapted to the material.

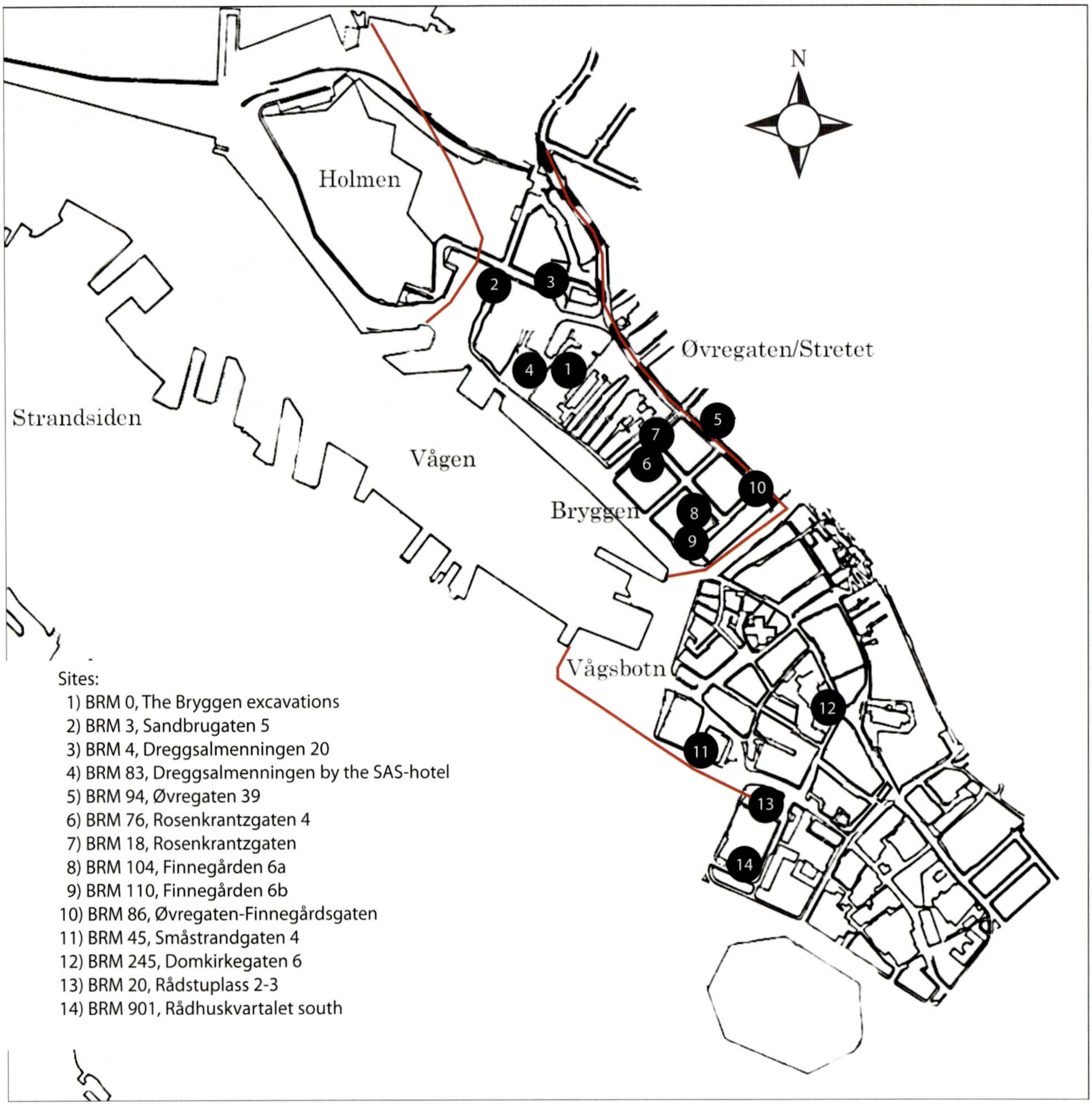

FIGURE 7.1: LOCATION OF SITES WITH FINDS OF QUERNSTONES AND MILLSTONES IN BERGEN. THE SOCIO-ECONOMIC ZONES ARE INDICATED BY LIGHT GREY LINES (MAP BASED ON MYGLAND 2007:14)

researched by Signe Hvoslef (2009), but without focusing on the provenance and trading of stones.

Bergen was exposed to several extensive fires during the Middle Ages, reflected as burnt layers in the substrata. Medieval documentary sources record the dates of several fires, thus giving a chronological basis for fire-layer chronology and absolute dating at Bryggen, which can also be applied to archaeologically documented finds, such as runic inscriptions, pottery and coins (Herteig 1985; 1990; Helle 1998). The archaeological contexts of the finds are therefore most often described in relation to fire layers. A problem remains in that the fire layers are not found across the entire excavated area at Bryggen, giving rise to problems of chronological reliability for some parts of the site.

Quernstone dating is primarily based on Herteig's fire chronology (Herteig 1990), supplemented and corrected by Gitte Hansen for the oldest fire of c. 1120/30, based on dendrochronology (1998 and 2005). In her analyses, Hansen also corrected the dates of some of the quernstones in relation to original documentation.

Quernstones and millstones in Bergen

A total of 133 quernstones and millstones have been studied according to geology and provenance. Nearly 50% of the stones originate from Hyllestad quarries, whereas c. 20% seem to have been extracted from Saltdal in Nordland. There are also a few quernstones of unknown provenance and not of mica schist (e.g. tuff, basalt and sandstone). 55 stones can be dated to the Middle Ages and most of these

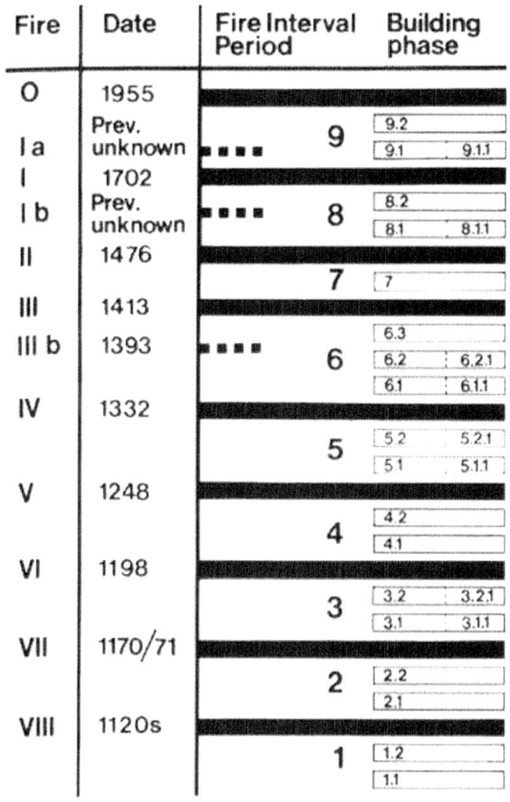

FIGURE 7.2: THE BRYGGEN FIRE-LAYER CHRONOLOGY (SOURCE: HERTEIG 1990:12, WITH CORRECTIONS FOR THE YEAR OF FIRE VIII)

come from Hyllestad, while c. 33% of the medieval stones were extracted from Saltdal (Baug *et al.* in prep. (a)). Stones of rock types other than Hyllestad material seem to have been of minor importance for querns in medieval Bergen, and have rarely been found.

Quernstones are found in most of the medieval urban area, and can generally be connected to the commercial area of international trade at Bryggen. The finds suggest that hand querns were ordinary tools in households, especially in the early and high Middle Ages. The quernstones have, however, only rarely been found *in situ* in fire layers inside the buildings where they can be directly related to domestic activities. The vast majority of these quernstones is from secondary deposits. The find contexts may thus in many cases be of a somewhat younger date than the actual use of the tools, as the finds may be earlier than the cultural layers in which they were found.

Several millstones and a few quernstones (34) were found in the sea, off Bryggen. All the stones were lying on the sea floor in the bay and their original contexts were disturbed.[41] Many were spread along the Bryggen side of the bay and some appeared in small clusters (Hansen 1997:62; Nøttveit 2010:10-13). It is therefore likely that the stones were lost accidently while loading and unloading, and some may represent sunken cargoes. It is also possible that quernstones, or perhaps more likely millstones, were sometimes used as weights for mooring.

Only one of the stones found in Vågen is from Hyllestad, while 24 of them derive from the Brønnøy quarries of Nordland. Geochemical analyses have been carried out on two stones, but the results are not yet available (Hansen 1997:62; Nøttveit 2010:10-11; Baug *et al.* in prep. (a)).[42] In spite of the considerable number of millstones from Brønnøy, none of these have been found in medieval contexts on land. This may indicate that the Brønnøy stones are post-medieval, which would explain their absence from medieval cultural deposits in the town.

The quernstones found in Bergen that have been measured (26) vary in size and shape, with diameters from 19-58cm; 20 specimens measure from 35-44cm. With two exceptions these stones come from Hyllestad and Saltdal. The millstones (7) are between 89-122cm in diameter,[43] and all except one come from Hyllestad. Only a few quernstones found in Bergen have traces of use (Fig. 7.3). A few stones appear as unfinished and are coarsely shaped, with off-central eye, which may explain why they were discarded. As most of the stones were found in fill material and foundations there is no indication that they were meant for further sale. Damaged quernstones and millstones may have been acquired for uses other than in a quern or mill, perhaps as foundation material. Stones found in fills and foundations of both buildings and *caissons* were most likely intentionally deposited there. Whether they were considered as unfit for purpose after being obtained, or were specifically acquired as fill material is not known. However some of the unused quernstones and millstones found in fill masses at Bryggen seem to be of a good quality. Bergen's role as a trading centre may then have led to an abundance of quernstones and millstones in the town, and therefore some of the stones could also be acquired for purposes other than milling. Their appearance in fills and constructions indicates a flexible and pragmatic use of the stones, and also that there were selection and quality control processes for stones in Bergen.

The stones retrieved from the extensive Bryggen site (site 1) date to period 2 (c. 1120-1170/71) and are represented throughout the Middle Ages and into the 17th century. Periods 4 (1198-1248) and 7 (1413-1476) stand out as the periods with most quernstones and millstones, with more than 50% of the datable finds. As many of the stones found at Bryggen could not be retrieved, the chronological representation of the stones may not give a fully representative overview. The oldest quernstone from Hyllestad relates to period 2 (1120s-1170/71), while stones from Saltdal occur from period 3 (1170/71-1198) onwards. Two rimmed quernstones were found, most likely from

[41] About 20 millstones were found and landed by divers in the 1980s. Ten (B.Sj. 4717/9) are today stored at the University Museum of Bergen, while some were kept by the divers. Additionally, 14 quernstones were found during a marine archaeological survey in Vågen Bay in 2010 (Nøttveit 2010:11).

[42] The provenance analyses of the stones found in the 1980s were conducted by the geologist Helge Askvik, University of Bergen.

[43] The millstones found in Vågen bay are not included.

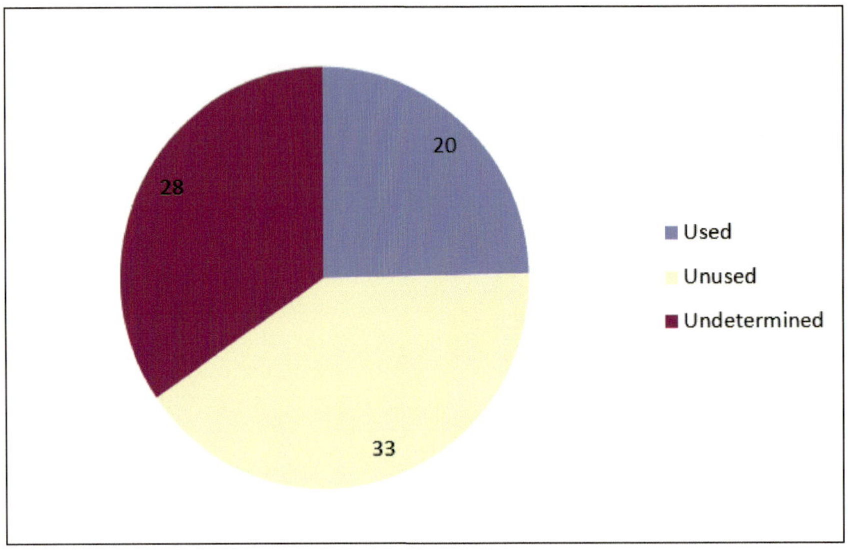

FIGURE 7.3: QUANTITIES OF USED AND UNUSED STONES FROM BERGEN

Hyllestad and related to the 12th century, periods 2 and 3. The oldest datable millstones are linked to period 7 (1413-1476), nearly all from Hyllestad, while one is of unknown origin (Baug *et al.* in prep. (a)).

Outside the Bryggen site, quernstones from 15 other smaller archaeological sites have been retrieved and are included in the analyses: ten sites from the northern town area of Bryggen (sites 2-11) and five sites from the southern town area (sites 12-16) (cf. Fig. 7.1). These excavations are all of minor scale compared to the main Bryggen excavations, and consequently fewer quernstones have been found. Dates and contexts of the stones from these sites are generally less certain than the Bryggen finds, but also here several are dated to the Middle Ages (Baug *et al.* in prep. (a)).

7.1.2 Borgund

The abandoned medieval urban centre of Borgund is located on a peninsula in present-day Ålesund (Sunnmøre). It may have begun as a seasonal market place as early as the 11th century, but traces of denser and permanent settlement first appear from the 12th century (Helle 2006:49). It was gradually abandoned by c. 1500 (Lossius 1977:10, 36, 49).

Four quernstones have been retrieved from Borgund. Their exact locations and dates are not known, but apparently from 1100-1500, judging by the assemblage as a whole. Hyllestad and Saltdal are suggested provenances for two of the stones. An upper quernstone from Hyllestad is 36cm in diameter and another upper stone, possibly from Saltdal, measures 43cm. All stones bear signs of being used at the site (Baug *et al.* in prep. (a)).

7.1.3 Stavanger

Compared with Bergen, there have been far fewer urban excavations in the smaller town area of medieval Stavanger further south, and quernstones have only been identified exceptionally. Stavanger was established as an episcopal see, with a small urban settlement in the 12th century, and archaeological finds indicating trade and craft activities are rare (Helle 2006:68-70). The ecclesiastical institutions, with their needs for commodities and services, as well as the income they brought to the town, created an economic basis for the modest levels of urbanization, rather different from development at Bergen (Helle 2006:68).

Three fragmented quernstones have been found at the site of Skagen 3 in the centre of Stavanger, covering an area of 150m² close to the medieval cathedral. All of them were upper quernstones with traces of use, coming from Hyllestad and Saltdal. The stones were found under the stone paving connected to a building that was destroyed in a fire in 1272 (Lillehammer 1968:12; Hauken and Anderson in press.), giving a *terminus ante quem* date for the quernstones, being older than the fire.

7.1.4 Oslo

The medieval town of Oslo, today's Gamlebyen ('Old Town'), developed on the eastern side of Bjørvika, a small fjord arm furthest along the inner part of Oslofjorden. Here, Oslo emerged as a town in the first part of the 11th century, but traces of craft and trading activities first appear in the 12th century (Nordeide and Gulliksen 2007:20, 23; Molaug 2008:77-79, 86-87). After the medieval town was ravaged by a major fire in 1624 it relocated to the other side of the bay, Bjørvika, where the new town Christiania was established close to Akershus castle. The medieval town was then to a large degree deserted, and mainly used for fields and meadows (Lidén 1977:11). Therefore the quernstones found may cover a period of some 500 years, from the early 1100s to c. 1600.

72 stones for hand querns were found and studied. They originate from eight excavations: Mindets tomt

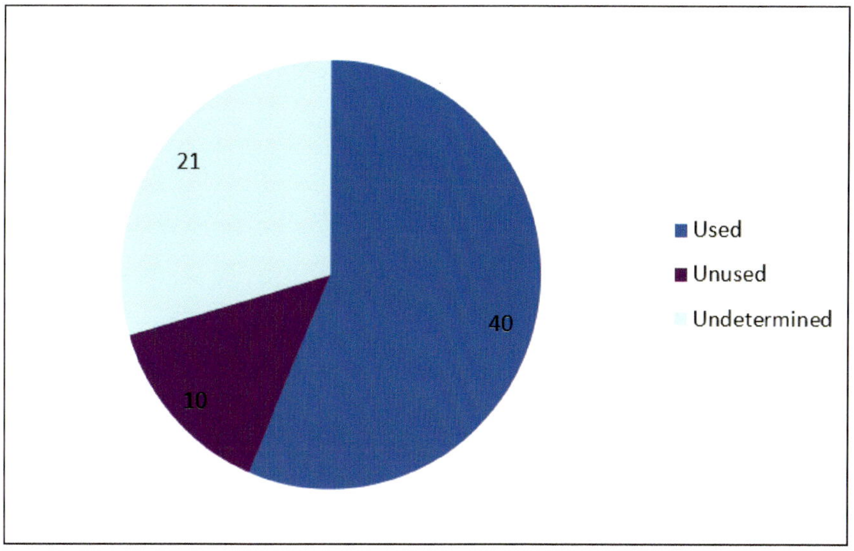

FIGURE 7.4: QUANTITIES OF USED AND UNUSED STONES FROM OSLO

(Klementsgate 2); Søndre felt; Nordre felt II; Oslogt 6; Schweigårdsgt 55; Kanslergt 10; Olavsklosteret; and Bispegårdsruinen. Their provenances differ from those of the Bergen assemblage. In Oslo nearly 50% of the stones come from Saltdal quarries, far more than from the Hyllestad quarries, and stones from production areas other than Hyllestad and Saltdal are also represented (Grenne *et al.* in prep. (b)).

The measurable quernstones range from 20-53cm in diameter (15 examples), most of them over 40cm (11). Compared with the Bergen finds the quernstones are more fragmented, as the majority here seems to have been used and broken through usage (Fig. 7.4). Three unfinished quernstones, two from Hyllestad and one from Saltdal, have also been found. One of the finds from Hyllestad and one from Saltdal have unfinished eyes, which is relatively uncommon (Grenne *et al.* in prep. (b)). Most of the quernstones from Oslo reflect domestic use, but it cannot be excluded that the quernstones were further redistributed. Stones stored for further distribution may possibly have been located elsewhere, undetected by the excavations. Several millstones have recently been found in the medieval harbour area at Bjørvika. These, unfortunately, could not be included in the analyses.

7.1.5 Tønsberg

South of Oslo, where the Oslofjord opens towards Skagerrak, is the town of Tønsberg, between the outlets of the Numedalslågen and the Drammen waterways. Archaeological investigations have, so far, found no remains of any urban settlement with traces of crafts and trading activities prior to the 1100s (Ulriksen 2008:102-103).

17 quernstones (no millstones) were found in the medieval town of Tønsberg, where stones from Hyllestad and Saltdal dominate the assemblage, with about the same number of stones from each, but also with a few stones from other quarries (Grenne *et al.* in prep. (b)). The stones measure 26-51cm in diameter (5 stones), and, as at Oslo, most of the stones reflect domestic uses at the site (Fig. 7.5). It cannot be excluded however that they were traded in that condition. Two unfinished specimens may represent commodities meant for further distribution.

7.2 Quernstones from rural sites in western Norway

Another issue for consideration is whether rural areas constituted a market for the products from Hyllestad and Ølve/Hatlestrand, and, if so, were they integrated within the contact networks of the towns? As many of the quernstones found in the rural districts are stray finds they are difficult to date. The distribution is nevertheless indicative as to where and how the quernstones were distributed. The finds from western Norway are commonly presented as two regions, one fairly close to the quarries in today's counties of Sogn og Fjordane and Hordaland, and a more distant region, the county of Rogaland, further south. Their locations in relation to the towns of Bergen and Stavanger respectively are also clearly significant.

23 stones for hand querns were found at different rural sites in the counties of Sogn og Fjordane and Hordaland. Measurable stones (7) vary in diameter between 38 and 45cm. Most of the stones derive from Hyllestad and Saltdal, with a few more quernstones from Hyllestad than Saltdal. Approximately 33% of the material is of unknown provenance, probably representing more sporadic and small-scale extraction (Baug *et al.* in prep. (a)). Despite the rather low number of stones represented, Hyllestad and Saltdal seem to have dominated the quernstone market within the counties of Sogn og Fjordane and Hordaland, and other rock types only rarely appear.

Extraction of other rock types for hand querns is most striking in the early Viking period, or perhaps earlier. Four quernstones of unknown provenance and dated to the early Viking period have been found at Bjørkum in Lærdal, Sogn

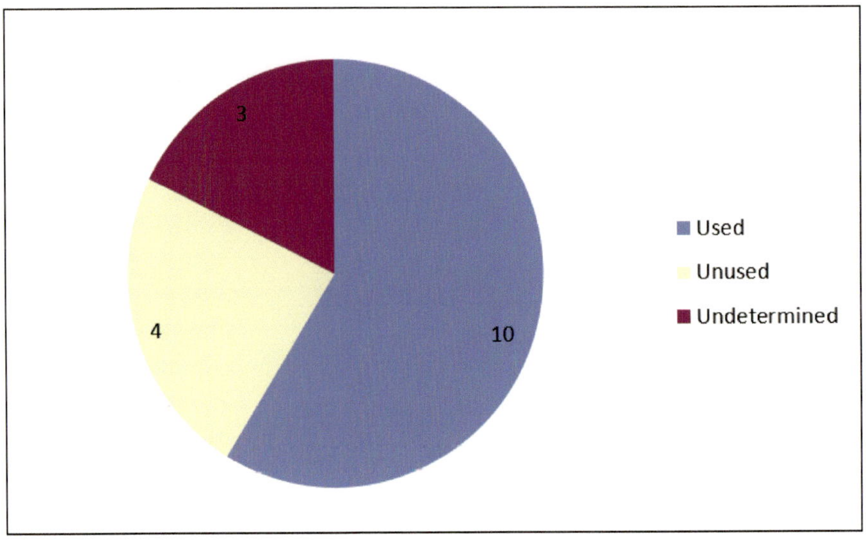

FIGURE 7.5: QUANTITIES OF USED AND UNUSED STONES FROM TØNSBERG

og Fjordane, a site where different crafts and productions have been uncovered. The main phase of settlement is dated to c. AD 700-850 (Ramstad *et al.* 2011), also indicating the date of the quernstones (Ramstad in prep.), and most likely prior to the large-scale Hyllestad production. All four stones are made of different types of rock, and none from the same quarry. Most likely different blocks of stones from the vicinity were used as raw material (Baug *et al.* in prep. (a)), indicating smaller-scale production. All have traces of use and were not commodities for sale.

The only quernstone from Hyllestad possibly dated to the Viking period in this region was found in a probable grave, most likely a male burial, in Flåm in Aurland, Sogn og Fjordane.[44] The stone shows traces of use on both sides and is interpreted as an upper quernstone (c. 40cm in diameter), reused as a lower stone. The stone was found with a spearhead considered to date to the latter part of the Viking period, c. AD 950-1000s (Petersen 1919:31-33, 138-139, fig. 21; Solberg 1984:94-99, 159, 206), which may also indicate the date of the stone and that production started prior to this. As the finds are old ones, collected locally and with no provenance, there is a chance that they may not come from the same context. The stone has slots for *millrynd* (Norw. *sigl*), a device for regulating the distance between the two stones, as well as switching rotation to the upper stone.

In addition, 18 quernstones broadly dated from the late Iron Age into the Middle Ages were found in the rural district of Rogaland, but only one quernstone with an uncertain date seems to originate from Hyllestad. A few stones were extracted in Saltdal, but more than 50% are of unknown origin, probably coming from small-scale production, unrelated to any of the large medieval quarry areas of Norway (Hauken and Anderson in press.). It is likely that households produced their own quernstones from suitable rocks in the vicinity, and were in this way self-sufficient. Compared with Bergen, Stavanger seems to have played a minor role in the distribution of quernstones and millstones.

[44] Information from www.unimus.no, find no. B9475b. Quernstones are also known from other Viking graves (such as B10578, B7831, B2651-2655), but these stones could not be retrieved. Two quernstones found in the Oseberg grave at Vestfold were not from Hyllestad (Grenne *et al.* b).

7.3 Quernstones from rural districts of eastern Norway

33 quernstones have been retrieved from rural districts of eastern Norway, the present counties of Telemark, Buskerud, Oppland, Vestfold, and Hedmark. Both Oslo and Tønsberg may have served as markets for these areas. The stones measure 18-61cm in diameter (22), while most (14) measure 39-51cm. For the largest stone (61cm) it is difficult to determine whether it was a millstone or a quernstone (cf. Chapter 2.4).

Only a few examples can be identified as products from the known quarry sites, and most of the stones are of unknown origin. This is a striking difference compared to the finds from medieval Oslo, seemingly not a supplier of quernstones to the rural inland of eastern Norway. The quernstones from eastern Norway also differ from Sogn og Fjordane and Hordaland, where extractions from Hyllestad and Saltdal dominate. As at Rogaland, a more random production system seems to have developed, with no sign of the organized production and distribution of quernstones. Stones from the large quarry areas are only exceptionally represented, indicating local, small-scale production. Only a few quernstones from Hyllestad and Saltdal have been found. One of the Hyllestad finds comes from the fill deposits of a well, dated to AD 783-970, from Hesby farm (Vestfold); it most likely also dates to the latter half of the Viking period (Gollwitzer 2012:128-134; Grenne *et al.* in prep. (b)).

7.4 Wreck cargoes of quernstones and millstones

18 quernstone and millstone cargoes have been found along the Norwegian coast (Nymoen 2011:68). Provenance

studies have only been carried out on six of these, indicating that they all derive from Hyllestad, therefore shedding light on maritime trade and seafaring connected to the quarries (Fig. 7.6). The finds clearly signify the importance of sea routes for export. No remains from the shipwrecks, or any datable objects on board, have been found, and it has not been possible to date the cargoes of Hyllestad stones more precisely (Fig. 7.7) (Hansen 1992; 1997; pers. comm.; Baug 2002:75-77). Dating therefore has to be based on the time framework known for this kind of production. So far, only millstones have been found in quarries from the early modern period in Hyllestad, and as all the cargoes Hyllestad products contain more stones intended for hand querns than millstones, they are likely to be dated to the Middle Ages.

Measurements have been taken for two of the quern cargoes, the finds from Alverstraumen in the present municipality of Lindås, north of Bergen, and the one from Bukken in the municipality of Sund, south of Bergen. Ten quernstones from Alverstraumen were measured, and are between 44 and 48cm in diameter (Løland in press), indicating a relatively standard size. The Alverstraumen and Bukken cargoes also contained half-finished whetstones. Potassium-argon dating, a method for rock dating, indicates that the whetstones were produced at the Eidsborg quarries in Telemark (Ganerød 2008), the largest production site for whetstones in Norway. Like Hyllestad, a very long production period can be seen here, from the Viking period until about 1950.

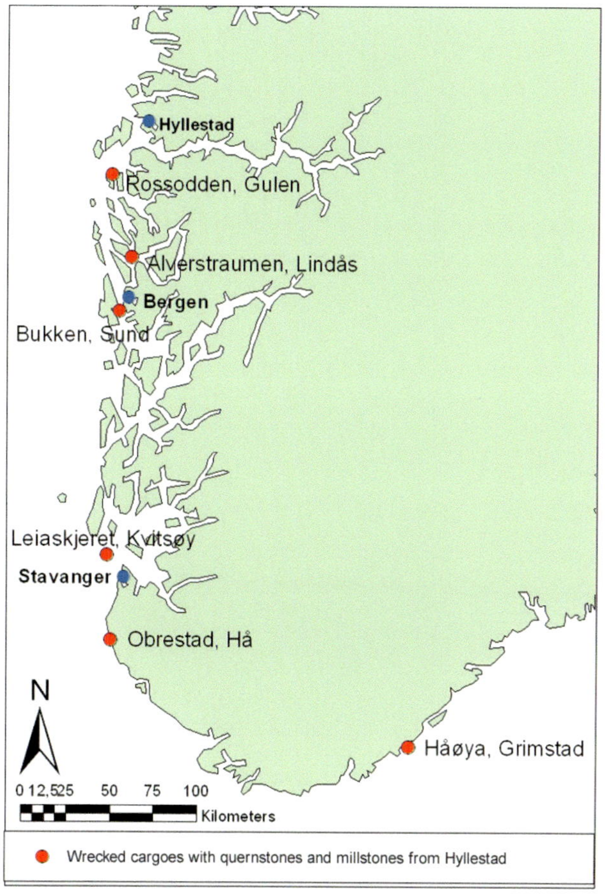

FIGURE 7.6: WRECK CARGOES WITH QUERNSTONES AND MILLSTONES FROM HYLLESTAD (MAP: IRENE BAUG, 2013)

Two of the cargoes with Hyllestad stones were found near Kvitsøy and outside Obrestad (Rogaland), both important harbours on the main sea route along the western coast. Vessels anchored up here overnight, or in bad weather, and crews could change over. The harbours are known from written sources from the 1800s, but are suggested as stages of an infrastructure that may go back to the Middle Ages, perhaps earlier (Elvestad 2005:12-13). The wreck cargoes show that not all ships managed to reach shelter in time.

7.5 Distribution of Norwegian quernstones outside of Norway

An important consideration is long-distance trade and the markets that extended beyond local and regional contexts. Earlier research has shown that quernstones of garnet mica schist, dated to the Viking period and the Middle Ages, have been identified in Denmark, Sweden, northern Germany, Iceland and the Faroe Islands. Geological analyses and provenance studies have, however, only rarely been carried out (Eldjárn 1964; Arge 1989; Elfwendahl and Kresten 1993; Schön 1995; Carelli and Kresten 1997). Together with geologists from the Millstone Project, I have therefore studied the geology and archaeological provenance of the quernstone material from Iceland, Shetland, the Faroe Islands, a few late Iron Age and medieval sites in Denmark, and in Hedeby and Schleswig in present-day Germany. Hyllestad stones are also suggested at Wolin in Poland (Carelli and Kresten 1997:124; Baug 2002:80), but this has not been fully confirmed.[45]

7.5.1 Denmark

As only a few sites in southern Scandinavia have been studied previously (Elfwendahl and Kresten 1993; Carelli and Kresten 1997), more information regarding quernstone exports to this area was needed. These new investigations were concentrated within north-eastern Demark, where the possibility of finding quernstones from Hyllestad was considered more probable than further south-west (cf. Elfwendahl and Kresten 1993; Carelli and Kresten 1997). A few Viking and medieval sites, offering likely contexts and dates, were selected in both Jutland and Zealand in Denmark.

One fragment of a well-used quernstone (41cm in diameter) from Hyllestad was found c. 100m from the Viking fortress of Fyrkat in northern Jutland, dendrochronologically dated to AD 979/980. The quernstone, found in a deposit (dated to c. AD 1000-1100) beyond the

[45] Quernstones of garnet mica schist found in Wolin do not come from Norway, according to geological descriptions of the finds (Majerowicz and Skoczylas 1983:63-73). There is a garnet mica schist quarry at Kemieniec Zabkowicki in Poland, and Polish finds of this material may come from this quarry (Lisowska *et al.* 2014).

FIGURE 7.7: A SMALL CARGO WITH QUERNSTONES (32) AND MILLSTONES (8) FROM HYLLESTAD, FOUND AT ROSSODDEN (GULEN, SOGN OG FJORDANE). THE MILLSTONES MEASURE 1M IN DIAMETER AND THE QUERNSTONES C. 40CM IN DIAMETER (PHOTO: ATLE OVE MARTINUSSEN)

fortress, was in all likelihood used for grinding within the fortress (Dobat 2013:111-112; Meyer *et al.* in prep.).

Seven quernstone fragments from the island of Mors in Limfjorden (north-west Jutland) appear to come from four different rural settlements, dated from the late Viking Age, c. 1000s to c. 1350. They are found in secondary deposits, reused in constructions at the site, and they all show traces of use. Their use as hand querns may therefore be somewhat older than the actual dating. All the stones are of mica schist, and with one exception they all came from Hyllestad. The two measurable stones are 40 and 50cm in diameter (Grenne *et al.* in prep. (a)).

13 quernstones from six different rural sites have been studied from the region of Thy, north of the island of Mors (north-western Jutland); they dated from the late Viking Age and the high Middle Ages, c. 1000-1300. 11 of the stones could be measured and varied from 34 to 50cm in diameter. Two stones had rim and *rynd* slots, and one of them dated to 1155-1220 at the latest. As the stones were found in secondary deposits, several in connection with stone pavements within buildings, they must be somewhat older than the find layer. Nearly all were extracted in Hyllestad, and only one stone, also of a garnet mica schist type, seems to have a different origin, but the provenance is somewhat uncertain (Grenne *et al.* in prep. (a)).

Quernstones of gneiss and mica schist are also found at Århus in eastern Jutland, with 21 finds dated to the Viking period and the Middle Ages (Meyer *et al.* in prep.). Århus was established as a market place c. AD 770 and developed into a fortified town during the 900s, although international trade was less prominent here than at e.g. Ribe and Hedeby (Skov 2008:215, 221). The geology of the quernstones from Århus has been earlier identified as granite gneiss and basalt (Andersen *et al.* 1971:162). As Århus was the first Danish town of some importance to be reached when sailing from Norway, and close contacts between Norway and Århus are documented in the tenth century through the presence of whetstones, soapstone vessels and walrus tusks from Norway (Skov 1999:603, 606), a more thorough study of the quernstones was regarded as important.

More than 50% of the quernstones investigated date widely to the Viking Age, and nearly 75% come from the quarries of Hyllestad. Quernstones from other areas are only sporadically documented. Most of the stones can only be roughly dated to Viking times (c. AD 850-1050), but three may date to the first half of the 800s. One of these early stones comes from Hyllestad, while the other two are of unknown origin. Three of the Hyllestad finds were found in a fire destroyed building with coins from King Harald Bluetooth (dated to AD 980), thus also giving an *ante quem* date for the stones (Meyer *et al.* in prep.). The Viking quernstones (7) range in size, from 35-46cm in diameter. One medieval stone is 48cm in diameter, and one stray find, probably late medieval c. 1300 or early modern period, is 23cm in diameter. Eight of the investigated quernstones from Århus are dated to the Middle Ages, most of them to the 1200s, and all the medieval quernstones of mica schist seem to have been produced at Hyllestad. One upper stone dated to the Viking period has a rim, but no stones were found with *rynd* slots (Meyer *et al.* in prep.).

Basalt quernstones from the Rhine area are also found in Århus (Andersen *et al.* 1971: 162-163), but according to Hans Skov from the Museum of Moesgård, the basalt finds constitute approximately 33% of the total, and Hyllestad stones c. 75% (Skov pers. comm. 2012), even though a few stones of different origin were also found. A certain competition between Hyllestad and the Rhine area seems, then, to have been present at Århus, but with Hyllestad as the dominant production area.

Additionally, quernstones from two mills, Thors mill (3) of the 1600s in Århus and Grundfør mill (6) (c. 1250s-1950), not far from Århus, have been studied. Here also the Hyllestad stones dominate the finds. Three quernstones from different medieval and post-medieval rural sites near Århus were also examined, but none of these were extracted from Hyllestad (Meyer *et al.* in prep.).

35 quernstones and fragments of quernstones found at the rural village of Tårnby, in Zealand near Copenhagen, were also studied (Baug *et al.* in prep. (b)). The settlement dates from the late Viking Age/early Middle Ages and continued into the beginning of the 1800s (Kristiansen 2005:13, 384). As well as agrarian use, the Tårnby finds reflect crafts, fishing and trade (Larsen 2005b:406).

The provenance study showed that the majority of the quernstones were extracted at Hyllestad, and only a few stones may have had a different provenance (Baug *et al.* in prep. (b)). Two of the stones from Hyllestad seem to date from the late in the Viking period, but most stones from Tårnby (30 specimens) date to the early or high Middle Ages (c. AD 1050/1150-c. 1300) and later (Kristiansen 2005; Baug *et al.* in prep. (b)). The measurable stones vary between 36 and 55cm in diameter (10 finds), where most (6) are between 36 and 42cm. Two stones have rims, one of them from Hyllestad and dated to c. 1300-1400. Six stones with *rynd* slots are widely dated from c. 1050/1150 to the 1800s. The quernstones at Tårnby are seen in connection with agrarian production in the village, and the large numbers of quernstones from Hyllestad can be explained by Tårnby's location on an easterly distribution route. Most of the stones were found in secondary deposits, many reused in constructions such as buildings and fireplaces (Larsen 2005a:378-379).

Additionally, 19 quernstones and millstones from Lejre mill, near Roskilde in Zealand, earlier studied by Carelli and Kresten (1997:132 -134), have been investigated. As only a few of the Norwegian quarries had been identified in the 1990s (Carelli and Kresten 1997:116), a re-examination was carried out. The study confirmed with certainty that nearly all the stones were produced in Hyllestad, with only two definitely not originating from Hyllestad (Jansen *et al.* in prep.). The re-examination, therefore, largely concurs with the Carelli and Kresten study. The mill has been dated by dendrochronology to the latter half of the 12th century (Carelli and Kresten 1997:118), and is therefore clear evidence that Hyllestad produced millstones for further export by the second half of the 1100s. Where diameters could be measured (5 examples), these varied between 80 and 110cm.

Carelli and Kresten's assertion therefore that all quernstones of garnet mica schist found in Denmark come from Hyllestad (Carelli and Kresten 1997) has largely been confirmed through the new investigations. However the picture is somewhat more complex than indicated in the Carelli and Kresten study. Hyllestad clearly dominated the market at the investigated sites, and most likely also in most parts of north-eastern Denmark during the Viking period and Middle Ages (Fig. 7.8). However stones of garnet mica schist with different origins have also been found and Hyllestad was clearly not the only Norwegian site producing and delivering quernstones to this area.

7.5.2 Northern Germany: Hedeby and Schleswig

Quernstones of garnet mica schist have been examined from the Viking Age town of Hedeby and the medieval town of Schleswig, as well as from two rural sites in Schleswig Holstein. The assemblages of quernstones from Hedeby and Schleswig were first examined by Volkmar Schön in 1995. He found that 11% (12 out of 88 specimens) of the finds from Schleswig, and only 1% (49 out of 5875 specimens) of the finds from Hedeby were of mica schist, while the amount of basalt finds from the sites was 82% (88 finds) and 99% (5875) respectively (Schön 1995:12, 22). Without a geological analysis of the mica schist stones, Norway and Central Europe were suggested as likely origins of the material (Schön 1995:106-107). To decide the issue more accurately a geological analysis of the finds was carried out. The re-examined stones from this area amounted to 73 finds, representing stones with a large degree of fragmentation. 45 of the fragments are smaller than 10cm (Baug and Jansen in prep. (d)). The number of finds does not, however, reflect the number of quernstones, as several of the fragments may come from the same quernstone. Only three stones could be measured, having diameters of 21, 38 and 46cm. Two stones have *rynd* slots and two have ribs, but for most stones it was impossible to determine these features (Baug and Jansen in prep. (d)).

The important Viking Age trading centre of Hedeby, located on the border between the Danish Kingdom and the German Empire, is mentioned in contemporary written sources in 804, and was abandoned after an attack in 1066. The site is well documented archaeologically (Clarke and Ambrosiani 1995:63; Hilberg 2008:101), providing a temporal framework for the finds. Of the 49 quernstones mentioned by Schön from Hedeby (1985:12), 46 could be retrieved, and altogether 40 of these are of garnet mica schist. Except for one stone that seems to have come from Saltdal, all the stones of garnet mica schist were extracted from Hyllestad. The remaining six stones are deformed pebbly quartz conglomerates of an unknown origin (Baug and Jansen in prep. (d)). Even though it is difficult to provide exact dates for the quernstones, they most likely derive from the latest phase of Hedeby, probably from the AD 900s onwards (Schön 1995:107).

Quernstones from Schleswig are dated to the centuries between AD 1000 and 1200, and the mica schist type is also found in the earliest phase (Schön 1995:70, 107). Eight fragments of mica schist were retrieved and re-investigated.[46] The fragments originate from two different sites in Schleswig, all of from Hyllestad quarries. Only two stones could be measured, with diameters of 43 and 46cm. One stone had *rynd* slots and none had rims (Baug and Jansen in prep. (d)).

There are no indications that the quernstones from Hyllestad in Hedeby and Schleswig were meant to be traded further, but this cannot be totally excluded. The finds have clear traces of being used, meaning that the Hyllestad stones were imported to be used in the town, which is also suggested by Schön (1995:103, 107) (Fig. 7.8). Quernstones, most likely from Hyllestad, have, however, also been identified at a few other sites in Schleswig-Holstein, as well as at Mecklenburg further east. Most finds are not dated, but a few stones may date to the 11th and 12th centuries (Schön 1995:106-107). They may have been distributed via Hedeby or Schleswig.

7.5.3 The North Atlantic region

During the Viking Age and the Middle Ages there was a close connection between the North Atlantic islands and western Norway. Iceland, Shetland and the Faroe Islands were settled from Norway, and contact was maintained throughout the following centuries. Various products from Norway were distributed westwards (cf. Arge 1989:115; Forster 2004; 2009:58-65; Øye 2008:246-250), and especially during the Middle Ages large volumes of commodities were transported between the areas (Helle 1982:165, 360-365; Øye 2008:244). Was this also the case with quernstones from Hyllestad? No urban centres existed in the North Atlantic region during the period under study, a situation quite different from southern Scandinavia. How did this affect the quernstone distribution?

Shetland

Quernstones of garnet mica schist have been found in Norse settlements in Shetland, and are documented in, e.g., Underhoull and Sandwick in Unst and Da Biggins in Papa Stour. Quernstones from these sites have been thought to be of Norwegian origin. 15 quernstones from Shetland were studied, but none of these was produced in Norway. The analysis showed that they were all locally made in Shetland, coming from several different quarries.[47] Here production seems to have started in the Viking period and lasted into the 1900s (Baug and Jansen in prep. (a)).

The Faroes

Quernstones of mica schist have also been found in the Faroes, where 40 stones for hand querns have been found and investigated. The measurable stones (11) are between 38 and 50cm in diameter. Seven stones have *rynd* slots and three have ribs, but none come from Hyllestad. Only four of the 40 stones are identified as Hyllestad stones, and additionally three more may come from Hyllestad, however the geological analysis has not yet finished. Most of the stones from the Faroes, therefore, are considered to have been of local origin, while a few may have been produced in Shetland (Baug and Jansen 2014; in prep. (a)). The four quernstones thought to be from Hyllestad are found at Durhus i Eiði and in Skarvanes, both dating to the Middle Ages, and two stray finds that may derive from the early modern period. The three stones that most likely come from Hyllestad were found near the old churchyard in Sandur at Sandøy (late Viking Age), at the churchyard in Sumba (late Middle Ages), and a stray find possibly from the early modern period (Baug and Jansen 2014; in prep. (a)).

Iceland

Quernstones and millstones of garnet mica schist have been found at six sites in Iceland, amounting to 30 finds.[48] Most of the quernstones studied (17) were found at Stóraborg, a farm mound dated from the 12th to the 19th centuries in south-eastern Iceland (Colic and Snæsdóttir 2010:1, 3). Five stones, including a small toy quernstone of soapstone, were found at the *landnám* farm at Bergþórshvoll, which has been settled up to the present day (Eldjárn and Gestsson 1951:5-6, 12-13), complicating the date of the quernstones (Fig. 7.8). A few stones lack information with regard to context and date (Baug and Jansen 2014; in prep. (b)). The majority of stones found in Iceland are, however, made of basalt, and these have been omitted with regard to identification of geology and provenance, but were obviously locally produced in Iceland.

Provenance studies indicate that most of the quernstones of garnet mica schist type come from Hyllestad quarries, with the exception of two quernstones and two millstones that were produced in Selbu, and seemingly are of younger date. All quernstones of garnet mica schist with a possible date to the late Viking period and the Middle Ages were produced at Hyllestad, and constitute about 25% (26 finds) of all the quernstones, whereas the rest are made of basalt. As the quernstones from Hyllestad in Iceland are very fragmented some of the fragments may come from the same quernstone. The representation of Hyllestad stones

[46] According to Schön, 12 finds of mica schist were found in Schleswig (Schön 1995:22), of which eight could be retrieved for this present study.

[47] Only quernstones from the Shetland Museum in Lerwick have been studied. Finds excavated before 1960 are to a large degree stored in Edinburgh, Scotland (Murray pers. comm. 2010).

[48] Quernstones and millstones from the National Museum of Iceland, as well as the Akureyri Museum, were studied in 2010. Quernstones excavated more recently (over the last five years) may be located with different excavation companies but due to the limited time and resource of this project these finds could not be investigated.

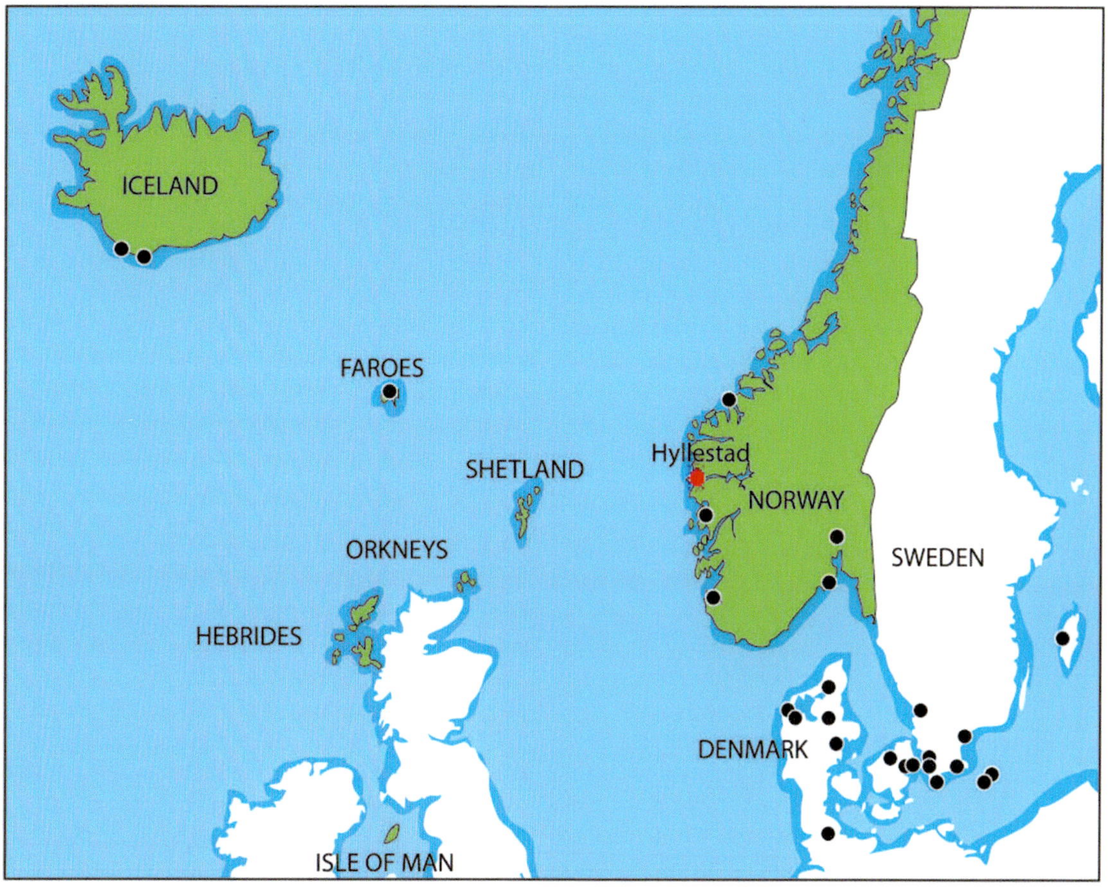

Figure 7.8: Distribution map of quernstones from Hyllestad. Black dots indicate finds of Hyllestad stones. In Norway only finds from medieval towns are indicated (Map: Irene Baug, 2013)

may therefore be somewhat lower than indicated. Only one of the quernstones from Hyllestad could be measured, having a diameter of only 16cm. Remains of ribs are found on three stones (Baug and Jansen 2014; in prep. (b)).

Within the North Atlantic region therefore, Shetland seems to have been self-sufficient in quernstones, whereas stones from Hyllestad were distributed to Iceland and the Faroe Islands. These finds are, however, few, and in the Faroe Islands they only occur exceptionally, and Hyllestad does not stand out as a dominant supplier even in Iceland, where most stones are made of lava. Only a few of the Hyllestad stones are possibly dated to the late Viking period, and it is striking that except for Bergþórshvoll, none of the *landnám* farms in Iceland seem to have quernstones from Hyllestad. Even though other Norwegian products are found at the *landnám* farms, the quernstones were of basalt and most likely locally made (see e.g. Gestsson 1959: 7172; Magnússon 1972:68, 72-74). Imports mainly from Hyllestad seem to have arrived from the Middle Ages onwards, but no significant importing is indicated in the archaeological material.

7.6 Spatial and chronological distribution of bakestones

Not only quernstones and millstones, but also bakestones were distributed to, and redistributed from Bergen, although never referred to in written sources. Large volumes of bakestone fragments have been found during the excavations at Bryggen and other sites in Bergen, as well as in other Norwegian towns. Within the five northernmost rows of dwellings at Bryggen, later known as Gullskogården, nearly 1,600 fragments of bakestones have been found. Approximately 90% of this material is made of chlorite-rich talc-bearing green schist, making Ølve/Hatlestrand the most probable production area. The earliest fragments of bakestones (15) found at Bryggen are dated to c. 1100-1120s (Hansen 2005:179).[49] From period 2, c.1120-1170/71, more than 250 fragments occur, most of them of green schist, but nearly as many were of soapstone (Tengesdal 2010:36, 46-47). An increase in bakestones of green schist is documented relatively late, in the period between 1248 and 1413 (Tengesdal 2010:31-36), which indicates an increase in distribution, and also production, from Ølve/Hatlestrand in the high and late Middle Ages.

Many of the fragments have traces of use, reflecting the extensive use of these tools in the town. Bakestones possibly intended for further distribution are difficult to identify through the archaeological record, but evidently appear first as traded commodities, or secondarily as easily transported personal possessions. They have been found in several places in the rural areas of western Norway.[50]

[49] One bakestone fragment may date 1030/30-c. 1070, but this is very uncertain (Hansen 2005:178).

[50] Cf. www.unimus.no.

Large numbers of bakestones have also been found in urban contexts other than Bergen (Weber 1989; Weber 1990:63). In Oslo some 300 bakestones were identified in the excavations at the sites of Mindets tomt and the southern site of Gamlebyen, and c. 49% of these (132) are considered as coming from Ølve/Hatlestrand. In Trondheim, from the large site of Folkebibliotekstomten, nearly 400 bakestones were found. Approximately 35% of these (141) are considered as coming from Ølve/ Hatlestrand, whereas the rest are described as local stones (Weber 1989:18). The bakestones found in Gamlebyen (Oslo) and Trondheim have been dated from c. 1050-1100 onwards (Weber 1984:159). The quarries at Øye (Melhus, Sør-Trøndelag), containing green schist, are, however, also a possible provenance, especially for the finds from Trondheim, which are closer to Øye than Ølve/ Hatlestrand. As in Bergen, an increase of green schist bakestones compared to other rock types seems to have taken place throughout the Middle Ages, from the period c. 1300-1400 (Weber 1989:18).

No study of the c. 200 bakestone fragments from Stavanger has been carried out.[51] Two of the fragments have the remains of iron nails and one has traces indicating that iron shafts were attached to them.[52] Two fragments have also been found at the farm of Store Eige, Eigersund, in the southern part of the Rogaland. They also have the remains of iron nails and traces of an iron shaft[53] (Petersen 1933: pl. III, fig. 8 a-b). A shaft is also indicated on a bakestone from Bergen, where the corroded remains of some sort of metal were found.[54]

In Stavanger, too, the geology of the bakestones seems to vary. Some are considered to be of soapstone and others have different schistose compositions. These stones may in part come from the quarry site at Ertenstein (Rennesøy), in the northern part of Rogaland, where building stones were also extracted during the Middle Ages. As at Ølve/ Hatlestrand, bakestones were produced with building stones; fragments of bakestones from the quarry have been found under a rock shelter at the quarry site[55] (Jansen pers. comm. 2011), evidently used as a worksite for finishing bakestones produced at the nearby quarry, a situation similar to Kvitafjell. The Stavanger bakestones could well have come from sites other than Ølve/Hatlestrand.

Bakestones were also recovered from farmsteads and farm mounds in northern Norway, but these are more frequent in coastal rather than inland areas (Reiersen 1999:47-48). Their distribution in these rural sites suggests that exports were limited to central locations, or the so-called hubs along the sea route, i.e. those sites located within the catchment areas of medieval towns, or otherwise connected to town networks (Øye 2009:232). It seems, therefore, that bakestones were not commonly used in northern Norway.

[51] Cf. www.unimus.no, find nos. S12198, S12257, S12259, S9454.
[52] Find nos. S9454 IXf15, IXf16 and IXf43.
[53] Find no. S4899h.
[54] Find no. BRM0/27091.
[55] Find no. S10390.

No study of the geology and provenance of these products has been conducted, and whether or not they originated from Ølve/Hatlestrand quarries is not known.

Iceland, Shetland and the Faroe Islands are the only areas where bakestones, most likely from Norway, have been found in larger quantities (Hamilton 1956:183; Arge 1989:119; Smith 1999:127; Weber 1999:134-139; Forster 2004). In Shetland locally produced bakestones are known from several sites in early Viking times, and thus are older than Norwegian production. It has been suggested therefore that Shetland bakestones were innovations transmitted from west to east (Forster 2009:65), and may have influenced Norwegian production. As well as at the archaeological sites of Jarlshof, Biggins, and Papa Stour in Shetland, two types of bakestones have been identified: one is interpreted as an import from Norway, possibly Ølve/Hatlestrand, the other from local sources, made from a rather homogeneous, heavy steatite of the kind used in cooking vessels. The bakestones extracted in Shetland were rectangular or sub-rectangular in shape (Weber 1999:134), and somewhat differ from Norwegian ones. Norwegian bakestones are found in Shetland from the 1100s (Weber 1999). Spectrographic analyses of stones from the Faroes indicate that they come from Ølve (Arge 1989:119). It remains difficult to separate Ølve/Hatlestrand from the quarries of Øye (Sør-Trøndelag) by this method (Jansen pers. comm. 2011). However, based on the close contact with Bergen and the Faroes, it is more likely that the bakestones derive from Ølve/Hatlestrand rather than Trøndelag.

Bakestones seem only to have been distributed rarely to Sweden and Denmark, where only a small number has been identified (Campbell 1950:14; Larsen 2005a:377; Bergström 2007:134). Bakestone fragments of green schist, in all likelihood from Ølve/Hatlestrand, have for instance been found at Århus, dated to the 1300s (Andersen *et al.* 1971:108-109; Larsen 2005a:377), and two fragments, probably from Ølve/Hatlestrand, were found at Tårnby and dated to 1300-1400 (Larsen 2005a:377). Four fragments were also found at Lund, most likely dating to 1200-1300 (C. Johansson Hervén pers. comm. 2012).

7.7 Distribution of stone crosses

Stone crosses have been found in several places in Norway, with the majority coming from western parts of the country, in the present-day counties of Sogn og Fjordane, Hordaland and Rogaland (Birkeli 1979; Gabrielsen 2007; Nordeide 2009:165-166), and this is also where the tall freestanding crosses are found. The 40 crosses here have been examined with regard to rock type: 22 are of garnet mica schist, most likely from Hyllestad,[56] and they are found in the counties of Sogn og Fjordane and Rogaland (Fig. 7.9). The crosses from Hyllestad vary in size and shape, from c. 1m-high crosses placed in churchyards, and freestanding ones standing 4m above the ground. The total

[56] The crosses from the Hyllestad quarries are not included in this figure.

heights are not known apart from one example: a cross, 2.65m above the ground, was sunk 1m into the ground, giving a total height of 3.65m. Based on these data, a 4m-high cross is estimated to have a total height of 5.7m (Gabrielsen 2007:146).

When the crosses were erected is much debated (cf. Birkeli 1973; Baug 2002; Gabrielsen 2007; Nordeide 2009; 2011). The C14-analyses from Hyllestad indicate that crosses were produced here sometime between the early and high Middle Ages, within a date range from AD 1050 to 1280 (cf. Chapter 5). The excavated crosses, however, come from the latest production phase at the quarry, meaning that the extractions may date back further. Large and small crosses were extracted simultaneously, but the tallest type of freestanding cross was not found. This may be a coincidence, but it is also possible that these crosses have an earlier date.

Previous studies of crosses in western Norway have tended to date the tall, freestanding crosses from 900-1000, and the phase of Christianization, while the small crosses placed on medieval churchyards are considered to date from the Middle Ages (Birkeli 1973:222; Baug 2002:87; Gabrielsen 2007:154-175). The suggested dating is to a large degree based on runic inscriptions and the locations of the crosses, as well as the production of crosses being an influence from the British Isles, where the oldest crosses may date to the 700s (Henry 1964:59-60; Lang 1988:9). Most of the British crosses, however, date to the period between 800 and the first half of 900, while some are placed in the 1100s (Henry 1964:60; Gabrielsen 2007:158). Such relatively early dates for the Norwegian stone crosses has recently been questioned by the archaeologist Sæbjørg Nordeide, who argues for a much later date for at least some of the crosses. According to Nordeide, cross production in Norway should be linked to the later, German tradition of erecting stone crosses in the late Medieval/early modern period. German crosses are often interpreted as memorial stones, as well as boundary markers, and Nordeide argues for a similar function for Norwegian crosses (Nordeide 2009:174-175). German crosses are generally 50-150cm high, a similar size to Norwegian churchyard crosses, and the smallest freestanding crosses (Nordeide 2009:166, 175-177; 2011:135, 139). The tallest freestanding crosses in Norway, c. 4m above ground level, are much higher than German examples.

The crosses from Myklebust indicate that contemporary crosses varied in shape and size. All crosses have V-shaped cutting below the arms, one also with possible traces of being semi-circular in shape. Except for two of the crosses, BRM 928/5 and BRM 928/6, they all vary in size. The complete cross head of the smallest find (BRM 928/5) is 35cm, and only the top part of the cross head of find BRM 928/3 measures c. 45cm (cf. Chapter 5). Cross size should therefore not be used as indication of date. Different crosses may have been intended for different purposes, indicated also by the different find contexts/landscapes (Gabrielsen 2007). The size and shape of a cross was most likely of

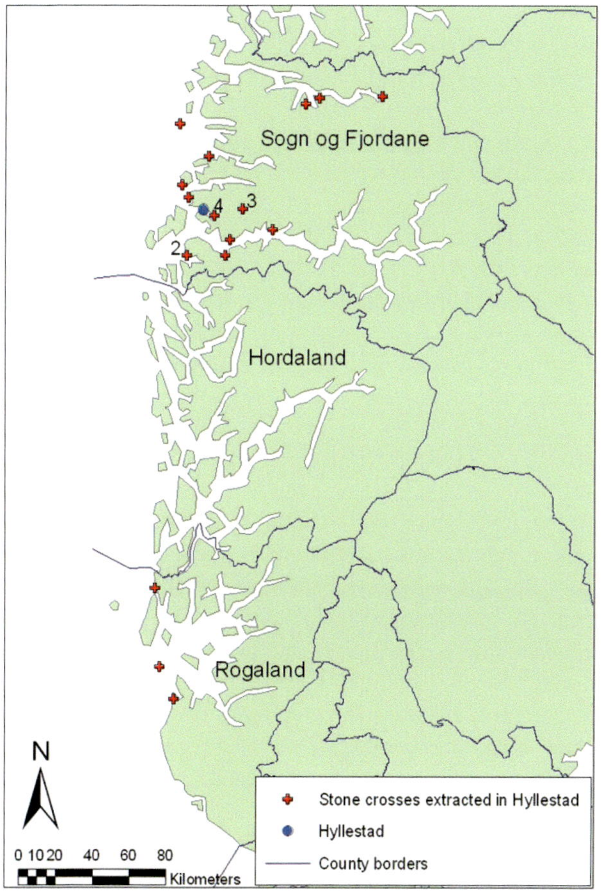

FIGURE 7.9: GARNET MICA SCHIST CROSSES, MOST LIKELY FROM HYLLESTAD, WHERE SEVERAL CROSSES OCCUR AT THE SAME SITE. THE QUANTITIES OF CROSSES ARE INDICATED ON THE MAP (MAP: IRENE BAUG, 2013)

crucial importance for those who commissioned it. There were also technological differences involved when carving a tall cross, rather than a short one. Tall crosses were more time consuming, difficult, and presented greater risks, than short ones; in addition more organisation and work was involved in the transporting and erecting of tall crosses.

There are few clues with regard to date on the crosses themselves, but in some cases runic inscriptions can give indications. Runic inscriptions on a find from Hyllestad (Svanøy, Sogn og Fjordane) have been dated to the beginning of the 11th century (NIyR IV: 223; Johnsen 1968:203). This concurs in time with runic inscriptions on a stone cross from Stavanger, interpreted as referring to the death of the nobleman Erling Skjalgsson († 1028) (Liestøl 1954:75). Dating based on runic interpretations is, however, considered to be relatively uncertain (Hagland 1998:10-11), and should be used with caution. A date to the late Viking Age or early Middle Ages is also supported by a document of 1394, referring to a large cross from Eidsskog raised in honour of St. Olaf († 1030), and pilgrims holding Mass by the cross (DN 1, 545). The stone crosses along the western coast of Norway may have had a similar function.

Runic inscriptions on Norwegian crosses, therefore, indicate that they were raised as memorial crosses.

However they also appear in other contexts, and to try and date them their contexts within the landscape is important. Several of the freestanding crosses are located on top of prehistoric burial mounds, while some are placed in medieval churchyards, although not as grave crosses, and others along sailing routes. Crosses on burial mounds have been interpreted as demonstrations against heathenism, and were therefore erected during the missionary phase or early years of Christianity (Gabrielsen 2007:177-233). Several of the freestanding crosses are located at important church sites of the Middle Ages, and they may have been used in connection with outdoor services before churches were built.

Crosses from the Hyllestad production may also shed light on those distributed along the western coast of Norway. The crosses found at Hylletad also indicate a possible date range for others. No remains have been recorded at quarries where gunpowder was used, and no quarry dates indicate production of crosses in late Medieval or early modern times. The Norwegian cross tradition, therefore, cannot be explained as a German influence, as Nordeide has proposed, unless the German monuments are earlier than assumed (Nordeide 2009:175-177). Most evidence points towards production and use of Norwegian crosses from the introduction of Christianity and the early Middle Ages, a tradition that continued into later medieval times. The crosses found in the quarry also substantiate that cross size and shape should not be used as an indicator of date, as this seems to have varied within the same period. Crosses and quernstones were extracted simultaneously from the same quarries. This is also indicated on one of the crosses from Eivindvik, where circular traces from quernstone extraction are visible on the beam of the cross (Baug 2002:69). Even though crosses were most likely commissioned products, they may have been produced for profit along with quernstones. These issues will be further discussed in Chapter 10.

7.8 The distributed material: an assessment

Most of the quernstones show clear traces of use, and, especially, those found outside Norway appear in a fragmentary state from heavy use. Quernstones within Norway are more complete, and Bergen, especially, stands out as an area with many unused stones, coinciding with its role as a distribution centre. The unused quernstones show a varied degree of finishing. In a few cases the eye has not been cut, as is the case with three stones found at Oslo, but for most stones it has been completed, and only final grinding surface and fittings for the second stone is lacking. The finds, therefore, indicate that quernstones with different degrees of finishing were exported from the quarries, most likely with the intention of being finished by the receivers, or perhaps by stoneworkers in the town before being distributed further. A similar situation is known from Selbu, where unfinished millstones were transported from the quarries in the mountains down to the farms, where the last finishing was carried out, in some cases also the cutting of the eye (Rollseth 1947:99-107). Some of the unfinished and unused quernstones and millstones found in the town probably represent commodities meant for further export, while some represent discarded stones, indicating that stricter quality control and selection of stones were carried out at Bergen than at Hyllestad. This may have more to do with the different agents involved in transportation to Bergen than to further export requirements.

Both quernstones and millstones can differ in size and shape, even stones found from the same quarries, the same period and the same contexts. The material from outside the quarries therefore supports the results from the Hyllestad excavations, in terms of product range and variation in size (cf. Chapter 5). Typology and stone size should therefore not be used as a chronological indicator. *Rynd* slots are found on a few of the distributed upper stones from the Middle Ages, but also from the late Viking period, documented on stones from Hedeby and on the stone from a possible burial in Aurland - even though the latter find should be treated with caution with regard to find context and date. Earlier assumptions that *rynd* slots first appeared in the larger watermills and were later transferred to hand querns (Stigum 1964:539-540) may still be correct, since watermills may have been used from late Viking times in Scandinavia. However most quernstones with traces of use do not have *rynd* slots. A rib around the eye of the quernstones also occurs on stones both in the Viking Age and the Middle Ages. Most stones however have no rib.

The stones from Bergen indicate large-scale distribution, where the town most likely also acted as a redistribution centre for the products, judging by the written sources. Quernstones from Hyllestad represent a less significant presence in other Norwegian medieval towns, and in Oslo they only represent a small percentage of the total numbers of stones. This is also the case from Tønsberg judging by the overall amount of quernstones found. Here, Hyllestad seems to have played a larger role than in Oslo. The Hyllestad quarries did not play an exclusive role in medieval Norway. Based on the finds from urban contexts, the towns functioned as important markets for two of the large medieval production areas, Hyllestad and Saltdal. Hyllestad, however, clearly controlled the largest commercial Norwegian centre (Bergen) and the stones also dominate abroad. The Hyllestad stones from Denmark and Hedeby/Schleswig do not, however, indicate large-scale distribution in themselves. These areas should be seen in total, as they were evidently parts of the same trading networks (see Chapter 10.1.1). The fact that Hyllestad stones were also more numerous at many of the sites signifies, as well, widespread distribution to the southern parts of Scandinavia from the late Viking period onwards, and where the main markets lasted for centuries. In the North Atlantic region, quernstones from Hyllestad seem to have been less important. As no towns or trading centres existed there, it was perhaps more difficult to make significant exports to these areas, where Bergen acted as their main commercial, as well as administrative, centre in the high Middle Ages. Local inhabitants seem to have been more or less self-sufficient in quernstones made of local rock. It is, however, interesting to observe that the early settlers had the skills and competence to produce their own quernstones at this early stage, seemingly bringing their skills with them from their homelands. This is also supported by the relatively low numbers of Norwegian soapstone products in the North Atlantic region, and no large-scale and organized exporting of these commodities seems to have taken place westwards (cf. Forster 2004).

The Ølve/Hatlestrand area evidently played a major, but not unique role in the production of bakestones; bakestones from other quarries are also found in the medieval towns. Products of green schist, in all likelihood from Ølve/Hatlestrand, are known from the early 1100s, and were transported to the North Atlantic islands more or less at the same time as they appear in Norwegian towns. The bakestones seem to become more common in the high and late Middle Ages (AD 1200-1400). At these sites, bakestone fragments are much more numerous than quernstones, indicating intensive use. Bakestones, however, had a somewhat narrower geographical range compared to Hyllestad quernstones and millstones, which is most likely related to different food traditions.

Tracing spatial patterns of exchange is a first step towards an understanding of how exchange took place (Earle 1982:6). Factors influencing distribution, the agents involved, organisation, and possible transaction forms for the different stone products will be further discussed in the following chapters.

8 Who Controlled the Quarries?

This chapter looks at the socio-political aspects related to the quarries, access and interests in terms of the resources, and rights of use connected to production, distribution and utilization. Central issues to be covered are whether the quarries represent economic specializations, and activities controlled and organized by elites, or whether they were situated on common land or commons exploited by the local populations. Is production the result of cooperative exploitation between local farmers, or were resources and areas subject to local or distant elites? The aim here is to come closer to the agents, indicated on the level of social groups, behind the production. The quarrying should therefore be understood both in relation to the local community of which they were a part, as well as in a wider geographical and socio-economic context. Emphasis is placed on reconstructing property structures in the late Iron Age and Middle Ages, as far as sources can reveal, taking into account ownership factors from the early modern period. Comparisons between the two production sites of Hyllestad and Ølve/Hatlestrand and their socio-economic contexts may help to assess the social and organisational relationships of this large-scale quarrying, and to reveal possible differences and similarities.

8.1 The concepts of ownership

The concept ownership and the form and function of manorial systems and tenures seem to have changed from the Viking period to the high Middle Ages (Iversen 1995:174; Øye 2004:87; Iversen and Myking 2007:296-297). Clarification and reflections on notions such as ownership (ON: *eign*) and property rights (ON: *eignarlýrittr*) is therefore necessary, as they most likely had different meanings in the Viking period and the Middle Ages. In both periods the control of land and estates constituted the political and economic power base. In the present study, an estate denotes a clustering of farms/parts of farms within an area owned by the same owner(s), either private individuals or institutions. However, form, function and organisation of the manorial system all appear to have changed from the Viking period to the high Middle Ages.

In Viking times ownership was intimately integrated within social relations. Control or power over land also seems to have included control over the people who lived on it. A system of personal dependency between superior landlords and subordinate land users seems to have existed in the early part of the period, as a kind of patron-client-relationship. This represented an asymmetrical bond between the parties, which also affected the possession and disposal of land. For a subordinate farmer it could mean protection. The return services provided may have varied, for instance according to the available resources, but would have included support and loyalty. This type of hierarchical tenancy would be flexible and dependent on the actors involved (Iversen 1995:171-174; Øye 2004:87). The different parties might have alternative rights, where social status would be decisive in terms of what rights they had: some may have had the right to live on the land, others the right to duties. Property was, in this way, embedded in social relations (Skre 1998:16-17). The relationships between the different actors and their disposal or rights to the quarries may thus have been of a social and economic nature. People had different rights to the land and the quarries - not ownership in the way we understand the concept today, but factors also relating to control and power (Skre 1998:18).

In a rural society of the Viking period, many farms seem to have been located around a main farm of an estate, also the residential farms of magnates, chieftains or petty kings, particularly in the coastal areas (Iversen 1995; Iversen 1999; 2005; Øye 2004:85). This kind of manorial system seems to a large extent to be based on personal bonds between land users and their landlords. Ultimately, power was based on ambulatory kings and magnates, travelling between a limited numbers of manorial sites (Iversen 1995; Øye 2004:87; Iversen 2005:133). In this first phase of the period there was, therefore, more direct contact between landowners and land users.

From around AD 1100, this system gradually changed. Considerable tracts of land had evidently been confiscated by the king from defeated opponents, and old manors were split up. Property must eventually have been transferred to different ecclesiastical institutions, resulting in large, supra-regional estates with dispersed property structures (Bjørkvik 1970; Iversen 1995:176-177; Øye 2004:85; Iversen 2005:133, 141). Landed property was gradually defined in a more impersonal way. From the 13th century onwards, tenancy was centred on economic factors rather than personal ones, with fixed land rents, based on an estimate of the farms' value and economic capacity. In the high Middle Ages it was normally paid in kind at each tenant farm, or in equivalent values (Bjørkvik 1965:278). By the 13th century, property rights, disposal and tenancy to a larger degree no longer involved the same kind of personal dependency as before; economic factors took precedence (Iversen 1995:171-174; Skre 1998:35-36; Øye 2004:87). The ecclesiastical institutions, as new landowners, represented this more impersonal form of property control, taking land rents from the tenants but without necessarily becoming directly involved in production at the farms. By the high Middle Ages some 70% of the farmers may have been tenants (ON: *leiglendingr*) who rented the use rights to land owned by others – either individuals or institutions

(Iversen 1995:169; Helle 2001:121). However it is often difficult to grasp these processes fully as sources are rather sparse. According to provincial laws (i.e. Gulathing's Law), representative of the 12th century in its written form, tenants had rights to use farm lands, both interior and exterior. In the outlying areas, however, exploitation of the resources was in principle restricted to the farm's own needs; the use of common land or commons was also restricted (G 73, 75, 95; L VII, 52; Helle 2001:112-114; Øye 2002:362-368, 379-391).

To recognize old property structures within an area, and to identify possible prehistoric estates, one methodological approach has been to look for coherent areas with farms owned by one individual, and clusters of such areas (Iversen 2008:89). An area where all farms have a common owner in the Middle Ages or early modern period would be denoted as a 'property cluster'. To define such an area, at least three or more medieval farms with a same owner should be clustered and border each other; or two bordering farms with a common owner should be more than three times the size of an average farm. Clusters of such areas may indicate the conscious policy of a landowner, while several clusters next to each other might indicate a division of a pre-existing larger estate into smaller units (Iversen 2008:89-90). The opposite situation would be an area characterized by a more diverse ownership structure, and areas of freehold known from the cadastres of the early modern period, which may indicate the presence of freeholders, perhaps as early as the late Iron Age.

Due to the length of time the quarries were in use, probably from the 800s until the early modern period, not only clustering, but also 'de-clustering' and 're-clustering' of such units must have taken place. Ownership that can be recorded for Hyllestad and Ølve/Hatlestrand lands can be dated from the late 12th century at the earliest, recorded at the monastery in Bergen, the Munkeliv cadastre, and the cadastre of the Bergen Bishop See in around 1360 and later diplomas from the 14th century onwards – but for most areas from the 16th and 17th centuries onwards. This makes it difficult to trace the processes more accurately, i.e. when the clustering started - when the gathering of farms by a common owner first began, and when estates were established. They may have been the acquisitions of available bordering farms by landowners in the Middle Ages, or they may represent older estate structures before Church and monasteries acquired the land, perhaps in the Viking period. At the same time pre-existing large units of ownership, or estates from the late Viking Age, seem to have been split up into smaller units or estates in the Middle Ages, judging by farm names and other information. The estates may also have been donated to, or acquired by, different ecclesiastical institutions. In the late Middle Ages the demographic crisis also made it possible for landowners to take possession of deserted farms. An area with a previously divided ownership structure and possible freehold may therefore have been included in an estate at different times. Such aspects complicate the question of the chronology of property units and estates, and how reliable reconstructions of prehistoric and early medieval property structures can be. They can therefore only serve as approximate, plausible reconstructions. However different types of ownership and use rights constituted important socio-political frameworks for quarrying activities and the organisation of quarry landscapes.

8.2 Private land or commons?

In Norwegian agrarian studies, the main production unit to have gained attention is the so-called 'named farm' (Norw. *navnegard*), a farm denoted by its own name (Øye 2004:96). This unit also seems an important structure in the present study. Even though the rural settlement pattern during the Viking Age and Middle Ages varied both in size and degree of subdivision into holdings (Øye 2004:96, 98), the 'named farm' represents a relevant unit when discussing quarrying and use rights, and when taking into account that their status and size may have varied over time (Øye 2011).

Quarries have only exceptionally been identified in the interior farmland, for example at Rønset in Hyllestad. More often quarrying took place in outlying areas, in Norwegian referred to as *utmark*. This is, however, a rather recent concept where the medieval terms denoting the areas within the farm were more differentiated, including different use rights. The term *utmark* today comprises forest landscapes, moorland, mountainous, and coastal areas, which can include the resources, use and production in areas on the periphery of core settlements. *Utmark* lacks an equivalent in English terminology, but is often translated as 'outfield', or outlying wasteland or 'outland'; a general equivalent for the Norwegian term is lacking (Øye 2002:361-362; 2005:9-10).

The use of different resources in these outer farm areas was vital for the farming economy in medieval Norway (Øye 2002:361-362; 2004:91). The outfield was an integrated part of the farming system which interacted with the more central areas of the farm: the interior area with the habitation structures and arable fields (Andersson 1998:6; Øye 2005:10). Viking Age and medieval farms consisted of two complementary areas, based on animal husbandry and arable farming. The farm interior or infield (ON: *innan garðs*) was located within fencing, consisting of arable land, meadows and enclosed pastures, whereas the outlying parts of the farm, the outfield (ON: *utan garðs*), were located beyond the fence. In the Middle Ages those outlying areas close to the settlement were an integral part of the farm, while the more distant outfields were often common lands, denoted by the ON term *almenningr*, meaning areas for 'all men'. Freeholders had the exclusive right of use to both the infield and the outfield. The tenants only had full rights of use to the infield, while the use of the outfield was for the household's personal use and not for profit, and otherwise belonged to the landowner (Øye 2002:362; Øye 2004:112-113; Øye 2005:9-11). Most of the quarries investigated in this study are located in outlying areas, at various distances from the historical habitation areas of the farms. Whether they were integrated into the farm territory, or belonged to what was defined as commons, therefore needs further study. The processes involved with quarry disposal are connected to the disposal and right of use of the outfields.

These outer areas were integrated parts of a wider economic, social and cultural system. From the early Viking period onwards, the use of different outfield resources evidently became more intensified, and a change from small-scale to large-scale exploitation can be seen. For example wood was exploited for timber and for charcoal and tar production; hunting was important for the provisions of furs and antlers. Deposits of bog iron were essential for all kinds of tools and weapons, and stone was quarried for various objects: quernstones, bakestones, building stones, etc. (cf. Øye 2002:362-393; Baug 2002; Tveiten 2012). Different resources offered possibilities to create surpluses for larger markets. Exploitation and distribution of these resources had a pseudo-industrial character that demanded wider organisation and contact networks. Consequently the outlying areas, and the resources located there, played an important role for landowners, and access to different resources located in the outfields was most likely of great importance for the development of farms and settlements.

Large-scale production is evident from the beginning of the Viking period. Both volume and distribution seem to be rather modest compared to the situation existing at the end of the period and further into the Middle Ages. From the latter part of the 1200s onwards, a decline in the large-scale exploitation and utilization of outlying resources can be documented. This may have to do with societal changes and also a lack of rural workers from the 1200-1300s, also changing economic networks and fewer entrepreneurs from the upper classes. The possibility that changes in trade systems, such as the development and growth of the Hanseatic League, had consequences for the native utilization of outfield resources should also be considered (Øye 2002:394). Despite these changes, quarrying in Hyllestad and Ølve/Hatlestrand continued throughout the Middle Ages into the 20th century, but a decline can be seen in the late Middle Ages/early modern period.

Although large-scale extraction of different stone products over centuries, which in Hyllestad and Ølve/Hatlestrand led to a marked change in the original topography, quarrying is hardly mentioned in medieval written sources. No legislation with regard to quarrying is known from the Middle Ages and early modern period.[57] Generally, information relating to the quarries and disposal of these resources is scarce (cf. DN I, 630; DN VIII, 428; DN XI, 131, 144; AB152; 154). Laws concerning use of outfields and commons are, however, also relevant for quarrying practices. These included medieval provincial laws such as the Gulathing and Frostathing Laws,[58] together with the Landlaw of 1274 that provide important information with regard to common rights in e.g. forest and mountain areas. Since the provisions of all three laws regarding common land are rather vague, it has been argued that not all aspects related to the administration of commons are included (Solem 2003:246).

The king's interests in the commons in the Middle Ages are expressed in the Landlaw, where the king has the right to freely lease the commons as he pleased (L VII, 62). Similarly it is claimed that the right to exploit other types of resources in the commons, such as quarrying, charcoal burning and mining, also belonged to the king, even though these activities were seldom specifically mentioned in the legal texts. To what degree royal disposal of resources was followed up in practice is also questioned (Herzberg 1913:8), as there may have been a discrepancy between norm and practice, between the regulations and administrative power, and the competence to enforce them.

In his doctoral project, Jon Solem (2003) suggests a duality in the administration of commons, based on what he denotes as 'inner' and 'outer' administration, the latter being the regulations from a central authority, while the former comprises norms and regulations developed on a local level. Solem's arguments are based on assumptions that royal authority, as manifested in the laws, may have worked on a different level than the local traditions (Solem 2003:258). Rights of ownership or disposal may comprise the judicial framework laid down from a distant central power, but they might also have depended on the actors' interests and capacity to defend such rights (Eriksson-Trenter 1998:169). Laws and regulations with regard to quarrying are, however, not referred to directly in the legal texts. A more locally based form of decision making should perhaps be considered with regard to quarry administration. Yet, the processes of decision making are seldom static, and may have changed with societal changes and shifting interests from the different authorities. Therefore the long period of activity at the two production areas studied, from around the 9th and 11th centuries into the early modern period, makes it probable that both the administration as well as decision making connected to production and distribution changed, but how and to what extent it is difficult to answer.

8.2.1 Quarrying: the documentary sources

The Hyllestad quarries studied are not mentioned in any of the documentary sources from the Middle Ages or early

[57] The geologist Amund Helland has interpreted legislation regarding 'earth-dug' goods (ON: fé) as a reference to minerals underground or ore deposits (Helland 1892:15). Gulathing's Law claims the landowner's right to the 'earth-dug' goods (G 148) and Frostathing's law gives certain rights to the discoverer, however, only if the discoverer has excavated with the landowner's permission (F XVI, 1). In the Landlaw the king's right to 'earth-dug' goods is claimed (L VI, 16). It is more likely that the expression *fé* refers to other valuable goods, including grave goods (Iversen 2008:78). The possibility that such laws may have relevance for other resources should perhaps be considered, and if so, these might indicate property rights on a more general level.

[58] Gulathing's Law in the form we know it today is considered to date from the 1160s, and the law expresses regulations valid from the latter half of the 1100s and first part of the 1200s (Helle 2001:17-18). Frostathing's law was in force in the Trøndelag-area in Mid-Norway. The law is now considered to have applied for the 1260s (Hagland and Sandnes 1994).

modern period, and only a few more recent documents exist. For Ølve/Hatlestrand there are documents from the 16th and 18th centuries (NRJ IV:474; Placat og Jordebog... 1744; Specification ofer Halsnøe-Klosters Goved-Gaard...1757). In the early 16th century, the nobleman Vincent Lunge owned a quarry site, as well as a forest, in the outlying areas of Fugleberg farm at Ølve (NRJ IV: 474).

These documents indicate that quarries belonged to landowners, partly the farm owner, but several other owners could own shares in the same quarries. In the Hardanger region, bakestone production is mentioned in the 17th century, and most likely it refers to the quarries of Ølve/Hatlestrand (Friis 1632:71-74). The quarries and their products were obviously known outside the local community, but according to Friis only within Norway (Friis 1632:71-74). In the 1700s at least some Hyllestad quarries were implicated in a dispute between two of the farms, Rønset (no. 17) and Berge (no. 70), over marine resources. The quarries mentioned (Furehågjen and Langemyr) were located in the outer areas and not considered part of the commons (Rønneseth 1977:9).

The only medieval regulation regarding quarrying appears in an agreement (1277) between Archbishop Jon Raude and King Magnus Håkonsson the Lawmender, whereby among other types of quarry, quernstone and millstone workings should be subjected to tithes, along with other outfield resources, e.g. hunting, tar, and iron (Ngl II: 475). No other regulations concerning quarrying are known and bakestone production is not mentioned at all, which may indicate a smaller industry. However, as with many of Archbishop Jon's claims from the 1270s, it is uncertain whether they were actually were carried out (NgL III: 117, 153; Bøe 1964: 300-302).

A few documents concerning medieval quarries in Norway do, however, exist; they are often related to the landowners' rights to the quarries, for instance the quarries of Saltdal (Nordland) and Vågå (Oppland). The quarries at Setså farm at Saltdal are mentioned in the cadastre of Archbishop Aslak Bolt of the 1430s. Here, one owner is mentioned, Eirik Raude, who some time before 1430 donated his share, which consisted of 15½ parts, to the Archbishop. The See then leased out 1 *spannsleige* for a yearly rent of 2 *pund* to Torberg Eirikson (AB152; 154), who may have been Eirik Raude's son. The term *spannsleige* is used to denote a land unit, in this case a quernstone quarry, which gave one *spann* of butter in annual land rent: one *spann* constituted normally 3 *bismerpund* (15.42 kg) (Steinnes 1982:25, 82). In this case, therefore, the quarry was leased for less, only 2 *pund* (10.28 kg). This reduction was evidently caused by the general decline in land rent in the late Middle Ages, but perhaps also a decline in quernstone production. Archbishop Aslak Bolt also owned some quarry sites in the counties of Sør-Trøndelag (AB 67, 91) and Sunnhordland, where slates were extracted (AB 181). It would appear that extraction at these sites was rented out. Additionally the Øye farm (Sør-Trøndelag), where the bakestone quarry is located, is listed in the cadastre of Aslak Bolt (AB 67; NRJ II 175). Here, only the farm is mentioned, not the quarries.

Production based on tenancy is also represented at Vågå (Oppland). In the 15th and 16th centuries, Tolstad farm and the quernstone quarries located there were both owned by landowners belonging to elites from outside the local community. Some 15th-century documents referring to the Tolstad quernstone and millstone quarries also signify that the workings belonged to private landowners, and that they were leased (DN I, 864; DN VIII, 416; DN XI, 144). The oldest document mentioning the quarries is dated to 1426, referring to a termination of a lease between a private landowner and his tenant (DN XI, 144). Another document of 1437 mentions that Tolstad was owned by a noblewoman, Fru Adalis Erlingsdatter, who leased out the farm (DN I, 760). A few years later, in 1466, the farm and quarries were sold to the prominent nobleman Herr Alf Knutsson and his wife Fru Magnhild Oddsdotter (DN I, 864, 865), and the quarries thus came under the control of one of the largest landowners in Norway in the 15th century. A document from 1488, concerning an agreement about compensation for an unpaid rent for quarries at Tolstad, shows that Fru Magnhild Oddsdatter still owned the quarries and rented out the right of use. Large sums seem to have been involved, and a share in a farm had to be given in compensation to the landowner for unpaid rent for quarry extraction, as well as one hide (ON: *hud)* annually as rent to the landowner (DN VIII, 416), amounting to 1/2 *laup* of butter (7.7 kg) (Steinnes 1982:69). This indicates that the tenants themselves carried out the production. There also seems to have been restrictions on the amount of stone the tenant was allowed to quarry; it was limited to the quantity he could cut himself, without the use of hired labour. This implies that the tenants themselves were the stonecutters, as well as the routine farming activities they had to attend to, but on a limited scale.

The Tolstad quarries belonged to the same noble estate in the 16th and 17th centuries. During this period Fru Magnhild Oddsdatter's granddaughter, Gyrvil Faderdotter (Sparre), is listed as landowner. In the 1580s the stonecutters who had no rights to extract stones after the 'established customs', had to give every third stone to Gyrvil (Sandberg 1986:6; Teigum 2001:268, 294; Teigum 2011), who owned land in Norway, Denmark, Sweden, and Shetland (Sandberg 1986:3, 5). In 1586 the king owned Tolstad and the quarries located here, but a hundred years later the farm and quarries again returned to private hands, and the landlord leased out both farm and extraction rights to the tenants (Teigum 2001:268, 294; Teigum 2011).

According to judgments from the 1600s, commoners from areas other than where the quarries were located, had the right to extract stones at Tolstad, after an old custom, in return for a fee to be paid in finished quernstones. The quantities varied. In the 16th century every fourth stone had to be paid as a tithe – the so-called *tithe-stone*. In the 17th century every third stone was required by the quarry owner, and every tenth stone had to be given by the owner to the king (Kleiven 1977:347, 390-391; Teigum 2001:268; Teigum 2011:52), indicating royal sovereignty with special rights over the stones. Whether a similar system existed in the Middle Ages is not known. It has been suggested that a judgment of 1677 reflects an intention to respect the old rights of the 15th century (Teigum 2001:268). A common right to extract quernstones seems to be similar to practice at the millstone quarries of Selbu, known from the 17th century onwards. In those quarries situated in the more distant outfields, areas that were obviously regarded as commons, inhabitants of Selbu could freely establish a quarry and extract millstones (Rollseth 1947). In contrast to Vågå, the state does not seem to have had any right over products from Selbu.

In the few medieval documents concerning quarrying in Norway, therefore, the quarries seem to have been regarded as the private property of the landowner himself. The quarries belonged to landowners, to a large degree owned by the elite and institutions outside the local community. In most cases these quarries were leased out, meaning that the landowners themselves were not directly involved in production - only indirectly through fees or land rent. The ownership seems thus to have been respected, and extraction was only permitted in return for a fee to quarry owners. However in some cases, extraction may have occurred on what was considered private land, not only by tenants, but also by others; a similar position is documented at Vågå in the 1600s. However, a fee to the landowners would also be expected here.

As the written sources regarding disposal and administration of quarrying in the Middle Ages and early modern period are rare, it is also valuable to look at later laws and customs concerning the disposal and exploitation of other outfield resources, as they may indicate general rules. Regulations from the Middle Ages and early modern period to do with outfield production refer in the main more to iron production from bog ore and mining than to quarrying (cf. DN VI, 238; RN I, 326, no. A86), but regulations may also have been relevant to quarrying at an earlier stage.

An indication of the king's interest in outfield resources appears in a document from 1246, when Harald Olavsson, from the Isle of Man,[59] donated all mineral deposits to the Furness monastery located here (RN I, 326, no. A86). Additionally, a few other documents give some clues with regard to the disposal of mineral deposits. A statute of 1358 from Østerdalen, a valley in the eastern part of the country, allowed the locals to extract iron from the royal commons and freely dispose any profits and ore 'after the old custom here in the valley' (ON: *eftir þeim siðuana sem aat forno hefuir veret her j dalenum*), as well as to freely distribute and exchange goods (DN VI, 238). The statute indicates that the king considered it a royal right to tax iron extraction from common land, but that the farmers still had authority to exploit this activity after the customary rights (Hertzberg 1913:8-9; Øye 2002:368). It has been suggested that this was an attempt to limit a previous privilege, in the hope of introducing royal sovereignty (Solem 2003:264-265), or that the statute was intended to revitalize declining iron production (Tveiten 2010:250). Whatever the motivation, the king at this stage obviously regarded these common lands as a royal prerogative. The king's interests in the outlying resources may have had ultimate precedence, indicating the importance of mineral production in the late Middle Ages.

This so-called 'mineral royalty' gave the king or overlord special rights to the minerals, and they could decide whether to use the rights to exploit the resources, or to lease them out to others. The mineral royalty would also involve the right to claim rent for exploitation, meaning that the king had both a judicial and economic right to regulate extractions. In Norway, the mineral royalty for mining, probably originating from German law, was imposed further and claimed by Queen Margrete in 1405 (Nagel 1991:41-42), long before the first mining law was enacted. Quarrying, however, is not mentioned, which may indicate less interest from the central authorities when it came to stone extraction.

According to the historian Anne-Hilde Nagel, three main agencies were central to these medieval and early modern Norwegian laws on mining interests: the landowner, the discoverer of the resources, and the state (Nagel 1991:41). There are three main theories to do with mineral disposal. The theory of *pars fundi* acknowledges the major claim of the landowner. It holds that all minerals on a property belong to the landowner, who assumes all rights to exploit these resources. The second theory, the *res nullius*, maintains that minerals below ground belong to no one, and therefore allows the discoverer of the resources the rights to exploit them freely, including the right of occupation. The third theory, the above-mentioned mineral royalty infers royal/ overlord rights to any minerals. This principle existed in Scandinavia since the Middle Ages, i.e. the statute of 1358 previously mentioned (cf. DN VI, 238).

None of these theories/principles preclude property rights to the land itself and only concerns any minerals below ground. According to these principles, the owner of the land was to be compensated for any surface damage to the land related to the mineral exploitation involved (Nagel 1991:41). As a consequence, the *pars fundi* theory, where the landowner's property rights are paramount, may be more important in terms of quarrying than mining. In fact, in the Norwegian mining legislation dating from 1842, minerals (e.g. mica schist) suitable for materials such as building stones and millstones are listed as resources belonging to the landowner, as opposed to metallic ores, and minerals subject to state control (Helland 1892:267, 279). The sparse documentary evidence from the Middle Ages and early modern period, however, does not provide

[59] The Isle of Man (off the west coast of England) was under Norwegian rule from the Viking period until 1266 (Kinvig 1975:54-66).

any information as to whether or not such a distinction was observed in earlier periods.

The first specific mining law in Norway is dated to 1540 (Beuther 1540),[60] and is in the main a copy of a German law, modelled after legislation in force in Annaberg from 1509 (Helland 1892:3; Nagel 1986:92; 1991:42-43). Under German mining law, the principle of 'mineral royalty' has a strong influence: however it is in combination with another principle, *mining freedom*, where mining rights belong to the discoverer. The principle in Norwegian mining law also incorporates the discoverer's right to exploitation, although the mineral rights belong to the king, or his designated authority, and were released to the discoverer. The mining law of 1540 was, therefore, a German law transferred to Norwegian conditions, and not necessarily representative of the contemporary Norwegian situation. The law was never actually translated from German and, consequently, very few people were able to read it (Helland 1892:3-4, 25, 29; Nagel 1991:42-43). It has therefore been claimed that the law never gained any significance in Norway, being little known, used and respected in the country, and therefore of minor influence on Norwegian practices (Helland 1892:4, 6, 29). The law was contravened several times during the 16th and 17th centuries, but the 'Royalty right' at least seems to have been respected (Nagel 1991:43). The legislation of 1540 grants the landowner some compensation for land taken over by mining activities in the form of a small share (Helland 1892:310-311).

The mining laws from the middle of the 16th century were in force until 1812 (Helland 1892:29), the same period as the extraction of millstones from Selbu in Sør-Trøndelag (Friis 1632:97; Rollseth 1947). All Selbu inhabitants had the right to exploit minerals in the mountains, establish quarries and extract millstones (Helland 1892:5). However, even though the legislation also includes non-metallic minerals, neither Helland nor Nagel view the Selbu millstone quarries in terms of discoverers' rights, or as a manifestation of the Royalty right, as far as exploitation of common land (Helland 1892:5; Nagel 1986:97). Slate quarries from Ål (Hallingdal) also seem to have been sited on common land in the 19th century (Helland 1893:43-44; Rynning 1934:451).

The 1540 mining laws had little relevance with regard to medieval production at Hyllestad and Ølve/Hatlestrand, but there was activity in both quarry areas into recent times. If the legislation was valid for quarrying, this may have affected the disposal of the two quarry areas and production from the sites, and most relevant medieval documents indicate that they were in the hands of different landowners. Based on indications from these few medieval documents (cf. DN I, 630; DN VIII, 428; DN XI, 144; AB152; 154) it has been suggested that the importance of different stone products, such as whetstones, quernstones and millstones, makes it likely that the rights to extract these products were reserved for the appropriate farms (Rynning 1934:446-447).

Based on current evidence, it is difficult to employ general models of quarrying, including organisation and development, carried over from the Viking period and the Middle Ages, but some indications can be traced. Most evidence points towards a situation where the quarries belonged to the farms on which they were located. Quarries are only rarely mentioned in medieval laws, probably because they were considered as linked to farm territories. This factor, together with other information from the few existing documents, suggests that quarry ownership should be further investigated with a view to understanding the quarrying activities at Ølve/Hatlestrand and Hyllestad. Quarries represent a resource directly connected to the different farms on which they are located, but the owners may not necessarily have been directly connected to the farms. The owners of the land may have been the agents with the appropriate resources - allocative and authoritative, to use Giddens' concepts - and thus had disposal rights and controls over the quarries and their output.

Written sources from both the Middle Ages and early modern period need to be analyzed to assess medieval property structure, According to written evidence, ecclesiastical institutions, churches and monasteries, held properties within the two quarry landscapes, and by the Middle Ages the majority of the farms formed components of larger ecclesiastical estates. Medieval sources from the 13th century onwards confirm that their property structure had roots in the early Middle Ages in the 12th century.

8.3 Property structure in Hyllestad in the Middle Ages

Nearly all the farms of Hyllestad formed parts of larger medieval ecclesiastical estates. They were divided between the local churches of Øn and Hyllestad, the ecclesiastical institutions of Munkeliv monastery and the Bishop of Bergen (Baug 2002:90-97). The farms with quarries within their boundaries were not owned by freeholders, either before or after the Reformation (Tab. 8.1).

Munkeliv monastery

According to a cadastre at Munkeliv monastery of 1175, the abbey owned four of the Hyllestad farms where quarries have been identified (DN XVI, 1). Later in the Middle Ages, a further two farms, together with parts of a third farm, were incorporated into the Munkeliv estates, all of them mentioned in a cadastre of 1427 (DN XII, 190).

Munkeliv monastery (Bergen) was founded by king Øystein Magnusson in the first part of the 12th century (DN XII, 12; Morkinskinna 254, 276), and belonged to the Benedictines until 1426, when it was given to the Birgittines (DN II, 1; Hommedal 1999:153). In 1175 the Munkeliv estate included 95 farms, and other parts of farms

[60] It is likely that the legislation was drawn up with a contribution from Paul Schmied, in Schneeberg, and Hans Glaser, however Anthon Beuther is considered to be the writer (Helland 1892:22).

Table 8.1: Medieval ownership structure at Hyllestad

Medieval farms with quarries	Medieval ownership	Date of oldest recorded evidence	Part	Complete	Ownership after the Reformation
Myklebust (No. 79)	Munkeliv monastery	1175		x	Munkeliv monastery
Hyllestad (No. 77)	local priest of Hyllestad	1300s		x	local priest of Askvoll (*mensalgods*[1])
Sæsol (No. 78)	local priest of Hyllestad	1300s	?	?	local priest of Askvoll (*mensalgods*)
Sørefjorden (No. 72)	Munkeliv monastery	1427	?	?	Munkeliv monastery
Berge (No. 70)	Munkeliv monastery	1175	x		Munkeliv monastery
Rønset (No. 71)	possible noble estate or Nonnester monastery estate	1535	?	?	Vincent Lunge from 1535
Skor (No. 23)	uncertain	1600s	?	?	local church of Øn and the church of Holmedal
Hatlem (Nos. 25, 26 and 27)[2]	1 *mmb*[3] belonged to the local priest of Øn	1300s	x		*Stigtgods*, former bishop-estate
Øn (No. 29)	Bishop's estate	1300s	?	?	*Stigtgods*, former Bishop-estate
Sørbø (No. 32)	local priest of Øn and the bishop	1600s	x		separated from the farm Ulvika (No. 31) in the 16th century. *Stigtgods,* former bishop-estate
Inner and Outer Gil (Nos. 33 and 34)	King, Munkeliv monastery	1175	x		Inner Gil was separated from the farm Ulvika (No. 31) in the 17th century: *Stigtgods*, former bishop-estate. Outer Gil belonged to Eide (No. 35) until the 16th century: Munkeliv monastery
Rutle (No. 10)	local priest of Øn, the local church of Øn, and from 1427 5 *mm* (ON: *mánaðarmatr)*[4] and 4 *teigar* belonged to Munkeliv monastery	11300s	?	?	Munkeliv monastery, but 4 *teigar* (smaller parts of land) belonged to the local church of Øn
Eide (No. 35)	Munkeliv monastery	1427	?	?	Munkeliv monastery
Åsmul (No. 36)	local priest of Øn bought a part of the farm for the local church	1300s	x		local church of Øn
Borsholmen (No. 37)	local priest of Øn	1300s	?	?	local priest of Askvoll

1 Mensalgods were estates belonging to the priest for his general needs an*d upkeep.*
2 Hatlem in the 17th century was split into three farms; Inner and Outer Hatlem (nos. 26 and 27), and Leirpollen (no. 25), which was separate from Inner Hatlem.
3 Land providing 1 mm in land rent was termed *månadsmatbol (mmb) (Steinnes 1982:84).*
4 The term *mánaðarmatr* was the amount of butter for one man for one month. *One mánaðarmatr = 1 laup = 15.43 kg (Steinnes 1982:70, 83).*

in western Norway, from Sunnmøre in the north, to Lista in the south. The estate had a concentration of farms in the rural districts close to Bergen, but also included properties further south and north. Behind this network may have been the objective of securing a diverse economic base for the monastery (Bjørkvik 1967:60-61), and this may also explain why the monastery owned farms within the Hyllestad quarry area.

Documentary evidence indicates that the monastery also acquired land with special resources in other areas. According to the cadastre of 1480-1490, king Magnus Håkonsson gave Halhjem (no. 44) farm at Os (Hordaland) to the monastery (DN XII, 190, 257), where two soapstone quarries have been identified.[61] These are not precisely dated, but are probably medieval. According to a cadastre of 1427 the monastery owned mills at Gravdal outside Bergen (DN XII, 190), which may be another example of the monastery acquiring areas with other, diverse resources. The mills were established by the early 13th

[61] Cf. https://askeladden.ra.no.

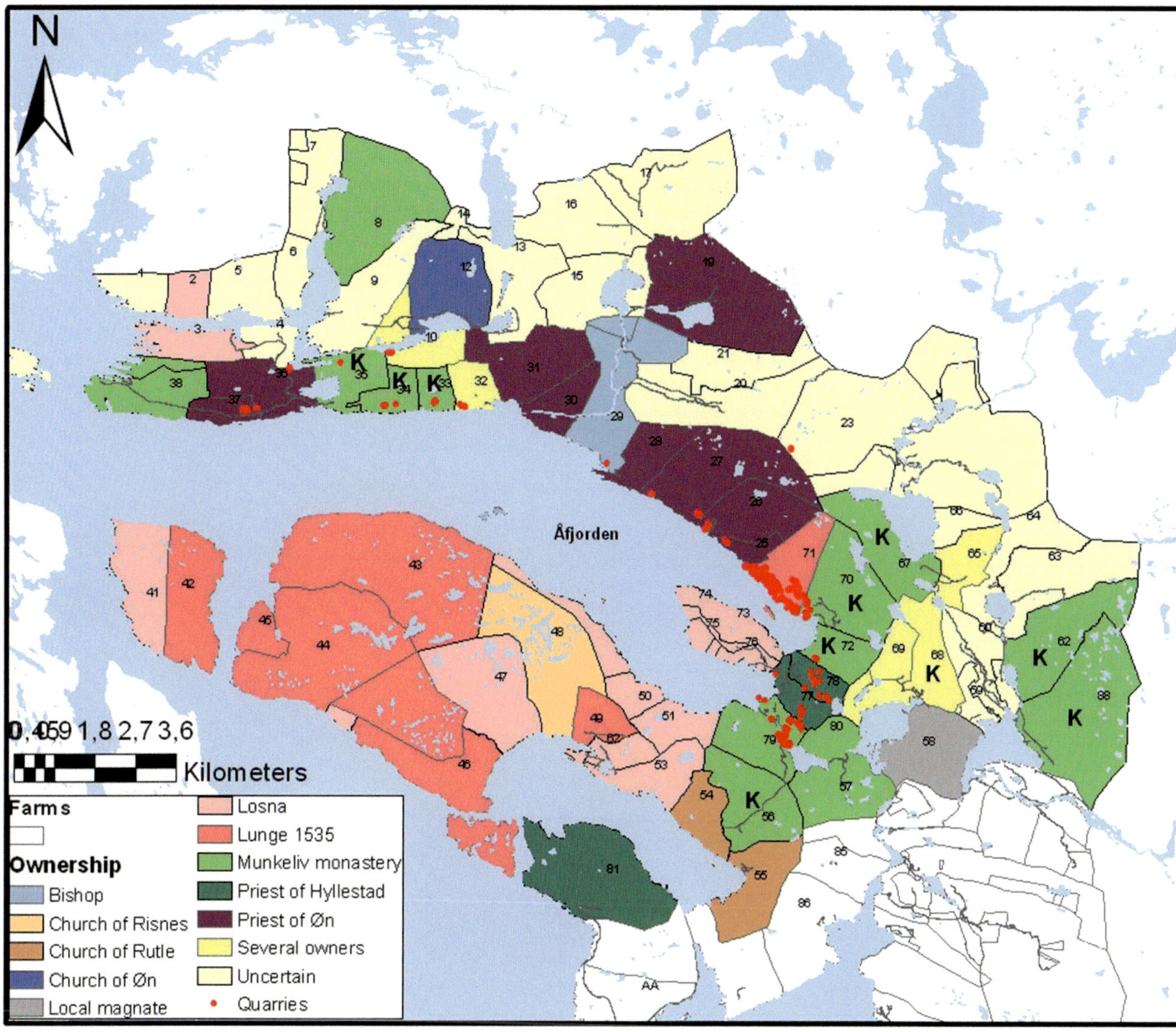

FIGURE 8.1: MAP OF THE MEDIEVAL OWNERSHIP STRUCTURE IN THE PARISHES OF ØN AND HYLLESTAD - ALSO INCLUDING TWO FARMS WITHIN THE PARISH OF BØ (NOS. 81 AND 88) MENTIONED IN THE TEXT. FARMS OWNED BY THE KING ARE MARKED WITH A K (MAP: IRENE BAUG, 2013, NORGE DIGITALT. FKB-DATA, 1413 HYLLESTAD KOMMUNE, 32_1413EIENDOM_FLATE, 28.02.2009)[62]

century (Helle 1982:430) but are not mentioned in the oldest cadastre of Munkeliv of 1175 (DN XVI, 1); this cadastre may be incomplete however (Bjørkvik 1967:53). After its foundation in the early 1100s, the estate expanded considerably over the next 100 years, with approximately 220 farms/parts of farms documented in 1328, and some 360 in 1427 (Iversen 2008:94-95). The monastery obviously had wide economic interests outside farming.

Munkeliv monastery stands out as one of the largest quarry owners in Hyllestad. Most of the farms belonging to the monastery in 1175 are considered to have been given by the king in connection with its foundation (Bjørkvik 1967:60-61). If this was also the case here, the Munkeliv estates at Hyllestad may have an older origin. It is also possible that the farms at an early stage represented confiscations in the unification processes, and following the conflicts of the time, and were also donated by the king to the monastery. More or less concentrated clusters of farms belonging to one institution often reflect larger donations and may indicate that the estate itself dates further back. The Munkeliv estate at Hyllestad shows, however, a diverse ownership structure, comprising only parts of the farms, which may indicate that the farms, and parts of other farms, were transferred from more random transactions, i.e. gifts or purchases, rather than being the result of a single, large donation by one owner.

According to the cadastre of 1175, the monastery owned ten farms in Hyllestad (DN XVI, 1).[63] Four of them,

[62] The map is based on an aggregation of single sub-plots: 32_1413eiendom_flate (polygon) to the cadastral units (named farm) by the dissolve-function in ArcGIS (Arne Solli, AHKR, UiB).

[63] In her study of the medieval settlement and ownership in Hyllestad, the historian and archaeologist Inger Kellmer seems to present an incorrect list of farms in Hyllestad belonging to Munkeliv monastery from 1175 (Kellmer 1963: appendix I). Three of the farms she mentions, *Ose*, *Vpsolum* and *Eiþi* are most likely not located in Hyllestad. *Ose* and *Vpsolum* are in all likelihood located in Ørstad (Bjørkvik 1967:54-55). Kellmer has mistaken *Ose* for the farm *Ass (Øvrås)* in Hyllestad, which was part of the Munkeliv estate in 1427 (DN XII, 190), and *Vpsolum* for the farm *Selsvik* (Kellmer 1963:14). The farm *Eiþi* was most likely located by the Aldafjord in Selje (Bjørkvik 1967:54), and is by Kellmer mistaken for the farm *Eide* in Hyllestad, which was part of the Munkeliv estate in 1427 (DN XII, 190).

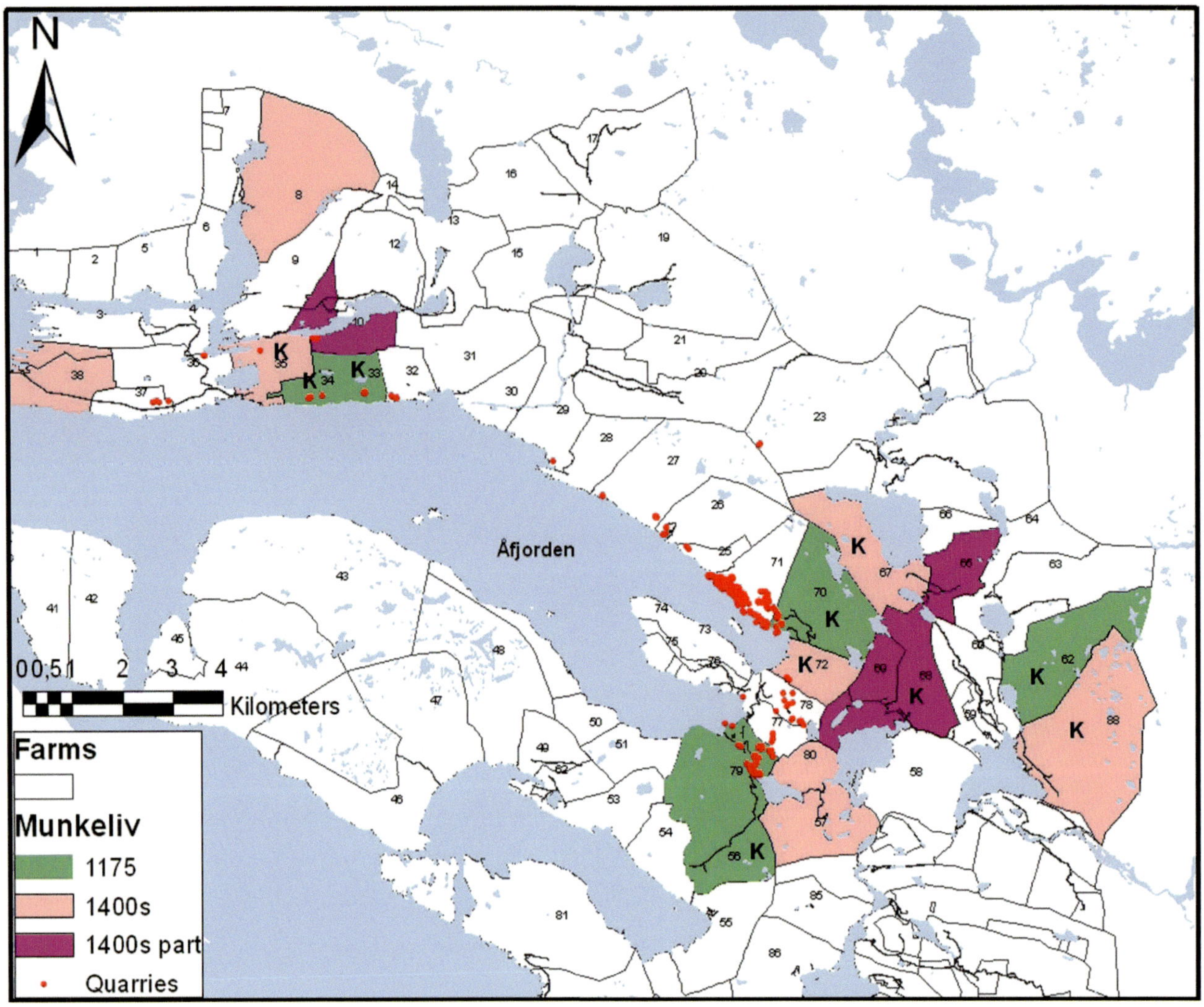

Figure 8.2: Map of the Munkeliv estate at Hyllestad. Farms shown in green are mentioned in the cadastre of 1175; farms shown in pink are recorded in cadastres of the 1400s. Farms partly owned by the monastery are shown in dark pink. Farms said to have been donated by King Magnus Håkonnson are marked K (Map: Irene Baug, 2013, Norge Digitalt. FKB-data, 1413 Hyllestad kommune, 32_1413eiendom_flate, 28.02.2009)

Bjarhastoþlum, *Stoþlum*, *Bruarase* and *Hiolmusætre* no longer exist and it is difficult to identify their former locations.[64] Consequently they cannot be included in the further analysis. Two different areas of the Munkeliv estate can be located, one in the parish of Øn, on the northern side of the Åfjorden, and the other in the parish of Hyllestad, on the north-eastern side of the same fjord (Fig. 8.2), but neither of them can be described as property 'clusters'.

The Munkeliv estate on the north-eastern side of Åfjorden comprised four farms in 1175: Myklebust (no. 79)[65] and Berge (no. 70), both located within the quarry area, and Tonning (no. 62) and Kolgrov (no. 56) (DN XVI, 1) (Fig. 8.2). At the latter two farms, the monastery owned only 1 *mm* (ON: *mánaðarmatr*), i.e. a land lease of about 1 *laup* of butter, corresponding to 15.43kg (Bjørkvik 1967:58; Steinnes 1982:83),[66] and indicating that only parts of the farms belonged to the monastery. This seems at least to be the case at Kolgrov, which is listed with a relatively higher rent in the land register of 1647 (Skattematrikkelen 1647 XII, 127). On the northern side of Åfjorden, only the Gil farm (perhaps referring to both nos. 33 and 34) belonged to the estate in 1175 (DN XVI, 1).

Since the farms owned by the monastery from the late 12th century onwards are dispersed, and do not represent clusters of ecclesiastical property (cf. Fig. 8.2), it may be less probable that the Munkeliv interests at Hyllestad were originally parts of older estate. The dispersed property may indicate that the monastery's ownership at Hyllestad

[64] The farm *Bjarhastoþlum* is supposed to be a strip of land north of the farm *Berge* in Hyllestad, which also belonged to the monastery. For *Bruarase* it is suggested that the location to Åfjorden is wrong. It is possible that the name refers to the farm *Bruaas* (no. 24) located in the parish of Lavik, near the parish of Bø in Hyllestad. The location of the farm *Hiolmusætre* may also be wrong, and it is possible that this was a deserted farm possibly located near Naustdal or Indre Holmedal. It is further possible that it belonged to a farm called *Hjelmeland*, possibly the farm belonging to Munkeliv from 1175, which is located at Holmedal. No suggested location for *Stoþlum* can be found (NG XII: 241-242).

[65] Two farms at Hyllestad are called Myklebust (no. 8 and 79). The oldest most likely refers to quarry area no. 79) (NG XII:236).

[66] For the Middle Ages, one cow is thought to produce 20-35kg of butter (Øye 2002:355).

reflect smaller gifts, donations and purchases. As for the monastic properties, and who donated lands to the monastery, the dispersed property structure of Hyllestad in 1175, with shared ownership of some of the farms, indicates that these parts were gifts and transfers given to the monastery in the 12th century, and that the monastery itself played an active role in the acquisition of farms (Baug 2002:94). According to a cadastre of 1480-1490, the Gil farms (it is unclear if it relates to both nos. 33 and 34), Tonning (no. 62), Kolgrov (no. 56), Berge (no. 70), Eide (no. 35), Sørefjorden (no. 72), Seluiik,[67] Markus (no. 67), Akse (no. 68), Espeland (no. 88), as well as the now lost farms of *Hiolmusæter*, *Bruaraas* and *Bergastaudul*, were donated to the monastery by Magnus Håkonsson (1263-1280) (DN XII, 257). These transfers must therefore have taken place in the latter half of the 13th century. As some of these farms are also listed in the monastery cadastre of 1175 (DN XVI, 1), it may indicate that only parts of the farms were donated to the monastery in the 1100s, and that the monastery acquired full property rights to the farms a century later. If this was the case, royal donations also in the 1100s should be expected. If the farms were given as a royal donation, a royal confiscation of farms from an earlier estate at Hyllestad should be considered.

During the 15th century a further expansion in lands belonging to the Munkeliv monastery seems to have taken place (DN XII, 190). Additionally, three more (today unidentified) farms are mentioned, *Haspaar*, *Klepp* and *Akranæss* (DN XII, 190).[68] As already noted, several of the farms were most likely donated to the monastery by Magnus Håkonsson (DN XII, 257). An expansion of the Munkeliv estate at Hyllestad must, therefore, have taken place in the second half of the 13th century. This later expansion of the Munkeliv estate is concentrated in the same areas as the farms acquired from 1175, indicating planned growth. Most of these farms are located outside the production area. It is therefore rather unlikely that quarrying was the reason for this acquisition, since only Rønset and Myklebust had large-scale production; all the other farms had only small-scale extraction. According to the cadastre of 1175, the farms within the quarry area had a relatively small value compared to many other farms owned by the Munkeliv estate. This is also the case at Myklebust (Bjørkvik 1967:58-59). Most of the farms of the quarry area remained within the Munkeliv estate, which was itself transferred to the state after the Reformation. An exception seems to be Inner Gil, which was transferred to the Bishop (Kellmer 1963:14-15).

Ownership structure before Munkeliv monastery control is blurred, but most evidence points to a royal confiscation of property during the unification process, followed by a royal donation to Munkeliv, shortly after the monastery was established in the 1100s, and even though the farms of this early donation were rather scattered. Further donations by the king took place in the late 1200s. The monastery most likely also played an active role in acquiring lands at Hyllestad (in the 15th century two clusters of properties in Hyllestad are attributed to Munkeliv).

The King

According to a cadastre at Munkeliv monastery, king Magnus Håkonsson (1263-1280) owned in the 13th century the farms of Gil (or parts of them) (nos. 33 and/or 34), Tonning (no. 62), Kolgrov (no. 56), Berge (no. 70), Eide (no. 35), Sørefjorden (no. 72), Selsvik (belonging to Sørefjorden), Markus (no. 67), Akse (no. 68), Espeland (no. 88), and the now lost farms of *Biarliastodlum*, *Hiolmusæter*, *Bruaraas* and *Bergastaudul* (DN XII, 257). As mentioned above, this ownership may have resulted from royal confiscations of land at Hyllestad during the unification processes, and thus older than the donation of the 1180s at the latest.

Local priest and church estate

The local priests, on behalf of their respective churches of Hyllestad and Øn, constituted the largest landowners within the quarry landscape in the 14th century (cf. Tab. 8.1 and Fig. 8.1). Altogether, eight farms/parts of farms within the quarry area then belonged to the local priests in the parishes of Øn and Hyllestad, and the estate was mostly concentrated around the two churches of Øn and Hyllestad (Fig. 8.1).

According to the cadastre of the Bergen diocese, Björgynjar Kalfskinn of c. 1360 (BK), six farms/parts of farms belonged to the priest of Øn (BK, 26), all located along the northern side of the Åfjorden, more or less clustered around the farm Øn (no. 29) with its medieval church.[69] Five of the medieval farms here have quarries within their boundaries: Hatlem (nos. 26 and 27), Borsholmen (no. 37), Åsmul (no. 36), Sørbø (no. 32) and Rutle (no. 10). The Øn farm itself was, however, not part of this complex, but belonged to the Bishop's own estate in the 14th century (Kellmer 1963:11); as several of the farms had shared control this points towards a divided ownership structure. The priest owned only 1 *mm* in Hatlem, indicating that the farm was shared between several owners (Kellmer 1963:9-10). At this time both the Munkeliv monastery and the Bishop owned parts of farms with the priest of Øn (cf. tab. 8.1) (DN XII, 190; BK, 26; Kellmer 1963:9–11), which indicate an earlier common heritage.

Two farms within the quarry area, Hyllestad (no. 77) and Sæsol (no. 78), belonged to the priest of Hyllestad in the 14th century. The priest also had income from three farms/parts of farms outside the quarry area (BK, 28). Two of these, Akse (no. 68) and Øvrås (no. 65), seem to have been shared between the local priest and Munkeliv (DN

[67] Seluiik (Selvik) belonged to Sørefjorden (no. 72) from the 16th century (GN XII, 243).

[68] Klepp is not mentioned in other cadastres, and it is suggested that the name refers to Kleive (no. 80) at Hyllestad (GN XII, 242).

[69] In the Middle Ages the church was located at Øn farm (no. 29). After WWII the church was moved to Sørbøvåg, Sørbø farm (no. 32), when the church at Øn (built in 1640) burnt down (Kellmer 1963, Kiledeanvisninger – merknader).

XII, 189, 190; BK, 25). The third farm, Sågnes (no. 81), is located within Bø parish, further southeast (cf. Fig. 8.1).

As was the case for the estate of Munkeliv at Hyllestad, parts of the farms belonging to the local priests were rather dispersed, located between the farms belonging to the monastery. These farms may originally have belonged to the same landowner (Baug 2002:95-96). Parts of a farm may have been given as donations for specific clerical services: the so-called *sjelemesse* (Eng. 'requiem'),[70] performed by the priests. These parts may have belonged to the same contributor who donated to the Munkeliv estate. It is also possible that some of the parts were planned acquisitions. According to Björgynjar Kalfskinn, the priest of Øn bought parts of Åsmul (no. 36), listed under the church of Øn (BK, 26),[71] indicating an active role in acquiring properties within the quarry area as late as the 14th century (Baug 2002:95-96).

Several studies of churches and their surrounding lands indicate a close relationship between larger farms and medieval churches erected by elites from early times. In the early Middle Ages, local magnates seem to have played an active role in building private churches (G 12; a.o. Skre 1988; Brendalsmo 1990; Ingvaldsen 1996, Brattekværne 2006; Skeised 2007; Haug 2009; Søyland 2009; Dahlberg 2012). This could also have been the case at Hyllestad and Øn. To assess connections between the churches and local communities, the locations of churches in relation to the contemporary built-up environment is important, as it may reflect a long-established relationship.

The location of the medieval church at Øn is considered to be identical with its later location and documented from the 1600s (Aaberg 1929:33), located on a plateau near the old farmyard to the west. The close connection to the farmyard indicates that it was originally built as a private church. This is also supported by its somewhat remote location in relation to the local community as a whole. Except for some steep and rocky footpaths, only the sea gave access (Aaberg 1929:42). Two burial mounds from the late Iron Age have been identified at the farm (Fett 1954:7), the bordering farms lack such burials, also indicate Øn as a central farm. The church may have been built by a local magnate situated here, and may have been a so-called *högendiskirkja* (ON).

The location of the medieval church at Hyllestad farm was probably also at the same site as the church from the 1600s. The church was built in the middle of the farmyard at holding 1, where the church and historic farm houses were (and still are) clustered (Aaberge 1929:17-18). Here the old churchyard is still preserved with stone walls, medieval/early modern burials, and four churchyard crosses produced in Hyllestad. This location may indicate that the medieval church was built as a private church in the first instance. Compared to the church at Øn, however, the Hyllestad church had a more central location in relation to the local community and parish, which is also named Hyllestad. It is therefore possible that the church was built as a common church, a so-called *heraðskirkja* according to the church categories mentioned in Gulathing's Law (G 12), perhaps by several local farmers. One of the landowners may have donated a part of his land for its building, whereas co-owners may have contributed with timber and other services. This kind of joint venture is known, *inter alia*, from Heddal in Telemark (Berg 2005). Within such joint enterprises, the obligations towards the maintenance of the church and upkeep of the priest were probably also shared, and eventually the priest may have received parts of the farm, or farms, for his maintenance (Skre 1988:10). In 1880 the church was moved to the later municipal centre of Hyllestad (Aaberge 1929:26).

According to BK, the Øn and Hyllestad farms belonged to the local priests and bishop in the 14th century, and thus represent secondary ownership. Local dignitaries may, however, have been the first owners of the farms and churches, at least at Øn, before they were transferred to the ecclesiastical institutions after the parishes were established from the 12th century. The church of Hyllestad may possibly have been built as a community project. It is also interesting to note that the two churches temporarily seem to have been without priests. In an account register for tithes in the Nordic countries in the decades between 1282 and 1334, tithes from priests to the Papal See in 1327 were entered for only one of the churches at Hyllestad and Øn, but both of the incumbencies seem to have been unfilled (PN:34; Aaberge 1929:13). A few decades later, both churches are mentioned in BK with their priests (BK 25, 26).

Vincent Lunge

Rønset is the only farm within the quarry area that belonged to the aristocracy in the 16th century, listed in 1535 as one of the farms belonging to the Danish nobleman, Vincent Lunge, in western Norway (NRJ IV, 484). The land register from 1647 shows a large concentration of land in Hyllestad in the hands of the aristocracy, of whom Vincent Lunge was a major landowner in 1535. Most of his properties, however, were located outside the quarry area, on the southern shores of Åfjorden (cf. Fig. 8.1), Lavik *skipreide*.[72] Bø farm (no. 85) in Lavik *skipreide* also belonged to Lunge. The farm has been interpreted as a possible medieval baronial centre (Norw. *lendmannsgard*)[73] (Iversen 1999:43, 45). Whether or not

[70] Land was donated so that a *sjelemesse* could be held after the contributor was dead. This ritual prevented the contributor from remaining in purgatory (Hamre 1964:310-312).

[71] The priest Sira Gæirmund, who is listed as the buyer, was most likely the same priest mentioned in a register of 1322-1323, and who was one of the persons responsible for the bishop's revenues from Askvoll (BK: Introduction by Finn Hodnebø, pp. 17ff.).

[72] A *skipreide* was an administrative-military unit attached to the naval fleet (Norw. *leidang*), and most likely dates back to the Viking period (Helle 2001:35). In the western part of Norway this was the fundamental administrative unit throughout the Middle Ages, but the medieval *skipreide* had a somewhat different extent and borders compared to the early Modern period and the land registers from 1647 (Bull 1920:108, 142-143; Iversen 1999:8-9).

[73] 21 centres in western Norway are known to have belonged to local

such an assumption is correct, a cluster of farms without burial mounds was gathered around Bø, indicating that it was once a major farm of an estate (Iversen 1999:44). In the 1500s, the ownership of Rønset formed part of the same cluster of aristocracy property at Hyllestad, further south (Iversen 1999:43, 45). It is possible therefore that quernstone and millstone production at Rønset should be seen in connection with this estate (Baug 2002:96).

Vincent Lunge came to Norway in 1523 as a royal vassal of the new Dano-Norwegian State, and through marriage and donations managed to gain a considerable estate. In 1529 he was given the Nonneseter estates by the king (Sandberg 1970: 93–97). Documentary sources do not, however, specify the extent of the Nonneseter property, and what was obtained through his wife. Whether or not the Lunge estate at Hyllestad previously belonged to Nonneseter, and thus part of an early royal donation, is not known for certain. If Bø farm was really a baronial centre in the 13th century and thus closely connected to the royal administration, it might explain why it was included in Vincent Lunge's estate (Iversen 1999:43-44). Consequently, Nonneseter monastery may also have possessed quarries in Hyllestad in the Middle Ages, including Rønset.

The other possibility is that the Lunge estate at Hyllestad came from his wife, the noblewoman Fru Margrete, whom he married in 1523. She came from one of the richest families in Norway (Gyldenlöve/Austråt) and her mother was one of the heirs to the Losna estate in Sogn, with its manor on Losna island at the entrance to the Sognefjorden. It has been suggested that Lady Margrete may have received parts of the Losna estate as a dowry when she married. Some of the land recorded in the cadastre of Vincent Lunge may possibly therefore have been previous Losna estate property (Kellmer 1963:44; Sandberg 1970: 93-94, 96=97). If this is so, Rønset may have belonged to Losna in the late Middle Ages. The spatial distribution of farms belonging to the two estates (Lunge and Losna) was also closely linked. The Losna estate is mentioned in documents from the 14th century (Kellmer 1963:39, 46), but may have an older origin. After the Reformation two of the Hyllestad farms, Tonning (no. 62) and Kolgrov (no. 56), which belonged to Munkeliv in the 12th century, became parts of the Losna estate (Kellmer 1963:39). The possibility that these farms were properties given in return for requiem liturgies, which after the Reformation were given back to the contributor, has been suggested (Kellmer 1963:39; Baug 2002:94).[74] There is, however, no evidence for such returns of property, although Erik Rosenkrantz in 1565 is known to have used his right to claim a mass in Bergen (Kellmer 1963:39). If this was the case at Hyllestad, the Losna estate may have existed before 1175, when these farms were transferred to Munkeliv. This is hypothetical, however, as there is no mention of the estate before the late Middle Ages (Kellmer 1963:39, 46). As parts of these farms were donated to the monastery by the king in the latter half of the 1200s (DN XII, 257), it is more likely that they were parts of a royal estate in an earlier phase.

Losna, however, may still have had an interest in stone production. In the 14th century, the Losna farms of Hyllestad were subject to the main Losna property (no. 4) on Losna island (Kellmer 1963:34). Communication between Åfjorden and Bergen further south went through the inlet of Storakersundet, where the eastern part of the Lihesten mountains partly belonged to the Lunge estate, and the western part to the Losna estate, after the Reformation. Traffic also went through the small inlet of Krakhellesundet, between Losna and Solund. In this area the Losna estate was sole owner. All communications out of and into the Sognefjorden were controlled by Losna (Kellmer 1963:45) (Fig. 8.3). At least from the 1300s, therefore, Losna may have had some control of, and probably benefited from, the distribution of querns and millstones, without necessarily owning the quarries.

Another factor to be considered is the status and extent of the farms related to quarrying at different stages. The Rønset farm should be assessed in relation to the neighbouring farms of Hatlem and Berge. Norwegian farm names ending with the suffix *-set* are considered to be secondary farms, from the late Iron Ages, often subject to other farms within the local community (Stemshaug 1976:91, 107; Sandnes and Stemshaug 1976:31). This may also have been the case for Rønset (Baug 2002:101; Foyn 2008:28-29), located between two larger farms, where the farm name had the suffix *-heim* and an uncompounded toponym, indicative of the early Iron Age. There are no signs of co-ownership between Rønset and Hatlem, but co-ownership of some kind with regard to the outfield seems to have existed between Rønset and Berge, where several outfield areas were shared (Foyn 2008:35). In 1751 there was a border dispute between Rønset and Berge over the right to use herring *seine* in the bordering zone between the two farms. The border was clarified and as a result the quarries furthest east at Rønset were transferred to Berge (Rønneseth 1977:9). C14-analyses from arable structures at Rønset may indicate that the farm was established during the early Iron Age at the latest (Foyn 2008:92). It is possible, however, that it was established as a subordinate holding subject to Berge before it became a farm in its own right.

For the medieval ownership of Rønset, therefore, both Nonneseter monastery and the Losna estate stand out as possible owners of the farm in the early and high Middle Ages. It is however more likely that the farm was part of a royal donation to the Nonneseter monastery, probably based on even older structures before being royal land and monastic property. The king may have confiscated land in Hyllestad during the unification processes, land that later was at least partly donated to the larger ecclesiastical

barons (Norw. *lendmenn*) before 1300. The barons were granted their powers by the king and had fiefs (Norw. *len*) and revenues (Iversen 2005:134).

[74] During the Reformation in 1536 the King confiscated catholic estates, and persons who had donated land in return for requiem masses (Norw. *sjelemessegods*) received their properties back (Kellmer 1963:39).

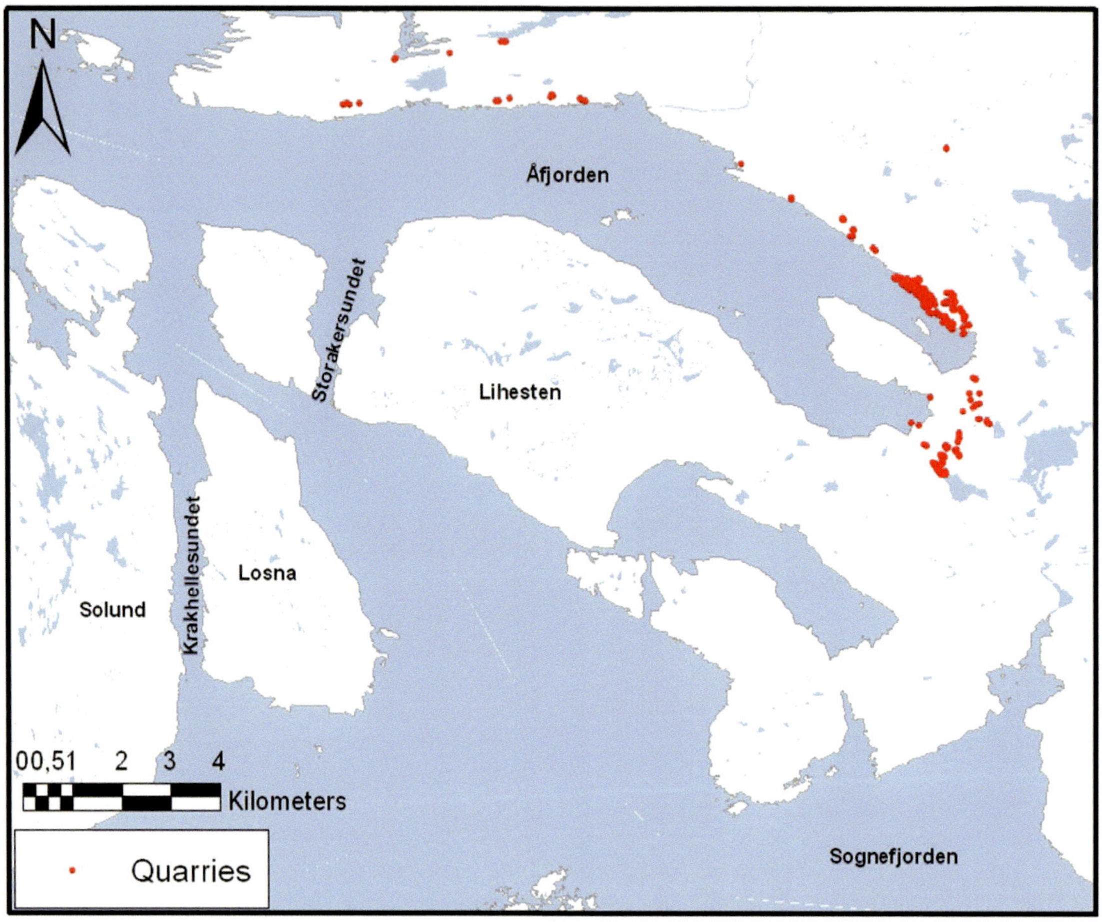

FIGURE 8.3: MAP OF THE INLET TO ÅFJORDEN (MAP: IRENE BAUG, 2013)

institutions. This may have been the case for Rønset (see Chapter 8.4).

Medieval disposal and quarrying

All the farms located within the quarry landscape therefore seem to have belonged to different estates in the Middle Ages, and also after the Reformation, and not to the local inhabitants. In the high and late Middle Ages the area was dominated by different ecclesiastical institutions, both local and central located in Bergen, but characterized by split ownership at this stage. It is possible that royal confiscations occurred at an early stage, as the king stands out as the major donator of farmland to the Munkeliv monastery in the 12th and 13th centuries. Production and distribution may in this way indirectly have been organized and controlled by landowners living outside the local community, but on a daily basis the work was undertaken by local tenants.

8.4 Viking Age estates in Hyllestad?

The earliest production in Hyllestad dates to the early Viking period, and quarrying went on prior to the ownership structure documented from the 12th century onwards. Production had started before the ecclesiastical institutions gained interests in the area. Is it possible, therefore, to discern earlier property patterns and shed light on the agencies who may have initiated production? From the middle of the 12th century, landowners obtained extended rights to donate land to ecclesiastical institutions (Bjørkvik1968:48),[75] As demonstrated above, documentary sources indicate a split ownership structure in the Middle Ages. However in some areas property structures indicate an early estate structure predating the medieval ecclesiastical estates. Here, archaeological evidence, especially grave mounds, may give further clues.

Since the Landlaw (and other records) refers to burial mounds as symbolic expressions of allodial privilege to ancestral farms, burial mounds may have reflected earlier rights to landed property. This view has been supported through several studies where the distribution of burial mounds indicates land rights (cf. Zachrisson 1994; Skre 1997; 1998; Iversen 1999; Iversen 2008; Ødegaard 2010). Frode Iversen's study of baronial centres and estates in western Norway in the counties of Sogn og Fjordane and Hordaland, substantiates a spatial connection between burial mounds of the late Iron Age and freehold farms in the Middle Ages and early modern period; the indication is of a general lack of burial mounds on medieval tenant farms

[75] Before 1152/35 landowners were only allowed to donate 10% of land they had acquired themselves; donations of allodium land was not allowed (F III, 17).

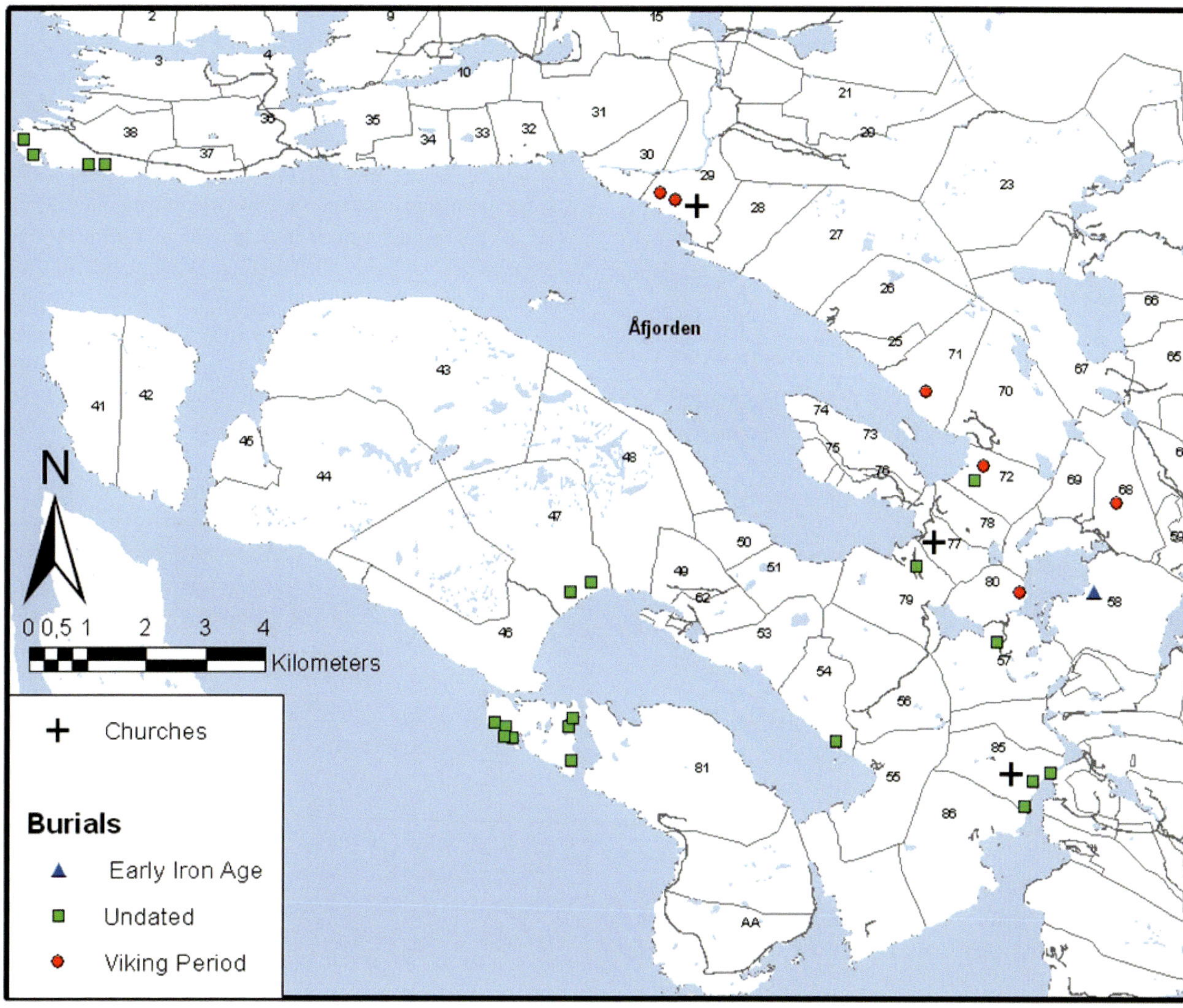

FIGURE 8.4: MAP SHOWING ALL IDENTIFIED BURIALS AND THE MEDIEVAL CHURCHES IN THE PARISHES OF ØN, HYLLESTAD AND BØ (BASED ON FETT 1954; FASTELAND 1983 AND HTTPS://ASKELADDEN.RA.NO). AS EXACT COORDINATES FOR MANY OF THE STRUCTURES ARE LACKING, THE SYMBOLS GIVE ONLY AN APPROXIMATE LOCATION WITHIN THE FARMS (MAP: IRENE BAUG, 2013, NORGE DIGITALT. FKB-DATA, 1413 HYLLESTAD KOMMUNE, 32_1413EIENDOM_FLATE, 28.02.2009)[81]

within estate areas. Furthermore, he shows that the core of an old estate, often marked by grave mounds, could be surrounded by farms with no mounds. This may, therefore, suggest a subordinate status related to the main farm where the landowner lived. According to this, analyses of burials and later property structures within this framework may be used as a method to trace prehistoric estates. This is a method used earlier in an attempt to trace early patterns of ownership structures in the Hyllestad production area (Baug 2002).

Obviously there was a difference between the medieval ecclesiastical dominance at Hyllestad from the late 12th century and the earlier ownership.

In the parishes of Øn and Hyllestad, burials and burial mounds from the Iron Age have been identified at 11 out of 48 farms (Fett 1954; Fasteland 1983), constituting c 23% of the total farms within the two parishes. Only at five of the farms have the burials been dated to the late Iron Age. This is the case for the farms at Øn,[76] Sørefjorden,[77] and possibly also Rønset,[78] within the quarry area (Fig. 8.4). Additionally, late Iron Ages burials were been identified at the farms Akse[79] and Kleive.[80] Except for the burial at Kleive, most likely all Hyllestad burials were marked with a grave mound (Dommasnes 1976:182-183), and may therefore be seen as indicators of freeholders in the late Iron Age.

[76] B 7649 and 7650.

[77] B 6642.

[78] A sword and an axe were found in a mound at Rønset in the 19th century (Top. Ark, Bergen Museum). The artefacts date to the late Iron Age, but the historian Ottar Rønnseth questions the find and claims the mound to be a spoil heap from quarrying activity (Rønnseth 1968:249). The objects are not included in Liv Helga Dommasnes' study of late Iron Age graves at Sogn (Dommasnes1976). I would suggest the possibility that the objects may be remains from a late Iron Age grave on the farm, as it is a more plausible context for such objects than a spoil heap. The location of the burial itself however is not known.

[79] B 9969.

[80] B 5866.

[81] The map is based on the aggregation of single sub-plots: 32_1413eiendom_flate (polygon) to cadastral units (named farm) by the dissolve-function in ArcGIS (Arne Solli, AHKR, UiB).

At Øn farm (no. 29), on the northern side of the Åfjorden, there are two burials from the late Iron Age and a medieval church (Fig.8.4). The surrounding farms have no visible graves or stray finds that could possibly signal a Viking estate. The neighbouring Hatlem farm (nos. 26–27) paid land rents to the priest at Øn in the Middle Ages (BK 29), while Salbu (no. 30), bordering Øn to the west, was owned by the church of Øn (BK 29), indicating that these farms were subject to Øn at an early stage. For the latter two neighbouring farms, Smilla (no. 28) and Sundbotten (no. 20), no medieval ownership is known, but in the land register of 1647 the farms are listed as properties belonging to the king and the priest of Askvoll respectively, and with no indications of medieval freehold (Skattematrikkelen 1647 XII: 157). At that time Øn was the second largest farm in Hyllestad and seems to have enclosed a wider area in the Viking period and in the Middle Ages, where the church looks to have been built as a private church. One of the burials at Øn, a Viking male grave, contained among other things an axe and 17.4cm-long chisel with part of a wooden handle, which may have been a quarrying tool.[82] However as it is doubtful that landowners actually worked in the quarries, the chisel and axe should rather be seen as symbols of status, and control, in this case related to quarrying in the area. The other grave at Øn, also a male, was provided with a sword, axe and *celt*, also symbols of status and rank.[83] There is, however, no certain information whether or not these two graves were ever marked with mounds. The archaeologist Bergljot Solberg has grouped Iron Age burials into three categories for social status, based on the number of weapons found in the graves. Graves furnished with one or two weapons, such as Øn, are considered to represent freemen and independent farmers (Solberg 1985:74).

On the eastern side of Åfjorden the ownership of land seems to have been different. Here the farms with burial mounds are more frequent, indicating ownership and property rights. Three of the farms have burials dating to the late Iron Age, located at Akse (no. 68),[84] Kleive (no. 80)[85] and Sørefjorden (no. 72),[86] all male graves with weapons (cf. Fig. 8.4). The burials at Kleive and Sørefjorden contained a sword and axe each, relating to freemen according to Solberg (1985:74). The burial finds at Akse were richer, containing two swords, two scabbards and one axe, indicating a high-status individual (Solberg 1985:74). All these farms belonged to the Munkeliv monastery in the 15th century, and a confiscation and later contribution, or donation, by the king seem likely (DN XII, 257). In the Middle Ages, the priest of Hyllestad also had a share in Akse (BK 25). The size of the farm, however, indicates that a part of the earlier complex may have had another owner.

[82] B7649.
[83] B7650.
[84] B9969.
[85] B5866.
[86] B6642.

A church was built at Hyllestad farm in the Middle Ages. As mentioned, the church may have been built either as a central community church or as a private place of worship by one, or several, independent farmers who may have given land to the priest for his livelihood. There is no burial mound at this farm although, as already mentioned, the neighbouring farms of Kleive, Akse and Sørefjorden have all revealed prominent grave finds. Burial mounds are identified at three other farms: Byrkjeland (no. 58), Myklebust (no. 79), and Atterås (no. 57). The burial mound at Byrkjeland most likely dates from the early Iron Age, but the burials at the other two farms have not been dated, and any artefacts found there are lost (Fett 1954:9). The farm names and burials do not signify subordinate status, and the area seems to have been inhabited by freeholders, with Hyllestad becoming a centre with its own church, and giving its name to the parish as well as the community as a whole. If the area on the eastern side of Åfjorden was freehold in the late Iron Age, it should be reflected in the property structure from the Middle Ages and the early modern period. This is not the case however. Byrkjeland farm (no. 58) was owned by a certain Orm of Folkestad in the 14th century. This character also had properties in the Vilnes/Skifjord area north of Åfjorden (Kellmer 1963:32), where he most likely lived at the eponymous farm. The extent of his property may indicate that he was a man of some importance. In the early modern period most farms on the north-eastern side of Åfjorden belonged to larger ecclesiastical estates, mainly Munkeliv (Skattematrikkelen 1647 XII; 159). A change occurred sometime before the late 12th century and probably later than the erection of churches, assuming they were not built by supporters of the king who gave the lands.

In the early Middle Ages, therefore, land within the quarry area belonged to local farmers, wealthy owners, and, at least temporarily, the king. It is plausible that local wealthy landowners or groups of freeholders may have built the churches at Øn and Hyllestad as private churches, which became parish churches during the 12th century at the earliest. The churches may therefore originally have been built by, and for, the upper levels of local society. From the middle of the 12th century the private landowners and founders of the churches gradually transferred their property rights to the churches as separate institutions (Molland 1968: 140). As a consequence of extensive land donations to ecclesiastical institutions from the middle of the 12th century onwards, new land was added and the property structure at Hyllestad changed dramatically. However, a possible royal confiscation of property during the unification processes complicates the picture. Whether or not the king also donated land to his own followers, who built the churches of Øn and Hyllestad, is not known. It is possible, and if so it may indicate that the ecclesiastical institutions represented at Hyllestad should be seen in connection with each other. It may also explain why different ecclesiastical institutions within and without the local community had property rights here.

Table 8.2: Medieval ownership structure at the quarry landscapes of Ølve and Hatlestrand

Medieval farms with quarries	Medieval ownership	Date of oldest recorded evidence	Part	Complete	Ownership after the Reformation
Ølve:					
Berge (No. 7)	King?	1563	?	?	freehold
Presthus (No. 5)	?	1563	?	?	the local priest of Skåla
Fugleberg (No. 17)	Halsnøy monastery	1535	x		shares were divided between freehold, Vincent Lunge and Halsnøy Monastery
Upper Tufta (No. 16)[1]	Halsnøy monastery	1600s	?	?	Halsnøy Monastery
Lower Tufta (No. 15)	local church of Ølve?	1600s	?	?	the local church of Ølve
Hatlestrand:					
Teigen (No. 26)	Skåla farm (no. 82), which may have been royal property	1600s	?	?	Skåla farm, Kvinnherad
Ask (No. 27)	local church of Skåla?	1600s		x	the local church of Skåla
Nettland (No. 29)	Bishop of Bergen, Olav Engelbregtson's estate	1500s	?	?	Crown in the 16th and 17th centuries
Hjortland (No. 30)	magnates, Lyse monastery?	1308	x		Lyse monastery, augmented by the Crown, and the cathedral of Bergen
Eikeland (No. 33)	magnates, Munkeliv monastery and the Bishop of Bergen, the King, Lyse monastery?	1300s	x		Lyse monastery, augmented by the Crown and secular aristocracy

[1] Upper and Lower Tufta most likely split from one farm at an unknown date, possibly in the early Middle Ages.

8.5 Property structure at Ølve and Hatlestrand in the Middle Ages

Several of the farms with quarries at Ølve and Hatlestrand were owned by different ecclesiastical institutions in the Middle Ages. Compared to Hyllestad, Ølve and Hatlestrand had a more divided property structure and no institutions appear as dominant owners within the quarry area (Tab. 8.2).

Halsnøy monastery

Halsnøy was an Augustine monastery allegedly founded by the nobleman Erling Skakke in 1163/1164 (Robberstad 1952:42), most likely in connection with the coronation of his son, king Magnus Erlingsson. According to the earliest preserved land register from the area (1614), the monastery owned two farms with quarries within their boundaries, Fugleberg (no. 17) and Upper Tufta (no. 16) (Robberstad 1952:52), with 46 and 7 quarries respectively. Some of the properties listed here may date back to the foundation of the monastery in the 1160s (Robberstad 1952:43), but whether this is the case for Fugleberg and Upper Tufta is uncertain. During the 18th century the former monastery property included only a part of Fugleberg farm and the quarries located there. This may indicate that the properties were not included in the foundation gift, but were probably acquired by the monastery at a later stage, either through purchases or gifts (Nysæter 2013:148).

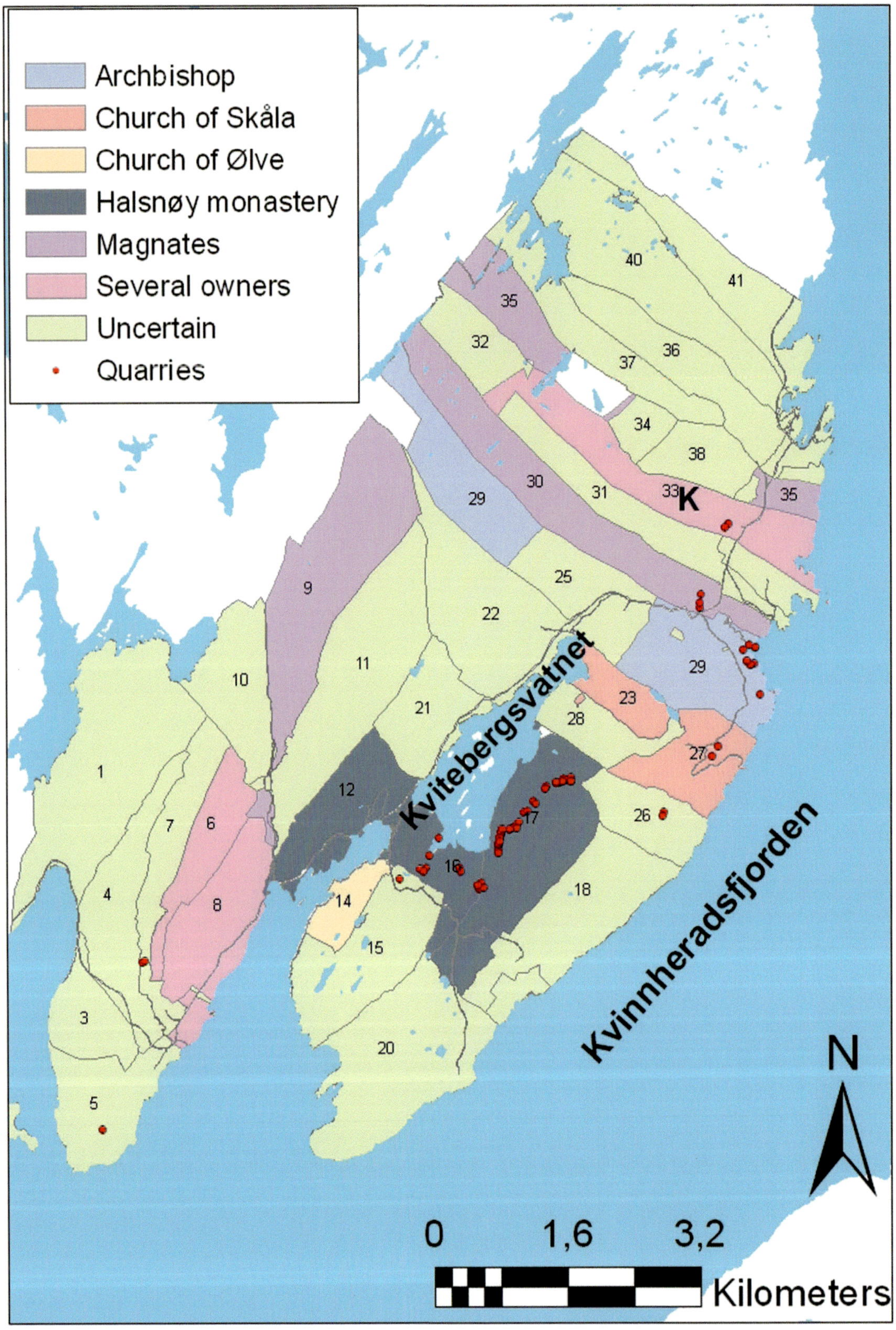

FIGURE 8.5: MAP OF THE MEDIEVAL OWNERSHIP STRUCTURE IN THE PARISHES OF ØLVE AND HATLESTRAND. THE FARM OWNED BY THE KING IS MARKED K (MAP: IRENE BAUG, 2013, NORGE DIGITALT FKB-DATA, 1224 KVINNHERAD COMMUNE, 32_1224EIENDOM_FLATE 21.11.2007)

In the 18th century one of the production sites at Fugleberg, *Helleberget*, occurs in two documents of farms and properties belonging to the estate of the earlier Halsnøy monastery, with an annual land rent of 1.2 *laupar* butter (18.5kg). Whether *Helleberget* refers to a specific quarry or a larger site with several quarries is uncertain. The name *Helleberget* may possibly refer to the largest quarry at the farm, Hedlebergshidlaren. The value of *Helleberget* was at that time estimated at 1 *riksdalar* and 48 *shilling* (Placat og Jordebog… 1744; Specification ofer Halsnøe-Klosters Goved-Gaard…1757), or about 15% of the value of the farm where the quarry/quarry site is located. The quarries at Upper Tufta are not mentioned in the written sources.

Vincent Lunge owned shares also in the Fugleberg quarry/quarry site of *Helleberget* (see below), which indicates dispersed ownership. In the 17th century a part of the farm was owned by wealthy farmers outside the local community. In 1614 Lars Tveit and Olav Munketun sold 12 *merker* of Fugleberg to Torstein Greipson Aga, all from Sørefjorden in Hardanger. This part constituted c. 1/6 *laup* or about 2.6kg of butter (Steinnes 1982:22; Vaage 1986:167), and represented a small share of land. Nevertheless it signifies that the farm and the quarries had different non-local owners but were run by local tenants.

There is little direct evidence with regard to the Halsnøy estate in the Middle Ages, and the information is mainly a series of reconstructions based on later documents from the 17th century onwards. Halsnøy monastery appears as one of the richest monasteries in the country, with properties concentrated in the southern part of Hordaland and northern parts of Rogaland: Sunnhordland, Ryfylke and North Jæren (Bjørkvik 1995b:148-149, 156). After the Reformation, when the estate was confiscated by the Danish-Norwegian King, it was organized as a fief. By the 18th century most of its farms were sold to tenants (Robberstad 1952:44).

Much of the rent from the farms at Sunnhordland was paid in animal products, butter, skin and leather, products that to a large extent were used as commodities for further sale. However, according to a document from 1626, rents to the old Halsnøy monastery fief also comprised six bakestones from a quarry in Kvinnherad (Bjørkvik 1995b:156-157). This shows an interest in quarrying from Ølve. Considering the large amount of bakestones that must have been extracted at Fugleberg and Upper Tufta, six bakestones represent a rather small amount. The general decline in land rent from the 1300s onwards may explain the low number, but it is also possible that the monastery only had small shares in these farms. Another explanation may be a possible decline in the extraction and use of bakestones at this time. In the 18th century, farmers at Fugleberg disputed over quarries within the farm and extracted slates for use on their roofs (Stuland 1924:536; Vaage 1987:177), indicating commercial activities in the quarries also in this period.

The place-name of Bussevika at the inlet at the farm Upper Tufta refers to the seagoing ship used in the Middle Ages, normally only owned by larger institutions, and may indicate that this was where bakestones belonging to the monastery were shipped from (Hansen 2013:187).

Munkeliv monastery

According to the cadastre of 1480-1490, king Magnus Håkonsson donated Eikeland farm (no. 33) within the quarry area to Munkeliv (DN XII, 257) in the latter half of the 1200s, meaning that the monastery gained land rents from the farm, or at least part of it. In 1369, a letter informs that the farm had been transferred to a wealthy farmer, who in return gave an equal value, that is to say 5 *mmb* (= 5 *laupar* butter = 77.15kg) from Haukanæs farm at Varaldsøy (Sunnhordland) to abbot Sigurd at Munkeliv monastery, with reservations that Eikeland should return to the monastery if the obligations were not kept up (DN XII, 104).

In 1338, Bishop Håkon of Bergen transferred the urban tenement of Fatten at Bryggen (Bergen), owned by the Bergen Cathedral to the law official Ivar Andresson, who paid from rural properties, among others, a value of 6 *mmb* (= 6 *laupar* butter = 92.58kg) for Eikeland farm (DN X, 33), presumably Eikeland (no. 33) at Hatlestrand. Munkeliv monastery cannot, therefore, have been the sole owner of the farm at this stage. The king may only have transferred a part of Eikeland to the monastery in the 1200s. Written evidence therefore reveals changes in ownership in this area, as well as a divided property structure during the Middle Ages. Eikeland, at least parts of it, changed owners several times and the King, as well as ecclesiastical and secular owners, held interests there. Nevertheless, none of the landowners mentioned were ordinary farmers, and the farm was shared between different larger and smaller medieval estates. In the 17th century the farm is listed as belonging to the fief of Lyse monastery, and two noblemen leased parts of the farm (Skattematrikkelen 1647 XI, 41). The quarries at Eikeland were evidently run by tenants during the whole period of production.

Olav Engelbregtsson

Netteland farm (no. 29) at Hatlestrand was part of Archbishop Olav Engelbregtsson's estate in the 16th century, listed under what is termed St. Olaf's land, allegedly *stigtgods,* denoting a former bishop's estate (OE: 9-11, 68).

Local priest and church estate

The local churches of Ølve and Skåla[87] are also represented among the landowners within the quarry area in the Middle Ages, according to the land register of 1647. The church at Ølve owned the farms of Lower Tufta (no. 15) and Presthus (no. 5) there, and the church of Skåla owned

[87] From the 1600s, Skåla church is referred to as the Kvinnherad church.

Ask farm (no. 27) at Hatlestrand (NG XI, 18-19; Kraft 1829:563; Skattematrikkelen 1647 XI, 55, 65). Since BK is not complete and the churches of Ølve and Skåla were listed on some of the missing pages (Hødnebø and Johannesen in BK: 14), it is difficult to be more specific. The three farms mentioned only had one owner in 1647 (Skattematrikkelen 1647 XI, 55, 65), the churches of Ølve and Skåla respectively, and evidently going back to the Middle Ages.

The local church of Ølve was built on Presthus farm (= house of the priest), obviously a secondary name. Presthus must have belonged to the church at least in the late Middle Ages (NG XI, 18), but probably earlier. As private landowners and founders of the churches gradually transferred their property rights to the churches from the middle of the 12th century onwards, this may also have been the case at Presthus. It has been suggested that the first church was built after 1640, and that Ølve earlier was part of the main parish of Kvindherrad (Haukenæs 1888:65). It is possible, however, that the church at Presthus was built as a private church in the first instance in the early Middle Ages. The farm and church have a remote location in relation to the local community as a whole, which may support such an assumption. Ownership of the farm in the Middle Ages is still somewhat uncertain, but was possibly the property of a local magnate. Five burial mounds have been found on the farm, indicating an early freehold (see Chapter 8.6). After the Reformation the farm belonged to the estate of the local priest, and the priest of Skåla is listed as the only owner (Skattematrikkelen 1647 XI, 64).

Skåla church, built in stone on the farm (no. 82), is first recorded in 1306 (DN III, 63; DN V, 159; NG XI, 21, 29–30; Kraft 1829:557). The style of windows and doorways indicates that it was erected approximately between 1275 and 1300, but probably with an earlier predecessor (Hoff and Lidén 2000:152-153), as has been documented in other central areas, i.e. Kinsarvik. Like Kinsarvik, the church at Skåla may have been one of four churches within the administrative *fjordung* in Hordaland county (Tryti 1994:35).[88] Teigen farm (no 26) and its quarries most likely separated from Skåla in the 18th century.

Vincent Lunge

In the 16th century Vincent Lunge (see above) owned shares in a quarry site at Ølve; a *pwndzboll* in Helleberget where bakestones were extracted (NRJ IV: 474). This is most likely the same site as the one owned by Halsnøy monastery. In the first half of the 1300s, one *pwndzbol* was equivalent to 1/3 *laup* of butter – 5.14kg or 5.4 litres (Steinnes 1982:76, 83). As large parts of Lunge's estate previously belonged to Nonneseter monastery, this may also have been the case for Helleberget. The Lunge property in Ølve also comprised part of a forest at Fugleberg and part of the now lost farm of *Arnwldz* (NRJ IV: 474). As there are no other traces of a previous Nonneseter estate in the area, and the fact that Lunge only owned parts of the farm, quarries and forest, it is perhaps more likely that these properties were acquired by Vincent Lunge himself, perhaps due to an interest in the quarrying and commercial logging that was expanding in the 16th century.

The King/State

The farm at Berge (no. 7) was held by the king in the 16th and early 17th centuries, but from the 1640s was owned by wealthy farmers outside the local community (Skattematrikkelen XI, 64; Stuland 1924:572). How far back the royal ownership stretched is uncertain. Royal ownership is also evidenced for Eikeland (no. 33) in the 1200s (DN XII, 257) and most likely also earlier.

Magnates

Two of the Hatlestrand farms, Hjortland (no. 30) and Eikeland (no. 33), seem, at least periodically, to have been owned by a secular aristocracy or magnates in the Middle Ages. In 1308 the 2 *mmb* (= 2 *laupar* of butter = 30.86kg) from Hjortland farm was sold by a Gudrun Thorkelsdatter (DN XII, 35). According to the same charter, she also owned farms in other areas and the extent of her properties indicate that she was a noblewoman, and evidently not living at Hjortland farm. In the 17th century Hjortland belonged to the Lyse monastery and Bergen Cathedral, and was farmed by tenants (Stuland 1924:507; Vaage 1987:344).

Lyse monastery

After the reformation the former Lyse monastery owned parts of Eikeland and Hjortland farms (Stuland 1924:496, 507; Vaage 1987:261). As the monastery was closed after the reformation in 1537, this ownership may go back to the Middle Ages. Lyse monastery was founded in 1146 and was a Cistercian institution. The monastery was closely connected to the Bishop of Bergen, who may have been involved in the foundation (Gunnes 1987:54; Gunnes 1996:135-136). The oldest cadaster we have for Lyse, however, dates to 1626 (Langgåt 2009:13), making it difficult to reconstruct the early medieval estate. As only parts of the farms were owned by Lyse in the 17th century, the land within the quarry area may have been transferred through random events, such gifts or sporadic purchases, rather than being the result of a larger donation by one individual owner.

Medieval disposal and quarrying

All the farms located within the quarry landscapes of Ølve and Hatlestrand belonged to different estates and not to the local inhabitants in the Middle Ages and after the Reformation. The king, a secular aristocracy, and

[88] Within the *Gulatingslag* (the area under Gulathing Law) the fjordung may have been the most important secular and ecclesiastical organisational unit. Fjording churches were administrative centres and important for communications between central and peripheral parts of the dioceses, as well as for an ecclesiastical system of tribute and duties paid in kind, involving large amounts of goods transferred to Bergen (Tryti 1994:35-36).

ecclesiastical institutions may have had properties here, but the ownership structure is unclear. As for Hyllestad, no clustered estate areas could be identified, and a split ownership structure seems to characterize the area. Landowners, both local and external, may indirectly have organized and carried out production and distribution, while daily activities were in the hands of local tenants.

8.6 Quarrying at Ølve and Hatlestrand

The quarries at Ølve/Hatlestrand date to the beginning of the Middle Ages, and here also quarrying began before any substantiated ownership structure. Quarrying activity seems to have started somewhat before the ecclesiastical institutions gained an interest in the area, i.e. from the middle of the 12th century onwards.

As demonstrated above, documentary sources indicate a split ownership structure in the Middle Ages, which may indicate smaller transactions, such as gifts and sales.

In the following sections the focus is on farms with quarries and a look into both archaeological and later property structures. Many prehistoric graves are undated, but most likely stemming from the late Iron Age, and are therefore included in the analysis. At Ølve and Hatlestrand, 13 of the 43 present farms have burial mounds (Fett 1956; Fett 1972) (Fig. 8.6),[89] i.e. about 30% of the farms, and relatively far more than Hyllestad. Nearly all the farms with marked burials are located along the fjords, indicating freeholds in this area. Four of the farms with grave mounds, Presthus (no. 5), Ask (no. 27), Netteland (no. 29), and Eikeland (no. 33) are located within the quarry area (Fett 1956; Fett 1972:122).[90] Most of the farms with quarries however have no associated burial mounds, which may indicate a subordinate status, as in the Middle Ages.

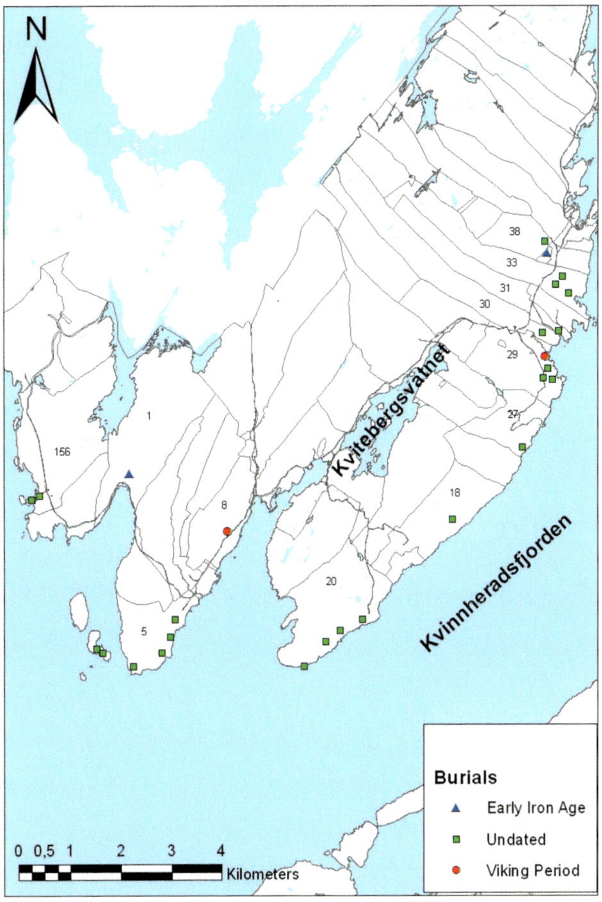

FIGURE 8.6: MAP WITH ALL IDENTIFIED BURIALS IN ØLVE AND HATLESTRAND (SOURCE: FETT 1956; E. FETT 1972; IVERSEN 1999:34 AND HTTPS://ASKELADDEN.RA.NO). AS EXACT COORDINATES FOR MANY OF THE STRUCTURES ARE LACKING, THE SYMBOLS GIVE ONLY AN APPROXIMATE LOCATION WITHIN THE FARMS (MAP: IRENE BAUG, 2013, NORGE DIGITALT FKB-DATA, 1224 KVINNHERAD COMMUNE, 32_1224EIENDOM_FLATE 21.11.2007)

Ølve and Hatlestrand are both located within the secular administrative area of Kvinnherad *skipreide* (Hatleskog 1997; Iversen 1999). By expanding the research area and including the whole *skipreide* in the study, it may perhaps be possible to come closer to the earliest property structure and therefore analyze the quarry area in a larger societal and socio-political context.

About 190 possible graves from the Iron Age have been identified (Fett 1956), with 40 that can be dated to the late Iron Age and 37 to the early Iron Age, located at 42 out of 87 possible medieval farms, representing c. 48% of all medieval farms (Iversen 1999:33-34). Here, most of the burial mounds are located at Rosendal (nos. 81-89), especially Skåla farm (no. 82) (Iversen 1999:34-35, 73).

The richest graves of Kvinnherad are found in the central farms (Hatleskog 1997:21, 25), but they are not surrounded by 'empty areas' without marked burials and archaeological finds (Iversen 1999:35), as at Hyllestad. This may indicate a different ownership structure from Hyllestad. According to the land register of 1647, most of the farms of Kvinnherad belonged to local aristocracy after the Reformation, with Axel Mowatt shown as the largest landowner. The estate appears as a cluster of property, but seems to derive from several earlier and smaller properties belonging to different noble families, the king and freeholders respectively. The cluster of property in the 17th century, therefore, does not date back to the Middle Ages. The local church at Kvinnherad, and churches in Bergen also, owned farms or shares in farms at Kvinnherad, but the royal land and property belonging to local churches and priests, as well as to freeholders, appear as relatively small shares (Iversen 1999:34–35). Overall, Kvinnherad seems to a large degree to have consisted of freeholds or perhaps small estates in the late Iron Age. According to the land register of 1647, Skåla, with its church, was one of the largest farms of Kvinnherad (Skattematrikkelen 1647 XI: 56; Birkenes 2004:122). The location of the church close to a grave field from the Iron Age, with burials from the early and late Iron Age, has been interpreted as continuity in cult location (Birkenes 2004:29, 117). The many burial mounds at the farm, most of them from the early Iron Age, indicate that this was a large and important farm, even at an early stage (Birkenes

[89] Cf. https://askeladden.ra.no.
[90] B 5775.

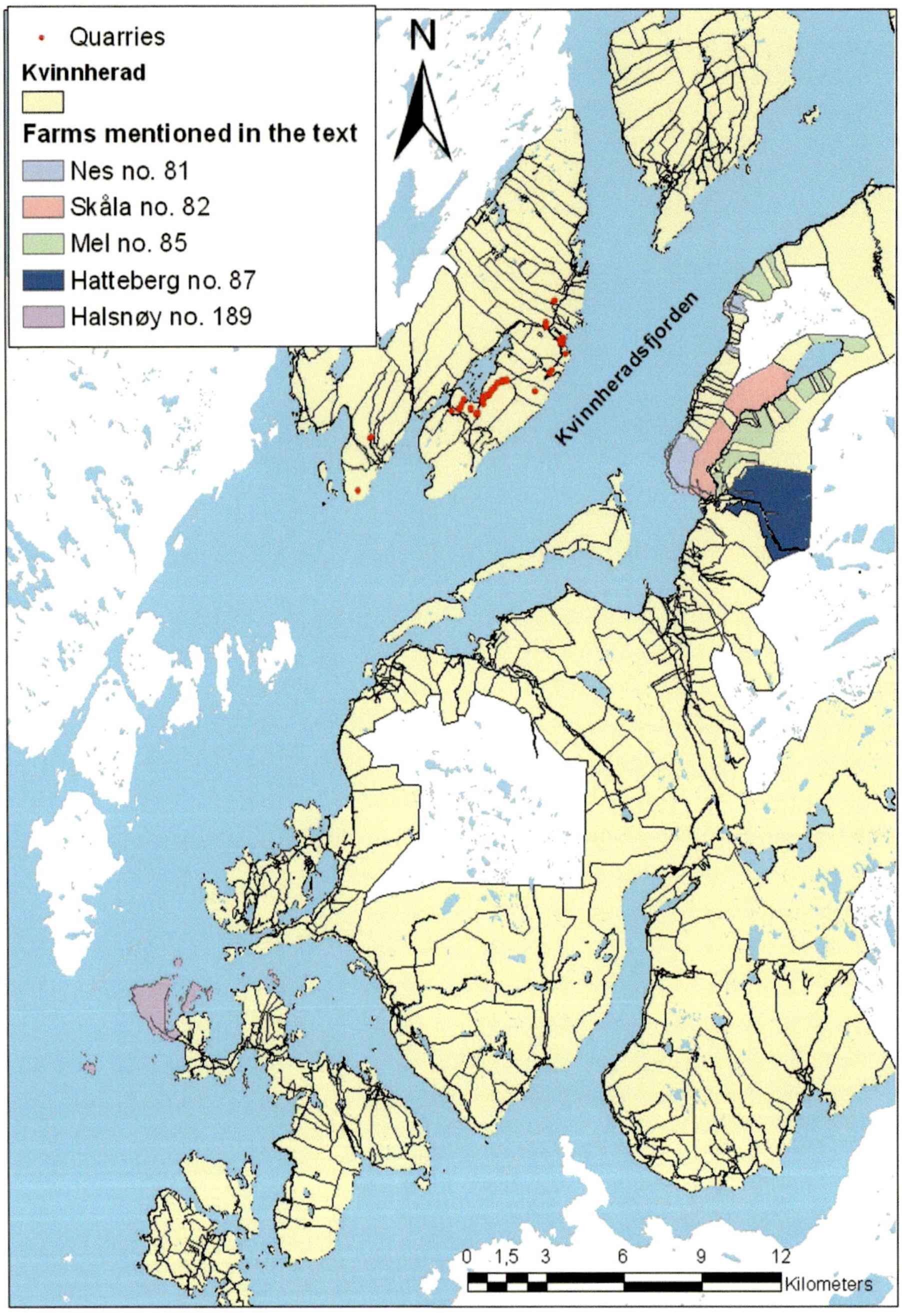

FIGURE 8.7: OVERVIEW OF KVINNHERAD SHOWING THE LOCATIONS OF FARMS MENTIONED IN THE TEXT (MAP: IRENE BAUG, 2013, NORGE DIGITALT FKB-DATA, 1224 KVINNHERAD COMMUNE, 32_1224EIENDOM_FLATE 21.11.2007)

2004:26–27). The name Skåla may refer to a guild hall (Norw. *gildeskåle*) where members of a guild had their meetings. It has been claimed that Skåla belonged to the king in the Middle Ages, because a burial mound called Årmannshaugen (årmann = royal deputy) is located at the farm (Bjørkvik 1995a:61–62; Birkenes 2004:120–121). This assumption has been questioned as no evidence can be found in later sources (Iversen 1999:35, 73). Most evidence points towards an environment connected to the elite. There is no clear evidence of a prehistoric estate structure around Skåla, and no connection to the quarry areas has been found, except for Teigen, which was most likely subject to Skåla in the Middle Ages and before.

Further southwest in Kvinnherad, on the island of Halsnøy (where there is a monastery), several burial mounds from the early and late Iron Age are located. One of the graves contained a sword, an axe and a hammer, representing a freeman and landowner (Solberg 1985:74).[91] The many and rich burials point towards a political and economic centre here from the early Iron Age onwards (Ringstad 1992:123–126; Iversen 2013). However, as the properties of the monastery within the quarry area seem to be later and smaller transfers and not the result of a foundation gift, there is no early link between the prehistoric centre at Halsnøy and the quarrying at Ølve.

[91] Cf. https://askeladden.ra.no; B1482-1482.

This indistinct ownership structure in the early Middle Ages, therefore, makes it difficult to distinguish all the agencies involved in the initial quarrying at Ølve and Hatlestrand. Local magnates and freeholders were present within the local community and quarry area. Most evidence points towards a situation where these agents disposed and controlled quarry areas. Compared to Hyllestad, more farms in Kvinnherad may have been freeholds prior to the Middle Ages, and local farmers or magnates may therefore have been those who initiated the quarrying in the first phase. However, for Hyllestad, it is possible that the king confiscated land in Kvinnherad and that later royal representatives contributed to the different medieval institutions. This is substantiated by the fact that the farms at Mel (no. 85) and Hatteberg (no. 87) on the eastern side of the Kvinnheradsfjorden became baronial centres, with lands given by the king in the Middle Ages (Iversen 1999:35), and that the king donated land to Munkeliv monastery within the quarry area.

8.7 Quarry control from the Viking period to the Middle Ages

Both quarry areas, particularly Hyllestad, therefore seem to have been subject to larger estates in the Middle Ages, as parts of the landowners' demesnes. Most evidence points towards a situation where political and economic elites – mostly different ecclesiastical institutions – had control of quarry areas in the Middle Ages from the 12th century onwards. Tenant farmers do not seem to have been able to use quarries for their own benefit in this period. The ownership structure at Ølve and Hatlestrand is more diverse than Hyllestad, but here also most evidence suggests a situation where ecclesiastical institutions dominated the area. For some farms, however, medieval ownership remains obscure.

At Hyllestad the different small estates seem to have been established as early as the Viking period, especially along the northern side of Åfjorden. Here the land seems to have been managed by farmers subordinate to Øn (no. 29), which may have been the residential farm of a local nobleman either in his own right or given land by the king. Local magnates may therefore have had direct control of quarry areas. On the eastern side of Åfjorden, marked burials indicate that farms to a large extent were based on a freehold system. At Ølve and Hatlestrand a more divided ownership structure seems to have been the norm in the area, based to a larger degree on freeholds compared to Hyllestad. Local nobles may still have had control of the quarries in the early Middle Ages before ecclesiastical institutions acquired lands, probably via the king in an intermediate phase.

The changes in control and disposal, and property structure generally, from the Viking period into the Middle Ages are difficult to understand, but different trends can still be traced in both quarry areas. In the initial stage, control and decision making seem to a large degree to have been carried out by local people within the communities where the quarries were located, although local elites probably initiated and organized activities. In the transition phase to the Middle Ages non-local agencies seemed to have gained access to land and income, probably through the king's intervention, during the turbulent phase of state formation and these lands were later transferred to ecclesiastical institutions. Quarrying activities may therefore have been controlled by social, economic and political elites, to a large degree located outside the local communities. The people who worked and lived at Hyllestad and Ølve/Hatlestrand were therefore not necessarily those who ran the work in the quarries, and who profited in these early centuries. How these relations affected the manner and administration of production in the Viking period and the Middle Ages, and to what degree the different owners and institutions were involved in production and trade/exchange of the products, is discussed in the following chapters.

9 Organizing Production

The aim of this chapter is to shed light on the work and organisation at the quarries under study. Landowners may have had influence upon the quarrying in various ways, both with regard to volumes and product types. Lack of written evidence of the activities makes it difficult to decide whether production was operated under a direct and dependent labour force or through different forms of tenancy. An issue to consider is whether ownership of the land involved more than the rights to collect rent, produce, and profits from those who ran the farms. Were the landowners themselves active in the quarrying with their own stonecutters? Was the production leased or in other ways rented to the land users; or was it to some extent a forced obligation? Were other workers involved?

9.1 State of control

Quarrying and those involved in it need to be understood in relation to the societal contexts of which they were part. The production era is decisive to help interpret the quarrying activities, as it was obviously not static. At Hyllestad production may have begun in order to satisfy local needs in the early Viking period. Most evidence points towards a situation where the actors involved were to be found within the local community. Hyllestad represents the oldest documented use of garnet mica schist for quernstones in Norway, and the oldest quarries for quernstone production; in this sense it was innovative. Most likely it was the local knowledge of the rock type and its properties, and local use, which later resulted in large-scale production and distribution. The output seems to be the result of local initiatives taken by local farmers or elites who had ultimate control over the land and its resources.

The increase in scale and character of production and distribution in the latter part of the Viking period bears witness to more developed activity and steady economic growth. Quarrying at Hyllestad evolved into a specialized proto-industry, most likely initiated by prominent men in the communities to increase production via the labour of subordinate farmers and perhaps, even, a captive workforce. A more or less direct involvement by the landowners may thus have taken place. As control over land during this period may also have included control over people, it is possible that quarrying was initiated and administered by the landowners and carried out by a more or less dependent workforce. Yet, not all of production sites needed to be parts of estates, and local farmers may have produced quernstones on their own lands, as part of their routine work. Only two of the farms, Myklebust and Rønset, reveal traces of more intensive exploitation, and quarrying on other farms was carried out on a smaller scale.

Similar organisation should also be expected at Ølve/Hatlestrand in the early Middle Ages. As the earliest dating from the quarries largely corresponds with finds of bakestones in urban contexts in the 11th century, the bakestones may have been produced for a larger market at an early stage.

A common assumption is that prior to state formation processes in the 9th and 10th centuries, the *veitsle*, where tributes and duties were paid in kind and services, was a well-established, institutionalized duty of the ambulant kingdom (Bjørkvik 1975:632; Iversen 2008:35).[92] Frode Iversen's study of royal estates in Hordaland in the late Iron Age and the Middle Ages may indicate a system where subordinate farms within the royal estates made contributions to central royal farms (Iversen 2008). This system is also referred to in Gulathing's Law (G 201, 278), but it was most likely completed and changed into a tax system by the 13th century (Bjørkvik 1964 (a and b): 425, 517). Iversen also suggests that farm names connected with birds, fish, sea-mammals, and grain within the royal estate may be traces of such contributions to the main farm with such products, suggesting a system where different farmers produced and delivered different products as tributes or duties paid in kind to the king, who was, then, indirectly involved in resource exploitation (Iversen 2008). This system of income from landed estates constituted the economic base for the king up until the beginning of the 13th century (Bjørkvik 1964a: 424–425). As the king in the early Middle Ages donated land to ecclesiastical institutions, duties also continued to be an important income for the new landowners, applied as land rents, tithes and other contributions.

The quarries may also possibly have formed an element of such a system, where quernstones and bakestones were produced as tribute for landowners. The system seems, however, to be an arrangement especially attached to older royal estates (Bjørkvik 1964a: 424–428; Bjørkvik 1975:632–633). Some quarries at both Hyllestad and Ølve/Hatlestrand may also have been held temporarily by the king. It is, however, uncertain whether or not common farmers were also obliged to pay such tribute to landowners at an early date.

[92] Two different forms of tribute and duty seem to have existed: one connected to accommodating the king and his men, and the other in the form of duties from farms and settlements subject to royal estates (Bjørkvik 1975:632–633). The latter form of duty may have applied to quarries.

During the high Middle Ages changes in the manorial system took place, from some form of personal dependency to a tenancy based on fixed rents and within an economic system. Personal contributions were replaced with land rents based on the size of land and output (Øye 2002:267). In the same way that land rent was normally paid in the form of products produced at the tenant farms, stones from the two quarry areas may have well have been accepted as rent.

A similar management of production and land rent is suggested for medieval mining activities at Näveberg in Södermanland (Sweden), where the whole activity has been described and explained as a feudal enterprise. Farms and smelting units were distributed among noble and ecclesiastical estates, where the landowners themselves were only indirectly involved. The mining may have been initiated by the nobility at a first stage, but organized and conducted by the tenants living at the farms. In the 13th and 14th centuries some of the finished products, including iron and copper, were given to the landowners as land rent. In the 15th century, however, a change seems to have taken place, with the transition to a monetary form of land rent. Tenants were now obliged to bring and sell their commodities at markets (Skyllberg 2001).

Germany has fuller early medieval documentary sources than in Norway, and one such source indicates that the monastery of Mendig at Eifel, studied by Meinrad Pohl, from the 13th century onwards owned several quarries, also some for quernstone production, that were leased out. Here, the stonecutters who ran the quarries were obliged to deliver a certain number of stones to the monastery, in amounts that seem to have varied. The documents also show that the monastery acted as entrepreneur, with quarry start-ups that were leased out even before they opened. The abbey was normally not directly involved in the quarry work but acted as the manager and owner and setting up the sites. In a few case the stonecutters seem also to have been taken into the monastery as lay brothers (Pohl 2012:77-78).

The monastery at Mendig was a Benedictine establishment, the same order as at Munkeliv and most likely Nonneseter too (Ommundsen 2010). The Benedictines were the first monastic order introduced into Norway, and the institution was based on far older traditions (Hommedal 1999:149). In the first phase, Norwegian monasteries seem to have been led by foreign brothers (Gunnes 1996:133). This makes it also relevant to look abroad when it comes to running and administration of their estates. Even though Halsnøy monastery close to Ølve/Hatlestrand belonged to a different order, the Augustines, there may have been similarities in the organisation of the quarrying. Quarries owned by monasteries may have been leased in the same way as seems to have been common practice in Germany. Production and distribution may well have been carried out by other agencies, with appropriate benefits for the monastery concerned. As mentioned earlier (Chapter 8.5), a document from the 17th century may indicate a similar running of the quarrying at Ølve, where rents were paid in different products, including bakestones. Such a system corresponds with what is known from medieval manorial systems, where the landlord leased most of the farms to tenants, who in return paid rent and fulfilled other obligations to the owner. It should be remembered that the owner normally retained control of resources in outlying areas. The tenants' use of such resources within a farm territory was normally limited to what was needed for normal agriculture production (Øye 2002:363, 366; Myking 2005:18-19).

Both at Hyllestad and Ølve/Hatlestrand the large-scale production from outlying resources have been examined, where the products formed part of wide-ranging trade networks. Bakestone production was rented out, at least in the early modern period (Bjørkvik 1995b:156-157). However, during this period bakestone production seems to have declined compared to medieval outputs, and changes in organisation of the work may also have occurred. As shown in Chapter 8.2.1, written evidence concerning the quernstone centres at Vågå and Saltdal also indicate leasehold of quarries in the 15th century. Production however appears at a smaller scale and based on more limited contact networks than occurred at Hyllestad. This also indicates different activities for production and distribution, where the agencies in Hyllestad operated within larger contact networks, demanding more control and a more developed system. It is therefore possible that the landowners of Hyllestad were more involved in overall production, and that people living on the farms worked in some sort of bonded system. Although quarrying may have been part of a tenant's duties, there is no documentation to confirm such a conclusion. Yet, to be able to run a large-scale production, a larger workforce would have been necessary compared to the normal agricultural activities of a farm, and also required agricultural surpluses to keep the workforce fed.

9.1.1 Landowner impact on quarries and products

Is it possible to see changes in production and quarry types after new institutional landowners had taken over, and a new impersonal system of tenancy developed in the early Middle Ages? If this was the case then it might best be seen at Hyllestad, because of the lengthy span of production documented there. The Hyllestad quarries are located in those areas with the most favourable geological conditions for extraction, and there are several examples of quarrying ceasing when inadequate garnet zones were reached. Systematic exploitation was conducted in the areas where the most suitable exposed rocks were found. Trial extractions are located along the shoreline of the Åfjorden and on exposed rock and in stream beds. The search for good quality resources must have been methodical, where quarrying was carried out within defined areas, mainly whenever geological conditions were most favourable, such as at Rønset and Myklebust. Some areas of similar quality further north-west were only exploited to a limited degree (Heldal and Bloxam 2007:23, 25, 59; Heldal and

Meyer 2011:328–330). The question remains therefore whether other factors, such as ownership and rights to land, were important conditions for quarrying.

Shallow quarries can date from the early Viking Age onwards, and represent an early activity represented over the entire quarry area. Production was mainly limited to stones for hand querns (Heldal and Bloxam 2007:49; Heldal and Meyer 2011:336). At some stage, however, a clustering of quarry areas characterizes the production landscape, with a strong focus on Rønset and Myklebust, where most quarries have been found (as wells as numerous shallow quarries and trial extractions). In the 10th and 11th centuries changes started to occur, with the introduction of a new quarry type (deep quarries) and an expansion of product types – millstones. These traces are very evident at Rønset and Myklebust. Deep quarrying involves more specialist operations and more intense land use, especially for the effective extraction of larger millstones (Heldal and Bloxam 2007:54; Heldal and Meyer 2011:335-336). These changes to the quarry landscape imply a change in activity (Heldal and Bloxam 2008:45) based on the switch from small- to large-scale exploitation. Could the change in production methods and the introduction of this new product type also be linked to changes in land ownership? The king and the church authorities seem, for instance, to have been involved in stone building construction and consequently in the quarrying of building-stone from the late 11th century, with a peak in the 12th and 13th centuries. It has been suggested that the deep quarry type was the result of a transmission of methods from other sectors (Heldal and Meyer 2011:335). This also accords with the traditional view that the technological development of watermills in Scandinavia was connected to the monasteries (cf. Fisher 2004:11, 25). Some of the oldest mills mentioned in medieval sources were owned by monasteries – arguably the earliest landowners engaged in milling by the end of the 12th century (Øye 2002:346).[93]

The millstone quarrying at Hyllestad should be seen as a response to growing demand, but extraction, however, began earlier than the establishment of the monasteries in Scandinavia (cf. Chapter 5). This process is further supported by studies of Danish mills. Previous investigations in Denmark suggest that mills were already operating there in the late Viking period (Kristensen *et al.* 1997:209-210). This has lately been substantiated by new investigations, where a mill at Ljørring has been dated to the middle of the 10th century, and wood used at a mill in the Viking Age settlement of Omgård has been dated by dendrochronology to AD 917, with a possible use-phase around 1000. Both mills employed vertically placed waterwheels (Fisher 2004:107; Jessen in press.). The stones found at the Omgård watermill were basalt (Jessen in press), and no millstones from Hyllestad have been dated this early outside the quarry area. Presuming the results from Denmark and the C14-dates from Hyllestad are correct, however, the earliest Scandinavian watermills documented, and the initial phase of millstone production at Hyllelstad, were not connected to monasteries. The introduction of watermills into Scandinavia represented a technological innovation based on specialized craftsmen, builders as well as operators, but who spread this technology, and how, in the Viking period is difficult to answer. During the Viking period, people from Norway and Denmark had close contact with both the British Isles and the continent, where the relevant technology had been used since before the end of the 2nd century AD and the early Roman period (Watts 2002:47). Therefore construction and use of early watermills was most likely connected to the upper strata of society, probably as mills close to their farms, and possibly on a communal basis. At a later stage the monasteries, as they gained foothold in Scandinavia, may have contributed to the dissemination of this technology.

The two Hyllestad farms, where deep quarries and the production of millstones are found, had different controlling interests. Myklebust was owned by the Munkeliv monastery, while Rønset ownership in its earliest phase is obscure: and may have belonged to a local noble, although Nonneseter monastery is also a candidate for ownership. It is interesting to note that two other farms within the Hyllestad quarry area (Gil and Berge) belonged to Munkeliv in 1175. However no deep quarries and no large-scale production have been identified here, only shallow quarries as well as some modern ones. At Gil, later farming activities may have destroyed the remains of a few quarries, and at Berge more recent production is likely to have disturbed the remains of some older quarries there. Nevertheless the absence of deep quarries and the few shallow types present indicate much smaller production here compared to Rønset and Myklebust. At Myklebust and Rønset the large-scale activities at the quarries have been documented in the late Viking period and the early and high Middle Age. The changes in ownership in the late 12th century do not seem to have affected the choice of production areas.

The output of stone crosses at Myklebust dates from the early and high Middle Ages (AD 1050-1245, AD 1175-1250 and AD 1225-1280), about the same time as the farm came into the possession of the monastery. As C14-analyses refer to the end-phase of activity at the investigated quarry site, the initial period of cross production may be somewhat older. As the extraction of slabs appears to precede the stone crosses, it also indicates that cross production began earlier – most likely before Munkeliv assumed ownership, which is also substantiated by the finds of standing crosses beyond the local community. Crosses also seem to have been extracted at Rønset, indicated by the find of a small and damaged, but undated, example (Baug 2002:67-68). Slab production at the farm in the high Middle Ages (AD 1285-1300 and AD 1220-1280) should perhaps be seen in

[93] Cf. DN XVII, 860 mentioning professional millers in 1209 from the monastery of Hovedøya. The oldest professional millers mentioned (1206) were from outside Bergen (Helle 1982:430). In the early 15th century four mills in this area were owned by the Munkeliv monastery (DN XII, 190), but how far back this ownership dates is not known. The mills are not mentioned in the monastery's cadastre of 1175.

connection with production, at least of small crosses, found at Rønset; but grave slabs may also have been extracted here. If Nonneseter monastery owned the farm for a period, two different monasteries may have been involved in the extraction of crosses at Hyllestad. The dating does not, however, indicate whether they were the party initiating this production. As discussed in the previous chapter, the king seems to have taken possession of land at Hyllestad before offering property to Munkeliv. This may have also been the case for Rønset and Myklebust, with the later royal donation of the farms to the Munkeliv, and possibly also the Nonneseter, monastery. The king may therefore have been involved, directly or indirectly through representatives, in the initial phase of cross production; this is also a plausible interpretation when considering their later distribution in the region and at strategic places along the coast.

Most of the evidence suggests a situation where the medieval landowners at Rønset and Myklebust acquired an already established activity, where quernstone and millstone production, as well as stone cross output, were already being undertaken. Large-scale production at the farms increased during the Middle Ages, but the medieval landowners did not expand activity beyond the borders of the two farms, even though the local geology would have permitted this.

9.2 Professional or seasonal workers?

When the quarries were worked and how many workers were engaged at them are important but difficult questions to answer, both for Hyllestad and Ølve/Hatlestrand. The quarrying documented in both landscapes must have been labour intensive. An important premise for major exploitation must have been access to an adequate workforce. Experiments indicate that a stone for a hand quern could take between one and three days, depending on factors such as the hardness of the stone, the skill of the stonecutter, and the quality of the tools (Løland in press.).[94] Bakestones may not have required as much time, as they are thinner than quernstones and the rock softer and easier to carve than the garnet mica schist of Hyllestad. The large amounts of broken quernstones and bakestones found in the spoil heaps also indicate that failures often occurred during production, which obviously would lengthen carving time.

The objects produced and their distribution from both quarry landscapes can be divided into two types: (1) serial production of food-processing commodities for daily household use and intended for specific markets (i.e. quernstones, bakestones and possibly millstones); and (2) production and distribution of more specialized and exclusive objects on special demand (i.e. stone crosses, grave slabs and building stones). The first group of objects was produced for a wide and unspecified market, while the latter seems to represent commissioned products for special end-users and societal elites. Production of these more exclusive objects may have been initiated by orders and instructions from high-status customers and produced on demand. The two groups of products must have been subject to different forms of transactions, and different levels of control can also be expected (see Chapter 10). The output of Hyllestad crosses may have been influenced externally, but those in control of the quarries must have contributed to the extraction, distribution, and final use of the crosses.

A common assumption is that stonecutters who produced commodities for an unspecified demand were independent specialists, whereas those who produced goods for social elites, or governing institutions, are regarded as associated specialists (Brumfield and Earle 1987:5). If the quarries were part of a system where quernstones and bakestones were tributes to the landowners, the actors involved in the extraction of these products may also have been attached specialists, obliged to produce the goods for the landowners. From the 12th century onwards, the different products may have formed part of the land rents that the tenants, or the others who ran the farm and/or the quarries, had to provide to landowners. A similar quarry system should also be taken into account for those quarries at both Hyllestad and Ølve/Hatlestrand belonging to the local priest and church estate. Study has shown that quernstones and crosses were extracted from the same site, as is clearly demonstrated at Myklebust (trench 6). Here the archaeological investigations indicate that quernstones were the main output, and crosses constituted a higher-status and more restricted product. This double production is also evident on one of the Hyllestad crosses erected at Eivindvik, near the old Gulathing site, where the foot of the cross has circles indicative of quernstone production (Baug 2002:68-69). This implies that the same stonecutters were involved in the extraction of both crosses and quernstones. The same situation is documented at Ølve/Hatlestrand, where building stone was extracted from the same quarries as bakestones.

Knowledge with regard to material, quality of minerals, and extraction techniques, were undoubtedly important, but probably not more difficult than any experienced person would be able to make, without being a specialist.[95] Consequently, workers with diverse backgrounds and positions – children and adults, farmers, tenants and slaves – could possibly be engaged in the quarrying. A certain operation and hierarchy or specialization of workers, including more skilled stonecutters and their assistants, would have developed. Others may have been responsible for different functions within the quarrying operation, where the most skilled stonecutters evidently oversaw the extraction of more complicated and special products, such as crosses. The skills of the workers would have been recognized and these individuals would acquire a certain

[94] In an experiment it took eight hours of continuous work to carve a quernstone for a hand-quern using Hyllestad rock (Løland in press.).

[95] Experiments at carving quernstones and stone crosses in Hyllestad indicate that no specialist skills were needed. 13-year-old schoolchildren can carve both quernstones and stone crosses by following some simple instructions and guidelines.

status within the community (Baug 2002:102, 104). Such a division of labour is known, for example, at certain quarries producing building stone in the Middle Ages (cf. Salzman 1967:119-127; Gardelin 2006:42).

Most workers at Hyllestad and Ølve/Hatlestrand were probably not full time stonecutters. Quarry work may well have been conducted along with the other farm tasks where they worked and lived. The marginal production sites within the two quarry landscapes never reached large-scale production, and the extraction levels here suggest that the quarrying was undertaken by people working and living at different farms, where they would have been tenants and semi-free workers, or slaves, in the first phase of the quarry as a seasonal or part-time operation, depending on the period and site. The medieval sources do not give a basis for assessing the size of an ordinary household. This would have varied according to the extent of the property, where larger farms involved more workers, and which also makes it difficult to further estimate the numbers of workers at the quarries. An average of four to six individuals in a nuclear family has been suggested, based on a household structure of the 1600s (Helle 2001:116); a situation where at least half of the household maintained production may have been sufficient within the marginal quarry areas. Before the farms were divided into smaller units and holdings, the households must have been considerably larger for running farms successfully. A large workforce in the quarries was probably not needed, and extraction may have been carried out during winter. Such seasonal production and semi-professional work probably reflects the situation at most farms of both Hyllestad and Ølve/Hatlestrand, but circumstances may have changed over time.

Within the central production areas, the situation would have been different. Extraction here was at a larger scale and demanded different organisation and a larger workforce. Rønset and Myklebust clearly stand out as the central production areas of Hyllestad, with nearly 70% of all quarries identified. Within the Ølve/Hatlestrand area, Fugleberg assumes a similar position, with more than 50% of the quarries (46 out of 71), even though large-scale production is also documented at Netteland and Upper Tufta. Lack of dating material from most of the Ølve/Hatlestrand quarries, however, makes it difficult to assess the temporal aspects with regard to quarrying and scale in this area. Whether a big quarry equates with large-scale production over a short period of time, or if there was a more restricted, but longer-lasting, extraction period, is not known for most quarries.

Large-scale extraction within the central areas of both production landscapes was evidently more intensive compared to the marginal areas. This indicates that some settlements of Hyllestad and Ølve/Hatlestrand had been, at least partially, more developed in quarrying. Here also those who worked in the quarries probably lived and worked on the farms. Their knowledge of stonecutting and quarrying operations, i.e. how to organize quarry work, must have constituted an important and transmittable asset. Although annual production volumes cannot be estimated, the numbers of quarries and sizes of the quarry areas indicate that the households would include hired workers, and perhaps slaves or semi-free farmers in the earliest Hyllestad period. Those with the rights to extract stones – whether they were hired stonecutters or tenants – would have been expected to deliver a certain amount of stone to the landowners – as is later known from Vågå (cf. Chapter 8.2.1). A large workforce also demanded suitable logistics and a certain level of operation at the quarry sites, and a presumed hierarchy within the workforce. Production in the central areas may have been conducted all year round, but more intensively in winter, as is witnessed by the large-scale millstone production at Selbu in Sør-Trøndelag (Rollseth 1947). This clearly shows that 12-month production was unnecessary for intensive exploitation aimed at larger markets.

It is also possible that large-scale production was rented out to agencies other than tenants, as is documented at Laach monastery in Eifel (Germany), where the quarries were leased to stonecutters who organized and operated the whole production process and paid rent in stone products in the 13th century (Pohl 2012:77-78). The written evidence does not suggest that such was the case at Hyllestad or Ølve/Hatlestrand, but it cannot be excluded. This would perhaps explain the concentration of large-scale quarrying at only two of the Hyllestad farms, since there may have been restrictions as to where the stonecutters were allowed to operate. A similar situation may also have been the case at Fugleberg in Ølve where the most intensive production took place, even though different owners here seem to have owned shares in the same quarries, at least in the later phases of production.

There are, therefore, several possibilities as to how production may have been carried out and managed. Most of evidence suggests seasonal production and semi-professional operations, where quarrying was conducted by tenants or subordinate farmers as part of routine agricultural work. Large-scale production at both quarry areas was aimed at larger markets, demanding diverse resources to plan, transport and distribute a range of products.

10 Forms of Transaction

The questions now arise as to what degree the stones (querns, millstones and bakestones) represented finished and traded commodities, and whether the distribution was the result of other modes of transfer. To understand how distribution and transactions were run, researches need to focus on shifting societal conditions within wider theoretical frameworks. Distribution and transactions of products were integrated components of larger economic systems, but necessarily of a different character before the emergence of Scandinavian towns in the early and high Middle Ages. During the Viking period distribution took place in a mainly rural and proto-urban society, although the process of urbanization had already begun in southern parts of Scandinavia. This chapter focuses on how distribution was managed in different geographical zones and within varied economic, social and cultural contexts, including rural and urban sites in Scandinavia and the islands of the north Atlantic. Seen in a long-time perspective, it may also be possible to trace both changes and stability in trade and trade routes. The possibility that varied objects were subject to different transaction forms should also be addressed.

Not all products were intended for market, and stone crosses and building stone are not found within market sites and towns, only as standing structures within the landscape. Whether or not they were included in the same trade category as the common products needs to be discussed.

Investigations at Hyllestad have shown that the area represents the oldest quarry landscape for quernstone and millstone production in Norway, revealing new production forms both in type and volume. Querns and millstones represent well known and common products used all over the world, and the large-scale production of these objects at Hyllestad may have been influenced by other areas in Northern Europe; the widespread use of querns and millstones led to the long-distance and bulk trade of these products. Bakestones, on the other hand, represent a product of far more limited use, and should most likely be seen as a more marginal commodity than querns and millstones, with a more limited group of users. Stone crosses and building stone represent relatively exclusive products, and consequently had a more limited range with regard to distribution and recipients.

10.1 A Viking Age trade?

As shown in Chapter 5, quarrying activity at Hyllestad dates back to the Viking Age, with a date range from AD 715-890 and AD 730-965, and a few quarries may have started in the early Viking Age. The latter date is from a spoil heap covering an older quarry, meaning that production may be somewhat older than was indicated by the C14-analysis. However, as the date range is wide, a more precise date is difficult to indicate. The distribution of Hyllestad stones is only exceptionally documented prior to the 900s, and the first quernstones extracted may primarily have been locally used. As shown, few stones are dated more or less securely to the early Viking period outside the community of Hyllestad. One from Vestfod is C14-dated widely to AD 783-970 (Gollwitzer 2012:128-129). One quernstone of garnet mica schist from Ribe has earlier been suggested to date from approximately 800 or early 900 (Carelli and Kresten 1997:120), but this date seems to be too early (Feveile and Jensen 2000:20; Sindbæk 2005:149). One of the Hyllestad stones from Århus is, however, considered to date before 850[96] (Meyer *et al.* in prep.), which may suggest distribution from Hyllestad already in the early Viking Age, but not within a process of large-scale, developed trade. Quernstones of garnet mica schist are also absent from the two proto-towns og Kaupang and Birka in the 9th and 10th centuries.

A fairly wide-ranging distribution of Hyllestad stones first becomes evident in the latter half of the 10th century. Not least the presence of quernstones in urban environments and market places indicates a more organized and commercially-based trade in trading centres such as Sebbersund (Limfjorden), the early town of Århus, and, to a certain extent, Hedeby and Ribe (Carelli and Kresten 1997:121; Birkedahl and Johanson 2000:29; Birkedahl 2000:36; Baug and Jansen in prep. (d); Meyer *et al.* in prep.). The stones found date mainly to the latter part of the 10th century in the towns/marketplaces. Quernstones from Hyllestad mainly appear in urban and pre-urban contexts dated to the late Viking period, and in towns that emerged in the decades around 1000, such as Lund and Roskilde. The emergence of large-scale production at Hyllestad and the wide-ranging distribution of stones therefore coincide with the second wave of urbanization in the Viking Age. The signs of increase in productions from Hyllestad may have been stimulated by this development. However, while Hyllestad stones dominate in centres like Århus, Sebbersund and Lund, they only rarely occur further south

[96] The quernstone was found in a building which has been cut into by a later house, where a skeleton is dated to c. 820-900. In all likelihood both buildings were out of use when the dead person was placed there, indicating that the building where the quernstone was found (as well as the stone) date to the first half of the 9th century (Skov pers. comm. 2012).

and south-west, in Hedeby and Ribe, indicating different levels of distribution within the areas. In Norway, a few stones from Hyllestad are dated to the Viking period, which may indicate that the products were primarily meant for markets in southern Scandinavia and not for local use.

As seen in Chapter 8, production and exports concur with the general development, from local, small-scale to large-scale exploitation in Viking Age Scandinavia for the wider distribution of resources from outlying areas. These are to a large degree focused on essential staples – such as wood and timber, furs and antlers, iron, whetstones and soapstone vessels. Increased production provided surpluses for trade and exchanges along wider-ranging trade routes and within new networks for exchange (cf. Øye 2002:361-391; Skre 2008b:340; Resi 2008; 2011; Baug 2011; Tveiten 2012:243). Here, production and distribution of quernstones fit into a general pattern of resource exploitation and trade during the Viking Age.

Hyllestad stones clearly demonstrate distribution over wide areas by the end of the Viking period. They reflect some forms of contact, but the spread cannot always be understood in terms of trade, as is the case with the Hyllestad quernstones in Iceland and the Faroes. The small amounts of quernstones from Hyllestad dated to the Viking period in these areas suggests that they were brought there by the inhabitants themselves, perhaps even early settlers, and cannot have been exported as commodities on a larger scale this early, judging by the few finds. Shetland seems to have been self-sufficient in necessary resources for making quernstones, mainly producing for a local market, but with a possible export route to the Faroes. Quarries have been identified both in Mainland and on Yell, all in lower numbers compared to Norwegian production, and there are no indications of managed and large-scale production. The querns and millstones seem to be made wherever suitable rock, like the mica schist, was found (Baug and Jansen 2014; in prep. (a)). Medieval Shetland, a purely rural community, was mainly relied on its subsistence economy, even though fish and wool were exported (cf. Helle 1982:308, 310). Whenever new querns or millstones were needed, people extract them themselves, or they called upon more skilled stonecutters for the stones from different quarries. The latter has, in fact, been common practice on Shetland until the end of the last century (Stummann Hansen and Larsen 2000:114). The long-lasting and widespread production has also left its traces in the Shetland language. Garnet mica schist is still called *möllgröt* by the locals, derived from ON *grjót*, meaning stone, and *möll* referring to the mill. Local quernstone production in Shetland was obviously sufficient, and no imports were needed from Norway.

Norwegian colonization of Iceland, which generally took place in the decades between 870 and 930, is verified both archaeologically and in the Landnámabók, recorded in the 12th century. In this, a certain Sigurðr Svinhöfði and his son Herjólfr are mentioned as immigrants from Fjaler (Førsund 2004:13 with references), one of Hyllestad's neighbouring municipalities.[97] Allegedly Sigurðr Svinhöfði lived in a place named *Kvernvágaströnd* (*kvern* = Eng. *quern, vág* = Eng. *bay* and *strönd* = Eng. *beach*) (Hauksbók: 29; Landnámabók: 70). This place was earlier identified as an area called *Vågestrand* in Askvoll, named after the farm of *Vågene* on the island of *Kinn*, most likely because watermills were located there (NG XII: 285). As this technology has so far not been documented this early in Norway, it is more likely that *Kvernvágaströnd* refers to Hyllestad and its quernstone production, as suggested by the historians Finn Førsund and Ottar Rønneseth (Rønneseth 1977:23-24; Førsund 2004). The name may indicate a direct link between Iceland and the quernstone extraction of Hyllestad in the Viking period. The value of the Landnámabók as a source of colonization information is much debated, however, and there is nothing in the archaeological record to indicate distribution of quernstones from Hyllestad this early. As mentioned in Chapter 7, Hyllestad stones have only been found on one *landnám* farm, and basalt quernstones seem to be far more common this early. The colonization of Iceland took place before the large-scale distribution of Hyllestad stones began in the latter half of the 900s, and the early settlers in all likelihood had to produce their own querns from the local rock.

A study of Norwegian soapstone vessels in the North Atlantic region indicates a somewhat similar situation as for the quernstones. Even though the finds are more numerous compared to Hyllestad stones, the relatively low number of these goods in this region does not support the theory of planned exports from Norway in the Viking period (Forster 2004:351, 360). The main mode of distribution of Norwegian stone products in the North Atlantic region in the Viking Age was probably the native settlers, who brought their necessities with them as personal belongings. Some stones may have been exported, perhaps as exchanges whereby products went from hand to hand through a network of personal contacts, but there are no indications that a recognized commercial trade of these commodities within this area took place this early.

The scale and range of quernstone distribution from Hyllestad to nodes in southern Scandinavia reveals a somewhat different picture. Hyllestad stones have been found over large parts of medieval Denmark (cf. Carelli and Kresten 1997; Sindbæk 2005:148-150; Baug *et al.* in prep. (b); Grenne et al. in prep. (a); Jansen *et al.* in prep.; Meyer *et al.* in prep.), but concentrated in the eastern parts of the realm, northern Jutland and the islands and towards the Baltic, as well as Scania and Bornholm. Here earlier investigations have identified both quernstones and millstones of garnet mica schist in the medieval towns of Lund, Roskilde, Lejre, Visby, Helsingborg, Ystad, Åhus,

[97] *Fjalir* was the name of the south-western part of Sunfjord, but the exact borders of the area are not known. In Hauksbók, which is a part of the Landnámabók edited in the 1300s, *Fjalir* may have reached Sogn, comprising an area that now includes present-day Gaular, Fjaler and Askvoll. Åfjorden in Hyllestad has been suggested as the southern border of *Fjalir* (NG XII: 285).

Figure 10.1: Quernstone regions in southern Scandinavia: (I) garnet mica schist from Hyllestad; (II) Mayen lava from Rhineland in Germany; (III) schistose sandstone from Malung in Sweden; and (IV) gneiss from Lugnås in Sweden (Carelli and Kresten 1997: fig. 18).

Trelleborg, Malmö and Halmstad, as well as at different rural settlements, such as Kyrkheddinge and Önnerup in Scania and on the island of Bornholm. According to Carelli and Kresten, all stones of the garnet mica schist type within these areas come from the Hyllestad quarries (Carelli and Kresten 1997:114-117, 123). This pattern coincides with my own investigations, even though they also show that a few stones of different origin are represented, but on a minor scale.

Carelli and Kresten point out four different and just overlapping stone regions, where quernstones originate from four different areas: (1) garnet mica schist from Hyllestad; (2) Mayen lava from Rhineland in Germany; (3) schistose sandstone from Malung in Sweden; and (4) gneiss from Lugnås in Sweden (Fig. 10.1). Figure 10.1 may not, however, give a completely correct representation. As already noted, a few stones of garnet mica schist, besides the Hyllestad material, are recorded in Denmark, and more quarries than the two mentioned are known from Sweden, for instance in Sala in Västmanland (Zachrisson 2014). It should also be mentioned that basalt quernstones were exported to Denmark from the 700s, about 200 years before Hyllestad stones appear. As Sindbæk has pointed out, this northern border for the basalt distribution was then not limited by competing export areas (Sindbæk 2005:149). Developed trade in quernstones in northern and eastern parts of medieval Denmark first appeared in the latter part of the Viking period with imports of stones from Hyllestad.

From the late 900s onwards, quernstones from Hyllestad seem to dominate the market in some areas, and at Lund c. 90% of the stones (140 out of 155) are from Hyllestad, while only 2% (3 stones) are made of local Scanian stone (Carelli and Kresten 1997:121, 123-124). Nearly all the quernstones found in Århus, Tårnby, Mors and Thy came from Hyllestad, and this is the only Norwegian quarry area where planned exports can be documented to southern Scandinavia – from the late Viking Age onwards. A few of the Hyllestad stones examined in this study are dated to the Viking period or the transition phase between the Viking Age and the Middle Ages. However the majority of the investigated stones are dated broadly to the Middle Ages. A more precise date has not been possible for most of the stones. In some areas Hyllestad stones were clearly preferred to others, while in western parts of Denmark stones of garnet mica schist are rare. Here the Hyllestad stones meet an area dominated by basalt from the Rhine area, with its western line through Jutland, from the outlet of the Limfjorden in the west, along the eastern coast of Jutland to Hedeby. To the east the distribution of Hyllestad stones meets an area dominated by schistose sandstone, produced from Malung in Sweden (Carelli and Kresten 1997:123-125; Sindbæk 2005:148-150).

Hedeby and Schleswig appear as the only nodes within the 'basalt region', where Hyllestad stones are represented to a certain degree. According to Schön, garnet mica schist is of lower quality than basalt, but was easier to produce and probably also less expensive to import than basalt, which could explain its distribution (Schön 1995:107). This also accords with Pohl, who has measured distances between quarries and markets and tried to calculate transportation times and costs, which are then used as explanations for distribution patterns of quernstones from Hyllestad and Mayen. Even though the distance is shorter from Mayen to Hedeby/Schleswig than from Hyllestad to Hedeby/Schleswig, the sea-route and cargoes from Hyllestad may have incurred lower transportation costs compared to the route between Mayen and Schleswig, with its longer transport overland and the need for more transhipments (Pohl 2011; 2012:138–143). Pohl's approach is rather common within a market-economic school and thus criticised by the substantivists. Distance and transportation costs, of course, need to be taken into account to make long-distance exports possible and profitable, but costs alone would most likely not explain the distribution patterns. The socio-political context must also have been important for contact networks, trade and trade routes. Pohl's models may represent an important tool for the understanding of quernstone and millstone distribution, but still do not capture the dynamics and complexity attached to the pre-modern economies that such commodities were associated with.

Whether or not the quantities of finds indicate large-scale distribution from Hyllestad is also dependent on how often the stones needed to be changed in querns in regular use. There are many factors for quernstone durability, such as the quality and hardness of the stone, the speed of rotation, and how much the quern was used every day. The Hyllestad quernstones found in the North Atlantic region and southern Scandinavia demonstrate nevertheless two different distribution spheres in the late Viking Age. Whereas the distribution to Sweden and Denmark signifies a commercialized and 'routinised' trade of quernstones, with refined logistics, stones from Hyllestad are almost absent from contexts on the North Atlantic islands, despite the close connection between the islands and western Norway, and also compared to other commodities of Norwegian origin. As stated above, soapstone artefacts from Norway seem to be far more common in the North Atlantic than Hyllestad stones (see Forster 2004), and in Iceland, soapstone is the most frequently recorded imported material from the Viking Age, but still relatively low in number.

10.1.1 Organisation of the Viking Age trade

Trade and exchange could take place in many and varying social contexts, as well as be variable in time and place (Skre 2010:131). Reliable information is lacking on who offered what, and who needed what goods before the emergence of towns and markets. Trade was most likely run on an intraregional basis, while long-distance trade was less common. An individual wanting to sell something would use already transmitted knowledge to try and find buyers (Skre 2011:77–79), which would favour local and regional exchange. This may explain the limited production in Hyllestad during the early Viking period, as well as the limited distribution. Quernstones were probably exchanged with other commodities within the local area, although there are few such traces.

This model does not, however, explain the long-distance trade of quernstones from the end of the 10th century. The widespread distribution to southern parts of Scandinavia demanded a wider network and routinised practices of sailing long distances. The establishment of such practices or networks depended on several factors. Both merchants and merchandise needed to be protected and an organisation able to secure and protect goods and trading routes as well as markets was important. Such protection could be guaranteed by an institutionalized political authority based on certain military presence, but this alone would probably not suffice. Trade between strangers required commercial transactions embedded in social and legal norms shared by people over a larger area than covered by regional law or social networks. This could be achieved at centres such as stable trading sites (Gustin 2004:243, 248, 264; Sindbæk 2005:272-273; Skre 2007:451). Power to secure trade and protect trading routes and markets was therefore important. This could be done through formal trading connections between different areas, as well as political alliances. The increase in political power within the early kingdoms, with expanding trade channelled more and more through ports and towns over the period, can be observed. According to Sindbæk, Viking Age trade was managed within networks, and routes between different strategic points, localities or hubs, where large cargoes could be loaded or unloaded for further distribution (Sindbæk 2008:154, 155). Within this system, the necessary protection of trade and goods could be provided. This may have also contributed to the long-distance distribution of quernstones from Hyllestad from late in the 900s.

Geological and contextual analyses of querns and millstones from Danish sites show that nearly all stones studied originate from Hyllestad. The quarries thus seem to have been a major supplier of quernstones for several hundred years in areas in northern and eastern Denmark, most likely the result of organized trade. This also accords with previous investigations in the region (Carelli and Kresten 1997). Only rarely are quernstones from other quarry sites found, which most likely represents more sporadic distribution rather than a developed trade. The distribution of quernstones does not appear as accidental. The question is, then, if this geographical pattern indicates exports within specific networks. The distribution demonstrates a shared artefact type, but not necessarily a connection between the sites where they were found. Based on the quernstone material alone, no clear networks for the Hyllestad stones can be drawn. In this pre-urban phase of Norway, ships may have sailed more or less directly from Hyllestad to the markets of southern Scandinavia. Many of the stones have clear traces of use and indicate imports for local use, but as many of these sites represent market areas, quernstones were most likely traded further on as well. Therefore there may not have been direct contact between Hyllestad and all the centres or marketplaces where these quernstones have been found.

Hedeby represents the southernmost site of Hyllestad stones, and an intersection point between the distribution of Mayen lava and garnet mica schist from Hyllestad. Most quernstones from Hedeby are lava examples, and, as stated above, only 1% (49 stones) of the total assemblage (5875) is identified as garnet mica schist (Schön 1995:12). Even though the number is low compared to lava stones, the querns from Hyllestad appear as traded commodities in this important trading centre. The relatively small numbers of Hyllestad stones found at Hedeby, and later in Schleswig, do not indicate a large-scale and managed trade in the late Viking Age and the Middle Ages. Hyllestad stones seem only exceptionally to have been distributed to this area, and there is little to indicate direct contact between Hyllestad and Hedeby/Schleswig. It is more likely that the quernstones found at Hedeby were distributed through other Scandinavian trading sites.

Ships may have sailed between Hyllestad and Hedeby without unloading their stones at other harbours. Pohl suggests that a seagoing ship would take c. 26 days to sail from Hyllestad to Schleswig/Hedeby, at an average of 30 nautical miles a day (Pohl 2012:137). Normally, however, 3 knots is the estimated average speed for rowing, meaning that an oared vessel would travel 36 nautical miles a day. By sailing, a ship would be able to achieve 72 nautical miles a day (Morcken 1964:22, 38). At this speed it would be possible to sail between Hyllestad and Hedeby in a minimum of 11 days, depending on weather conditions, but it would still have been a major effort.

Larger cargoes of Hyllestad stones only were probably not transported to Hedeby, judging by the number of finds. Trade at Hedeby operated as part of a nexus of wider trade networks, which also included several of the other contemporary Scandinavian towns and market places (Sindbæk 2007:67-68). It is therefore possible that quernstones may have been reloaded at some of these sites to be distributed at Hedeby, with other commodities. Århus clearly had close contacts with Norway from early times (Skov 1999:603, 606), and Sebbersund on the eastern Limfjord also received goods from Norway to be traded and transshipped onwards (Birkedahl and Johansen 2000:29). These centres may have acted as gateways for various products from Norway, including quernstones from Hyllestad. Århus, with maritime access to Hedeby, was closer to Hedeby than, for instance, Ribe, which was dependent on routes overland (Sindbæk 2007:70), and basalt quernstones found at Århus are considered to have been distributed from Hedeby (Andersen *et al.* 1971:165). The contact between Hyllestad and some of the market sites mentioned, such as Hedeby, may therefore have been indirect, based on transhipments and redistribution. As urbanization in Norway started late in the 10th and early 11th centuries, and the proto-town of Kaupang went out of use in the 10th century (Pilø 2007:177–178), none of the Norwegian towns could have acted as gateways for the quernstones this early.

Studies of Viking Age trade indicate that long-distance trading networks were vulnerable to political conditions. They expanded during peaceful periods and declined in periods of conflict (Sindbæk 2007:71; 2008:150), resulting in changes in both intensity and routes. Yet established routines and traditions were important for the distribution of Hyllestad stones. Already from the 700s, eastern parts of medieval Denmark appear as important areas for goods from Norway, as large amounts of soapstone vessels were exported here (Jensen 1990:126; Sindbæk 2005:149-150). 200 years later, the same area was still important for Hyllestad quernstone exports. Despite chronological and functional differences between vessels and quernstones, they indicate long established contacts important for the distribution of products from Hyllestad. The export of both these commodities reflects traditional patterns of contact between areas that appear to have neglected trade with the Rhineland (Sindbæk 2005:149-150). Nodes along the Carolingian-German border, such as Hedeby and Dorestad, generally seem to mark the southern border of commodities traded from Nordic countries (Steuer 1987:187). The distribution of quernstones seems to have been run within defined routes and areas from the latter half of the Viking Age and were maintained throughout the Middle Ages. There are no signs that the development of different quernstone regions was initiated by difficulties or obstacles to further distribution. As Carelli and Kresten have pointed out, the eastern extent of what they define as the Hyllestad region also reflects the political border between the medieval Danish and Swedish kingdoms, which may also have operated as separate trade regions (Carelli and Kresten 1997:126).

Maritime connections and the development of larger cargo ships throughout the period stimulated large-scale production and commercialized exchange of quernstones, as well as other commodities, from the late Viking period in south-eastern Scandinavia. Specialized cargo vessels most likely appeared in the 800s, indicated by the single find of Ralswiek 4. From the latter half of the 900s there is little doubt that such ships were in common use (Crumlin-Pedersen 1999:15; Bill 2008:174). Western Norway stands out as an area for ship-building in the Viking Age, and the two ships Skuldelev I and 6, found in Zealand (Denmark), were in all likelihood built in Sogn around 1020/30 (Crumlin-Pedersen 2002:65, 68, 323). The introduction of slower bulk-carrying vessels without armed crew for protection also testifies to improved marine security and a certain integration between trade, towns and royal security (Sindbæk 2008:152).

Little is known of those who were involved in production and trade. The landowners need not necessarily have been central agents in trade, but they would have ensured and somehow ran the distribution of any of their surpluses to market (Skre 2008a:353). Landowners may have used their own servants as agents for distributing and selling their commodities, perhaps in a similar way as is known from the Middle Ages (see below). It is also possible that ships, crews, and independent traders were all hired. Another possibility is that tenants and farmers living and working on the farms over the centuries managed the quarrying and distribution of surplus products, perhaps cooperating among several local farms, as a common *félag,* where they received any profits based on their contributions. Production may in this way have been largely organized and carried out by tenants, with compensation to the landowners. Over generations the different families involved may have managed to establish contacts and networks they could regularly operate within. In this way the landowners were never directly involved in production and trade of the stone products themselves, even though they obviously had a great interest in the enterprise, as well as a certain income from it. Nevertheless, the delimited and extensive distribution area for quernstones indicates logistics and management on a level that would be easier for elites to establish. Distribution from Hyllestad may have been initiated by individuals with political allies, and could gain access to markets and routes within networks of trade and exchange. Landowners could themselves be part of larger alliances with connections to wider Scandinavian regions. The relatively limited production of good-quality quernstones in Scandinavia from the 10th century may also have encouraged agents from Hyllestad to become important suppliers within certain areas. Eastern parts of medieval Denmark had no large-scale and managed production of quernstones of their own, and Hyllestad stones more or less took over from locally produced quernstones.

10.2 Medieval trade: changes in forms of transaction?

From the 11th century and into the high Middle Ages (1150-1350), multi-functional towns serving as ecclesiastical and royal centres, as well as centres for handling different commodities, developed in Norway (Helle 2006:24). In an attempt to control the Norwegian and international trade and exchange, the king in 1299 issued an embargo against trade outside the towns, an embargo that was repeated throughout the Middle Ages. Trade and exchange between farmers was allowed, but merchants were obliged to bring their commodities inside towns for resale (Norske middelalder dokumenter: 218). This must also have affected the distribution of querns and millstones, as well as bakestones, which became common products in households.

The large numbers of stones of all types found in the towns signify a more or less regular trade. That Bergen became a centre for quernstone and millstone trade in the Middle Ages is testified both archaeologically and by the Urban Code of 1276 (Bl VI:8). Millstones, however, had a more limited use compared to the smaller quernstones and are more rarely found in urban archaeological contexts. Most of the quernstones found in medieval Bergen come from Hyllestad quarries, unsurprisingly considering the relative short distance between Hyllestad and Bergen. Expanding commercial activity in Bergen from the 12th century onwards may have motivated and made possible an even larger trade in Hyllestad stones than occurred in the Viking period.

A majority of the stones found at Bergen comes from the quarries of Hyllestad, and also a relatively large amount of quernstones from Saltdal. The oldest distribution from Hyllestad is dated to AD 1120s-1170/71, but from 1170/71-1198 onwards Bergen apparently becomes a market also for products from quarries of northern Norway. The volume of quernstones from Hylletad and Saltdal indicates organized distribution to Bergen in the Middle Ages, and the two quarry areas were most likely competing in some of the same markets. The location of the Hyllestad quarries by the outlet of the Sognefjord afforded a relatively close sailing distance to Bergen by boat of 62 nautical miles, making it one of the most convenient trading centres for these quernstones with a travel time normally of one day. The distance between the Saltdal quarries and Bergen is much longer, 560 nautical miles and taking about eight days, depending on the wind and sailing conditions. More investment in transportation would be needed. It was more time-consuming and perhaps also involved larger costs and greater risks compared to the transportation from Hyllestad. Nevertheless stones from Saltdal were distributed and traded via Bergen, but in much smaller quantities compared to Hyllestad. A wider distribution of quernstones from Saltdal is further indicated in the archaeological material in Oslo and Tønsberg. It is reasonable to see the distribution from Saltdal in connection with the expanding stockfish trade to Bergen in the Middle Ages. Stockfish and quernstones were probably transported on the same ships, and opened up larger markets for the Saltdal stones, through Bergen. From the towns the stones were distributed further by agents from the rural districts who came to exchange commodities. The provenance of the quernstones from the rural areas of Sogn og Fjordane and Hordaland counties largely correspond with finds from medieval Bergen, and the town could actually have acted as a supplier of quernstones to the medieval rural districts. However with regard to local and regional distribution, direct exports by Hyllestad to other areas cannot be excluded. Compared to Bergen, the medieval towns of Stavanger, Oslo and Tønsberg seem, however, to have played a minor role in the distribution of quernstones, as most stones within the rural districts in Rogaland and eastern parts of Norway seem to be of local origin.

Hyllestad and Saltdal clearly stand out as the two quarry sites with the largest distribution within medieval Norway, and in some areas, such as Rogaland and eastern parts of Norway, Saltdal stones took precedence slightly over Hyllestad products. As seen in Chapter 2, extraction at Saltdal is dated to AD 900-1190 and AD 1040-1270 (Helberg 2010:114; 2010/2011:9), and activity may have begun in the late Viking period. Viking production in Saltdal, therefore, demonstrates a wider exploitation of stone products, but on a small scale prior to the Middle Ages. At Saltdal, querns were most likely produced for local, and perhaps regional, demand in the Viking Age, and large-scale production and distribution did not take place before the Middle Ages. The quarries were further from the markets of southern Scandinavia compared to Hyllestad, making it more complicated and resource demanding to establish routes, necessary networks and alliances for effective distribution. However it is more likely that limited distribution was the result of small-scale production and that surpluses were insufficient for interregional trade this early.

Quernstone quarries have been recorded at several locations in Norway with a type of rock very similar to that found in Hyllestad, most likely dating from the Middle Ages onwards.[98] These sites were nearly completely excluded from trade and markets both within and beyond Norwegian territory. The almost total absence of quernstones from quarries other than those at Hyllestad and Saltdal suggests that most Norwegian quarries produced quernstones for local markets in the Middle Ages.

The quality of Hyllestad products is not considered superior to quernstones from other quarries of garnet mica schist, and the rock does not seem to be easier to extract compared to other Norwegian quarries (Løland in press). The location of the quarries by the fjord at Hyllestad made them easily accessible, giving a competitive advantage to inland quarries such as at Vågå and Lierne, as well as access further north with longer sailing routes. However this alone cannot explain why quernstones from Hyllestad

[98] Cf. www.millstone.no.

dominated. Well-established tradition and knowledge of how to organize transport and exchanges may also have given an advantage to Hyllestad over other quarry sites. Hyllestad operated within well-established trade routes and markets, perhaps also networks, and increased exports within these areas are indicated during the Middle Ages. Transactions in Hyllestad stones changed during the Middle Ages as new agencies and larger towns and markets became involved in the activity. Allies, routes and networks may have changed but they can be difficult to trace in the archaeological record. Nevertheless, the trading routes and foreign markets established during the Viking period were largely sustained, giving an advantage to the new owners of the quarries of Hyllestad. Hyllestad was the only Norwegian quarry site producing for an external market and with an organized trade abroad; at some locations in medieval Denmark, such as Århus, Tårnby and Lund, Hyllestad stones seem to have dominated the market completely. Long-lasting trade contacts were important to maintain trust and security (Gustin 2004:176, 249, 264), and cultural affiliations and foreign contacts were of vital importance for distribution.

Trade between Norway and England flourished during the high Middle Ages (Helle 1982; Nedkvitne 1983:507), but quernstones from Norway are not mentioned as commodities nor are they found in the archaeological record in the British Isles. Quernstones from Mayen were imported to southern and eastern England from the 700s, continuing into the 1700s, while Scotland and Ireland seemed to have been more or less self-sufficient (Parkhouse 1997; Watts 2006:1-2). This is an argument for the lack of Norwegian quernstones in England, but also transport costs would have been higher than from Mayen (Pohl 2011:174; 2012.139-140). Since ships already sailed between Bergen and England, distance and transportation costs alone cannot explain the absence of Norwegian quernstones in England. Already established markets may explain the confirmed trends in Medieval quernstone export, making it difficult to compete in England with quernstones and millstones coming from Mayen and France. Production in Scotland and Ireland may also have prevented other producers from entering this market. There was also extensive trade between Bergen and towns in Germany and the Netherlands during the Middle Ages (Helle 1982:324-329), but seemingly without any influence on the quernstone trade from Norway. New markets for Hyllestad stones were not established.

Bakestones were generally distributed over shorter distances than quernstones from Hyllestad. As shown above, only a few bakestone fragments, most likely from Ølve/Hatlestrand, have been recorded in Denmark, for instance Århus and Tårnby, dated to the 1300s (Andersen *et al.* 1971: 108-109; Larsen 2005a:377). There is, however, nothing to indicate wider exports of these products to southern Scandinavia. From about 1180-90, at the latest, Bergen had direct contact with areas around the North Sea as well as the Baltic (Helle 1982:323; Nedkvitne 1983), explaining the distribution of both bakestones and quernstones in this area. In Shetland, bakestones of Norwegian provenance, most likely from Ølve/Hatlestrand, date back to the 1100s (Forster 2004:170, 176), not long after production started in Ølve/Hatlestrand.

Bakestones seemingly had a more limited range than quernstones, mainly connected to the Norwegian cultural sphere, including the islands of the North Atlantic, because of different food traditions and habits. Food processing represents conservative aspects of culture, where traditions and availability, and also strategic decisions and social roles, were fundamental (Øye 2009:225), aspects that may have made it difficult for bakestones to become important outside of Norway. To use bakestones, bread and other meals had to be baked or fried, normally over a fire. Baking plates or frying pans of other materials, such as clay, iron, or other types of rock, could also have been used. Bread was also baked using other methods, sometime directly in hot ash or in ovens (Bergström 2007:129-139, 155-162). Alternative ways of cooking would have limited the market for bakestones and they never managed to gain a foothold in foreign markets. Bakestones were, however, used at Bryggen, in Bergen, after the establishment of the Hanseatic Kontor at Bryggen about 1360, and the utensils were here used in an environment dominated by foreigners with other traditions of food processing. Residue analyses of three bakestones from northern Norway dated to the high Middle Ages, have revealed remains of vegetable material of oily plants, milk fat, peas or beans, largely corresponding to bread found in late Iron Age sites at Birka, Gotland and Grobin (Reiersen 1999:74-75). Bakestones may therefore been used for activities other than baking flatbreads (Øye 2009:232).

10.2.1 Organisation of Medieval trade

As Bergen was the closest town to the production areas of Hyllestad and Ølve/Hatlestrand, it may well have been the main distribution centre, as is indicated in the archaeological assemblage there. Direct distribution between the quarries and receivers outside Norway most likely did not take place in the Middle Ages. Producers and landowners organized the transport of stone products to towns that then traded them on further. Both Norwegian and foreign merchants were involved in these further exports.

Documentary sources that appear more frequently from the late 12th century onwards give more information about such Medieval trading agencies. In the Middle Ages, larger landowners, institutions and social elites brought those products acquired as land rents to market towns for consumption and further trade (Helle 1982:330-337, 346, 354). This would also have been the case for large-scale outputs from their quarries. As both monasteries and local churches were now owners of quarries at Hyllestad and Ølve/Hatlestrand, they would have been directly or indirectly involved in the distribution of such outputs. Stonecutters - tenants or hired workers - would also need to sell or exchange their share of the stones for other commodities. This would have been conducted within the local community, even though it was not allowed under the law (Norske middelalderdokumenter: 218). It is possible that tenants, or others working in the quarries, were obliged to bring and sell the commodities at the nearest market, most likely Bergen, so as to convert the stones into other products, or money, to pay their land rent, as is known in connection with 14th-century mining production in Sweden (Skyllberg 2001). The activity, production and distribution, would have been run and undertaken by tenants, after compensation to the landowners, largely represented by ecclesiastical institutions. Already established contacts and networks from the Viking period may have been sustained by the same families well into the Middle Ages. The ecclesiastical institutions in the urban centres, however, to a large degree would have been involved in the distribution of their land rent commodities themselves. As previously mentioned, the Urban Code of Bergen from 1276 states that quernstones and millstones were to be sold at Stranden, on the western side of Vågen bay (Bl VI:8). This is the same area where the Munkeliv monastery was located, and it would have been easier for the monastery to control and administer trade near its location. As Munkeliv, and perhaps also Nonneseter, owned land within the central production area of Hyllestad (i.e. the farms of Myklebust and Rønset) quernstones and millstones were probably important land rent products for the monasteries. Ecclesiastical institutions also acquired tenement sites in the commercial areas of Bryggen in Bergen (Helle 1982:368), as reflected in several place names. According to documents from the 14th and early 15th centuries, one of the Bryggen sites was called Åfjorden, apparently named after the Hyllestad location (DN II, 97; DN XII, 133, 134, 148, 167, 171, 180; Helle 1982:227-231, 239), but later became known as *Jakobsfjorden* (Helle 1982:239).[99] The name indicates some sort of contact with Hyllestad and the Bryggen trade activities. Åfjorden is first mentioned in 1309 (DN II, 134), and the tenement must predate this record. Some of the documents referring to Åfjorden concern a sale of a part of the property, and Munkeliv monastery in Bergen first obtained part of it in 1399 (DN XII, 133, 134, 148, 167) and further extended it in the 15th century (DN XII, 171, 180). Munkeliv monastery owned several farms within the production area (see Chapter 8), which also connects Bryggen and Hyllestad. As the name of the Åfjorden site at Bryggen is older than the monastery's interests here, it may imply that contacts with Hyllestad started earlier.

Regarding millstone output, ecclesiastical institutions may also have been important recipients of these as well as being active in the production and distribution of the stones. In Norway monasteries appear among the earliest landowners engaged in milling on a larger scale, and often connected to urban centres with professional millers. Mills owned by ecclesiastical institutions were normally leased. In rural districts also, larger watermills were often run by professional millers, while the smaller mills were largely owned by the farmers themselves (Øye 2002:346-347). The locations of mills in Denmark also indicate a close connection to ecclesiastical properties, especially monasteries (Fisher 2004:11, 25). Even though millstones are found in towns and marketplaces from the Middle Ages, it is possible that they represent commissioned products. Millstones were produced and distributed in smaller quantities than quernstones, which were used daily in households. Fewer excavations of mills compared to settlement sites, where hand querns were used, will influence the data, but it accords with excavation results (cf. Chapter 5). Stones for hand querns in all likelihood represent the main product type from Hyllestad. Although the introduction of deep quarries to Hyllestad was important for the extraction of millstones (cf. Heldal and Bloxam 2008:51-53; Grenne *et al.* 2008:57-58), the main product was still quernstones, which were as extensively used even after the introduction of water mills.

The Norwegians sailed over large distances and their presence is documented in the North Atlantic, eastern parts of England, Denmark, Sweden (e.g. Gotland), Belgium, and several places in Germany from the 1100s and into the 1300s (Helle 1982:360, 364-367). King, magnates and ecclesiastical institutions were all directly involved in foreign trade, with their own ships and commodities. In some cases they rented their ships to others (Helle 1982:367-368, 398, 402). From the 1250s and into the 1300s it was common to use agents, such as servants, ship-crews and independent traders. This could be organized into so-called *hjáfélag*, where one party handed over commodities to a second party acting on their behalf. Profits were divided between the two parties. By partnership in a common *félag* a number of agencies took space on a ship for their commodities and shared the profits after costs. In other circumstances one party would own the ship and another the cargo. Written evidence also provides information about those traders who acted as distributors, transporting and selling the commodities of third parties without sharing profits. The merchant would at the same time sell his own commodities (Helle 1982:370, 372).

During the high and late Middle Ages, the North Atlantic islands became more integrated into the medieval trade than was the case in the 10th century, because of close contacts with Bergen. The town was a cultural, economic and legal centre for the North Atlantic region, and in the first part of the 14th century, the North Atlantic islands were subject to the *féhirði* (Eng. tax-collector) of Bergen - meaning that taxes were paid to the Norwegian Crown (Helle 1982:331). Even though inhabitants from the North Atlantic islands, from the 1100s onwards, travelled to Bergen to buy and sell, trade was mainly organized by Norwegian traders and ecclesiastical institutions in Bergen, sending annually several ships westwards with Norwegian commodities (Helle 1982:165, 360-365), most

[99] The oldest document mentioning the tenement site is from 1309, and the name is here written *Jáfjorðr* (DN II, 97), which in Middle-Low German has been translated to *Jakobsfjorden* (Brattegard 1943-1946:272-273). Later documents, however, refer to the tenement as *Afiordhin* (DN XII, 133, 134, 148, 167, 171, 180), which is considered to be the original name (Lorentzen 1952:65; Helle 1982:239).

likely including quernstones and bakestones. In a charter issued by king Håkon Håkonsson of 1217-1219, Halsnøy monastery was granted the same rights as the canons of the Bergen diocese to ship their land rent commodities to Iceland for sale (RN I, 430). This may also have included products from the Ølve quarries belonging to the monastery at Halsnøy, as Iceland, Shetland and the Faroe Islands are the only areas where bakestones, most likely from Norway, have been found in larger quantities (Hamilton 1956:183; Arge 1989:119; Smith 1999:127; Weber 1999:134-139; Forster 2004). Bakestones, however, were not only regarded a commodity for sale. During the excavations at Halsnøy monastery in the 1960s, the bakestones found indicated that the monastery inhabitants used them when preparing food.

In the Middle Ages, querns, millstones and bakestones were distributed from the towns, however these products are not mentioned in contemporary customs records of commodities exported from Norway (Nedkvitne 1983). Exceptionally bakestones seem to have been exported and their modest distribution to Sweden and Denmark can perhaps be explained as personal possessions distributed by travelling individuals who had bought them at other markets. This may also explain their absence from written records. Quernstones, on the other hand, present another picture. Archaeological remains of Hyllestad stones in southern Scandinavia indicate organized trade within certain areas from the Viking period onwards. It is possible, of course, that cheaper commodities in smaller quantities on board were not fully documented. The list of dutiable commodities is not systematic. Despite the large amounts of imported pottery found in medieval Bergen, there is also no written evidence testifying to these types of imports. However these lists are the best contemporary accounts of Norwegian exports and imports during the high Middle Ages (Helle 1982:305, 320-321).

The majority of the wrecked cargoes of quernstones and millstones found along the Norwegian coast involve such small quantities that they cannot have constituted the main cargo on board, and a function as ballast has been suggested (Nymoen 2011:68). Exports of these stones therefore probably constituted ballast for resale, which has also has been suggested for soapstone vessels (Forster 2004:344). Ballast has also been regarded as a factor in stone distribution in medieval England, for instance in connection with the reuse of Roman building stones (Buckland and Sadler 1990). By defining and using quernstones and millstones as ballast, customs duties may perhaps have been avoided. Saleable ballast, however, was unlikely to have been a primary factor for their inclusion as cargo, as they obviously were distributed to markets within specific areas in southern Scandinavia throughout the Middle Ages. *Umlandsfart*, where ships loaded various commodities at several different harbours, is documented in Norway since the 1200s (Nedkvitne1983:33, 52, 57), and seems to have been common into the early modern period. Evidence points towards a situation where commodities were transported to harbours close to the quarries and from there distributed (Nymoen 2011:70, 72).

As referred to earlier in Chapter 7, whetstones from Eidsborg were found in two cargoes containing Hyllestad stones (from Bukken, south of Bergen, and Alverstraumen, north of Bergen). Whetstones were important Norwegian exports from Viking times, having a wider trade network than quernstones: Britain and the continent also seem to have received Eidsborg whetstones (Myrvoll 1984; Livland 1992; Resi 2008:65). Whetstones were transported along the waterways of Telemark to the coast of the Skien district, where they were distributed by sea (Tveitane 1981; Myrvoll 1985:31; Livland 1992:24-32), many of them to Bergen. Similar to the quernstone trade, Bergen played a central role in the further distribution of whetstones from medieval times (Livland 1992:33). In the 1800s whetstones were also transported to the western parts of Norway across the mountain plateau of Hardangervidda, either by individuals carrying the stones themselves or by loading them onto pack animals (Livland 1992:33). The market of Kaupanger in Sogn seems to have been a centre for local and regional exchange, with connections to the inland districts of eastern Norway and Bergen. Products were transported to Kaupanger across the mountains from eastern parts of the country in the Middle Ages (Øye 1989:159), possibly including whetstones from Eidsborg. Whetstones may have reached Hyllestad from Kaupanger, even though no stones have been found in Hyllestad, or nearby areas, that indicate such activity. From here they would have been transported further by ships, together with cargoes of quernstones and millstones from Hyllestad. With more than 500 examples, the Alverstraumen assemblage is larger than the other cargoes found. It is possible that this ship was on its way from Hyllestad to Bergen, and was carrying quernstones and millstones, with whetstones used to wedge them in some way. The cargoes found to the south were probably return loads from Bergen and containing a variety of commodities. This would explain the reduced quantities of stones in each cargo.

Beyond Norway, not only Norwegian agencies but also foreign merchants may have been involved in the distribution of stones. Contemporary sources indicate that most of the trade in Bergen from the 1180-90s onwards was conducted by foreign merchants, especially from England and Germany (Helle 1982:364). However, according to 14th-century regulations, foreign traders were not allowed to sail north and west of Bergen, and Bergen had the monopoly on foreign trade further north (Helle 1982:354). A reloading of quern and millstones evidently took place in Bergen, and different actors were most likely being involved in the transportation to Bergen than to the further export. Finds of unused quernstones in building and harbour constructions indicate an abundance of quernstones, perhaps due to a certain sorting of commodities in the town.

From the end of the 13th century and into the later Middle Ages, the Hanseatic League dominated foreign trade

(Helle 1982:378-385, 730-736; Helle 2006:85). A few quernstones of basalt and tuff, most likely from the Rhine area, have been found in Bergen (Baug *et al.* in prep. (a)), and it is interesting that these appear mainly after the establishment of the Hanseatic Kontor at Bryggen in the late Middle Ages, and were probably brought there as personal belongings. German influence and dominance in foreign trade had little impact on the Norwegian quernstone trade. Hyllestad and Saltdal stones were still used in Norway. Outside Norway no change in the markets for Hyllestad stones is found.

10.3 Commissioned products

While quernstones and bakestones, and possibly also millstones, represent commodities aimed at larger markets, other products from the two quarry areas, stone crosses, grave slabs and building stones, represent categories with a much more limited output; they were not distributed via market places and towns (Fig. 10.2).

Distribution of Hyllestad crosses to the counties of Sogn og Fjordane and Rogaland indicate that the recipients were linked to the western coast of Norway. Hordaland, Sogn og Fjordane and Rogaland represented the initial territory of the Gulathing (*Hordafylke, Sygnafylke, Firdafylke* and *Rygjafylke*) (Indrebø 1936:25; Helle 2001:76-77), the core area of king Håkon den Gode (Håkon the Good). The establishment of the Gulathing in the early 10th century has been connected to a royal attempt to consolidate power in western Norway, and Håkon was probably the first king to try and introduce Christianity (Helle 2001:30, 46). The exact location of the Gulathing is not known, but Eivindvik stands out as the most probable site and two Hyllestad crosses have been found here. The crosses may have been erected as part of the process, used in connection with outdoor services before the construction of a church, which is first mentioned in written documents in 1327 (Helle 2001:52-61). A block of stone from Hyllestad is also found here, today used as a font in the church. The initial function is not known, but it may have been related in some way with the crosses, perhaps as a vessel for holy water (Helle 2001:59). Kristine Holme Gabrielsen argues that most evidence links the crosses to social elites; the king, or his representatives, are suggested as sponsors of the Eivindvik crosses (Gabrielsen 2007:242-244). King Håkon den Gode (ruled c. 931-961) had a close connection to England through his foster father, Athelstan, and from which country missionaries were brought to Norway (Helle 2001:34). These agents would have been familiar with the British tradition of erecting crosses. A king is also suggested as being responsible for the cross at Kvitsøy (Rogaland), an important harbour along the sea trade/travel route. It has been suggested that the cross may have been connected to this route and had a practical function as a navigation sign, as well as representing an expression of political power (Gabrielsen 2007:230, 242, 244). The pilot books from the 1500s, however, make no mention of a cross used for navigation, and in any event the cross is too small to be seen from any distance. Any symbolic or

FIGURE 10.2: THE C. 4M FREESTANDING CROSS AT KORSSUND, FJALER (SOGN OG FJORDANE). FROM ITS LOCATION ALONG THE SEA ROUTE, IT HAS BEEN SUGGESTED THAT IT ACTED AS AN AID TO NAVIGATION (GABRIELSEN 2007:196). THE STONE FOR THE CROSS COMES FROM HYLLESTAD (PHOTO: ASTRID WAAGE, 2013)

political role for the cross may have been intended for the sailors arriving in the bay (Elvestad 2005b:89). The king may also have been involved in the erection of the cross at Gard, near the strategic inlet of Karmsundet, on the sailing route to the south. The farm possibly belonged to the king and the cross is also suggested to be an early demarcation of Christian territory (Fyllingsnes 2000:79; Gabrielsen 2007:238). However none of the crosses are directly located at royal residences, and there are no confirmed links between the king and the erection of crosses.

Crosses were also placed at churches connected to the *fjordung*, one quarter of a county, as mentioned in the oldest sections of Gulathing's Law, dating at least to the early Middle Ages. The secular division of the *fjordung* within Gulathing's Law most likely coincides with the ecclesiastical *fjordung* known within the Bergen diocese in the Middle Ages (Helle 2001:77-78).[100] Two of the Hyllestad crosses are located at churches that seem to have been connected to the *fjordung*, Svanøy (Bru) and Vereide within Firdafylke, which represented administrative aspects within the diocese (Tryti 1994:34; Gabrielsen

[100] Cf. Chapter 8.4.2.

FIGURE 10.3: TWO MEDIEVAL CHURCHYARD CROSSES IN HYLLESTAD (PHOTO: IRENE BAUG, 2011)

2007:230). The cross at Svanøy has a runic inscription that gives a clue as to its originator, reading: *'þórðr lét reisa kross þenna ept (ir..')* (NIyR IV: 223), translated as '*Thur erected this cross after...*' Thur is unknown, but he was probably a wealthy farmer or noble. Svanøy (Bru) (no. 117) belonged to the Bishop of Bergen in the 1400s (DN XVI 303), but the owner prior to the bishop is unknown. It has been suggested that the *fjordung* churches were built on a communal basis, but probably following a royal initiative. The cross may have been erected as a common activity (Gabrielsen 2007:242), however, only one person is mentioned on the inscription, which indicates that it was erected as a memorial. It is a large cross, standing c. 2m high and 1.26m from arm to arm (Birkeli 1973:196). It is considered to have been directed towards the farm and churchyard (Gabrielsen 2007:198).

Landowners and elites are thought to have been responsible for erecting crosses in the main, but in a few cases, for example Loen and Rygg, a collaboration of farmers and magnates – a so-called *bygdeting* (assembly within the rural district) is likely: crosses were a symbolic substantiation of conversion to Christianity (Gabrielsen 2007:230, 248-249).

The seven small Hyllestad crosses found in the medieval churchyards (four at Hyllestad and three at Guddal; Fig. 10.3) may have been erected in connection with the consecration of the churchyards, and thus represent a second generation of crosses with a different function than the larger, freestanding ones. The crosses may have been ordered by the church builders. Both Hyllestad and Guddal were parish churches in the 1350s (Tryti 1994:34), however, as mentioned in Chapter 8, the church at Hyllestad was possibly built as a private or common church in the first instance.

The precise dates of the crosses are difficult to ascertain, but most evidence - in line with the C14-dates from Hyllestad, as well as location and runic inscription - indicates production in the early and high Middle Ages. The size and shape of the crosses found at the Hyllestad quarry site indicate that they were intended for different uses. Most are small and intended as churchyard crosses. The two largest crosses (BRM 928/1 and 3) may originally have corresponded in size to the smallest freestanding crosses, but this is difficult to confirm as the stem is broken.

A few grave slabs from Hyllestad have also been identified, but they all lack proper contexts and dating. The shape and size of the slabs indicate that they date to the Middle Ages and were commissioned by high-status individuals.

FIGURE 10.4: GRAVE CROSSES FROM SKÅLA CHURCH PRODUCED IN ØLVE/HATLESTRAND (PHOTO: IRENE BAUG, 2006)

The production of building stones at Ølve/Hatlestrand during the Middle Ages would have been connected to the elites. Using stone for building material represented a new technique in Norway at the very beginning of the Middle Ages. The technique was introduced from abroad, and most of the stone buildings were churches and castles, with the church or king as owner and builder. In a few cases, however, examples of churches built as collaborations by farmers and magnates within a rural district are found, such as the *fjordung* church at Voss in Hordaland from the middle of the 13th century (Ekroll 1997:11-13, 104-106; Helle 2001:78). Building stones from the quarries of Ølve/Hatlestrand have so far only been identified at Onarheim church in Tysnes (Hordaland). The church is mentioned in written sources from 1327 (DN X, 25), but it would have begun in the latter half of the 12th century (Hoff and Lidén 2000:267-268). Geochemical analysis confirms that the building stones derive from the Ølve/Hatlestrand quarries (Jansen pers. comm. 2009), indicating extraction of the building stone in the early Middle Ages. Onarheim farm belonged to local elites in the Middle Ages, and was settled by nobles with a close connection to the king (Håkon Håkonssons saga, chapter. 23 and 87). The church was no doubt built as a private church, but was being used as a parish church at least from 1347, when it is mentioned as such in written evidence. There was also a guild at Onarheim in the Middle Ages (DN IV, 316; Hoff and Lidén 2000:267). This emphasizes Onarheim's function as a religious, administrative and cultural centre in western Norway, and therefore linking products from Ølve/Hatlestrand to the upper classes in the Middle Ages.

It was important to trim the stone as much as possible at the quarry site in order to reduce the weight before transporting it. Buildings may have been planned in detail before they were started, and measurements and layouts needed could well have been sent to the quarry before production, as was the case at several medieval quarries in England (Salzman 1967:123).

In the 1900s, stone crosses used as grave crosses were also produced at the Ølve/Hatlestrand quarries (Hoff and Lidén 2000:166). These are ring crosses, where the cross is encircled in stone, similar to ring crosses found in eastern Norway, where several are considered to date from the early 1100s and 1200s (Fig. 10.4) (Nordeide 2009:164-165; 2011:129, 133). Whether or not the production of crosses at Ølve/Hatlestrand also date to the Middle Ages is not known. They have only been found at the churchyard at Skåla, where geological analysis has confirmed them as Ølve/Hatlestrand products (Jansen pers. comm. 2011). No remains are documented in the quarries, indicating small-scale production.

The crosses, grave slabs and building stones from the quarry landscapes in this study had a limited geographical distribution, with limited groups of users. However, with regard to building stones large quantities were needed.

This material was commissioned by the upper strata of society in western Norway, where the church, nobles, and perhaps even kings, represent the agents who ordered and authorized production. The recipients would have had specific requirements in terms of form and size and this may have been important also with regard to the intended use of crosses. The largest cross found at Myklebust (BRM 928/1) was 1.2m high, with a complete head and the foot partly preserved; it would be possible to erect the cross despite the damage to it. However the cross was clearly discarded, probably because its height did not comply with the specifications ordered. The products would have usually been transported directly from the quarries to those who had commissioned them.

As referred to above, millstones may also represent commissioned products, with agents and institutions having the rights and resources to build watermills placing orders for the stones. This may well have been the situation until the 12th and the first half of the 13th century, when there were few watermills. During the latter half of the 13th century smaller watermills with horizontal wheel became more common on farms (Øye 2002:345-346), and this would have affected production of millstones at Hyllestad. There would have been a development from production on demand, at the earliest phase, to serial production with a more regular trade and commercial exchange of millstones from the high Middle Ages onwards, when the stones were distributed through towns and marketplaces together with quernstones. This has been substantiated through finds of millstones in medieval towns, for instance Bergen.

Different products therefore would have required different forms of production and transaction. A network of buyers would have been important for the products commissioned from Hyllestad and Ølve/Hatlestrand, and production itself would be the eventual result of actions taken by elites, often outside the local community, most likely on a regional or even national level.

10.4 The medium of exchange

More than 90 coins have been found at Hyllestad, on the headland between the arms of Hyllestadfjorden and Sørefjorden, most likely at Gjerde farm (no. 76). The deposit contained English, German, Danish and Cufic coins, which are thought to have been buried around 1030 (Skaare 1976:102; Gullbekk 2006:41). The value of the coins is estimated at c. 2/3 *mark* silver (1 *mark* = 214.32g in the Middle Ages) (Steinnes 1982:24), not a large value by the end of the Viking period (Gullbekk 2006:47). The find can be seen in connection with the quernstone trade, used as a means of payment for quernstones.

Silver currency is known in Scandinavia from the early Viking Age, from the 700s (Hårdh 2008:97; Skre 2008a:346), but a monetary system was not introduced into Norway before the end of the Viking period (Skaare 1976:58-60). Coins were regularly used in towns from the middle of the 11th century, and in the agrarian countryside from the second half of the 12th century. Commodities were still used for payments however. From the 14th century, the Norwegian currency collapsed and the use of coin became less common and commodities resumed their importance as the normal form of exchange, also for towns, a situation that lasted until the 15th century (Skre 2008b:330-331).

Coins and silver, however, may just have been one of several means of payment for querns, millstones and the bakestones, and it has been argued that during the Viking period and into the Middle Ages silver and coins should be seen from a commodity perspective, rather than as representing a coin-based economy. The medium used for payment carried the value within itself, and both during the Viking period and the Middle Ages silver and coins were used as payments alongside commodities (Skre 2011:68, 70-71, 74). The values of animals and agricultural products varied, but some rates were more or less fixed in the Middle Ages - at least with regard to payment of land rents and tithes. One of the most important means of payments was a *kyrlag* – the value of a cow with a defined worth as regards age and productivity (Steinnes 1982:66).

The values of the Hyllestad and Ølve/Hatlestrand products are not known. A few sources give some indications regarding Vågå millstones in the latter half of the 17th century, when prices seem to have varied. A millstone was said to be worth 24 *riksdalar* at market, which - presuming the amount is correct - was a very high value. Sold locally, however, the stones were much cheaper. On one occasion the value of two millstones, with diameters of 150cm, was the same as one horse (= two *kyrlag* (Øye 2002:354)). On another occasion a buyer gave a horse worth six *riksdalar* for the right to quarry stones for his own use for several years (Teigum 2011:52, 54-55).

10.5 Stability and change in transactions and trade

Trade in quernstones from Hyllestad can be documented from the late Viking period in southern Scandinavia, and the products would have been traded between different market places and towns in this period. The distribution pattern indicates an organized and routinised trade, where the actors involved - most likely belonging to the upper levels of society - had widespread contacts with southern Scandinavia. From the 12th century onwards, the medieval and multi-functional towns that developed in the early and high Middle Ages in Norway became important markets for Hyllestad stones, and the quernstones were increasingly channelled through the towns, especially Bergen, for further export. While the quarries in Hyllestad represented the only production area with large-scale production of quernstones in the late Viking period, they were no longer unrivalled in the Middle Ages. In medieval Norway the Saltdal quarries seem to have played an important role as quernstone suppliers, perhaps partly because of the growing stockfish trade to Bergen. Stones from Saltdal, however, were never able to compete with Hyllestad stones abroad. Here, the already established trade networks and markets for Hyllestad stones seem to have been strong. Despite new agencies and different distribution methods the markets established during the Viking period were sustained, with a slight increase in the volumes of commodities circulated. An expansion in distribution to the island of the North Atlantic occurred in the Middle Ages, but Hyllestad stones played a minor role compared to quernstones of different origin.

The many cargoes of quernstones and millstones found in wrecks along the coast underline the risks involved in distribution, where products, ships and sailors were lost regularly. Nevertheless wide-ranging distribution was carried on for centuries, indicating that large sums were involved in the trade of these products. Those involved in transactions changed after the Viking period but little is known of the different groups of entrepreneurs involved. For generation tenants and their families organized both production and distribution, but perhaps is it more likely that people or institutions belonging to the upper classes were those who controlled and organized trade in some way during both periods, and they were the ones who in the end profited from the activities.

Bakestones represented a new product in the early Middle Ages, and the easily broken stone plates were more numerous in the medieval towns than quernstones, indicating an extensive use in everyday households. The towns also acted as redistribution centres for bakestones, and distribution decreased in relation to distances from the towns. The elites at Ølve/Hatlestrand, including various ecclesiastical institutions, are also thought to have been involved in distribution. Different food traditions and cooking habits had a major bearing on bakestone exports and they had a far more limited distribution abroad.

During the Middle Ages both the quarry areas under study were also involved in on demand supply, whereby the elites commissioned the extraction of different product types. These items, such as crosses and building stones, never reached the markets or towns, but were transported directly between the quarries and their intended recipients. Extractions of this kind set the quarries in wider contexts linked to the King and Church, and to the major social upheavals of the Viking period and Middle Ages. Organisation of the distribution and forms of transaction for the two quarry areas changed over time, and also spatially, related to the type of products involved.

11 Conclusions

This final chapter presents an overall assessment of the results and conclusions of the analysis of two quarry landscapes in western Norway, Hyllestad (Sogn og Fjordane) and Ølve/Hatlestrand (Hordaland).

People have always adapted to their environments, and a precondition for any production was the intimate knowledge of different geological resources and their potential. Successful exploitation in both production areas was based on this knowledge, but it was triggered by new demands, new agencies and new organisations. The project involved three different levels of research, with the quarries being assessed at the micro level, their relation to the local communities at the medium or 'meso' level and trade and exchange within a wider geographical context at the macro level. Issues concerning production, distribution and the organisation of these activities have therefore been addressed, as well as control and disposal of quarries and their products. Some main questions have been addressed:

- How old are the quarries and the different products?
- What was the extent of production and what product types were extracted?
- What markets were central for these products and what was the nature of the transactions?
- Did production processes and product types, as well as transaction and trade, change over time?
- Who controlled production and how was production and distribution organized?
- How did societal changes in the Middle Ages (new institutions, church and monasteries) affect production?

The quarries of Hyllestad and Ølve/Hatlestrand were chosen as study areas for several reasons. Both quarry areas indicate large-scale and specialized production with a limited range of products. The main products - quernstones and bakestones - represent serial production where the products seem to have been distributed over large distances. These goods played an important role in daily life, representing everyday utensils and common tools for food processing. Both quarry areas also had a wider range of products, with special items being commissioned for the elites - crosses and millstones from Hyllestad, and slabs and building material from Ølve/Hatlestrand. Long periods of production are documented at both quarry areas, with peaks in the Middle Ages, when large-scale extraction also led to major changes to the landscapes. Seen together, the concurrence in production and product type at Hyllestad and Ølve/Hatlestrand - to a large extent household utensils - and production date made it possible to analyze the exploitation of different resources within the larger regional perspective. The long-term view, from the beginning of the Viking period to the end of the Middle Ages, has made it possible to study production processes and product types through time, as well as the distribution and trade of these commodities. The comparison of the two quarry areas and the geographical distribution of the products also opened up a better understanding of how this was organized and the agencies behind it.

Archaeological investigation of the quarries was essential for further analysis. Small-scale excavations were carried out to assess and date production and product types, and to be able to place the quarries and their products in spatial, temporal and societal contexts. Trenches were excavated in the spoil heaps at selected quarries, and radiocarbon-dating of organic material, mainly charcoal, combined with stratigraphical analyses were the main methods for dating. It was considered important to investigate quarries at different distances from harbours and transport routes, and quarries belonging to different farms and landowners. It was a central aim to study the quarries within their societal contexts and to assess aspects such as control, disposal and organisation of production. It was especially enlightening to discuss the quarries, production and distribution in relation to structuration theory, which was used for perspective when analyzing connections between human actions and social and physical structures. As far as possible it was also an objective to make the key actors visible. Little is known of the people involved in both the production and export of the objects; they can only be viewed at the level of social groups. In this way the perspectives and levels of analysis have been extended and social aspects of the landscapes taken into account.

Quarries, production and archaeological investigation

The main focus of this study has been on the production of quernstones, millstones, crosses and bakestones. The methods and approaches used in the fieldwork have been dependent on the resources available, and larger and more destructive intrusions within cultural landscapes have been avoided. The scale of the quarry areas and the methods involved made it difficult to date the beginning of production in both areas. The oldest dates in the quarries represent *post quem* dates where the production may have started earlier than the earliest date. However, the wide date ranges allowed the initial phase also to be covered within the dates provided. Finds from other contexts, graves and settlement sites have substantiated the dates achieved.

Hyllestad seems to be the oldest quarry area for quernstone and millstone production in Norway, representing new production forms in terms of both context and volume. Nearly 400 quarries have been identified but large-scale production has only been documented at two of the farms,

Rønset and Myklebust, representing the central production areas. Here production was of a different character and scale compared to activity at the other farms, with smaller and more widespread quarries. Rønset and Myklebust are also, with one exception, the only farms where deep quarries and a combination of shallow and deep quarries have been located. Forges seem to have been used at the quarry sites or nearby, most likely for sharpening tools on a regular basis and an important adjunct to quarry work.

A few Hyllestad sites producing quernstones could date to the early Viking period (AD 715-890 and AD 730-965), but both volume and distribution seem to have been modest at this early stage: production began to satisfy local needs. Greater output is documented from the 10th century and the late Viking Age, when extraction took place at three of the investigated farms, Rønset, Sæsol and Myklebust. The activity represents an innovation with regard to product form, scale and rock type used. Production seems to have increased in the Middle Ages, not only in scale but also in the repertoire of products with millstones and stone crosses as new form types. Precise dates for different items are difficult to give as C14-analyses provide broad parameters. Millstones c. 1m in diameter may have been extracted already by the late Viking period (AD 890-1015, AD 990-1160), as is indicated by previous investigations at Hyllestad, but at least from the 11th century Hyllestad millstone production was already taking place, as is suggested at Rønset (e.g. trench 5). Millstones seem mainly to have been extracted from deep quarries and sites that appear as a combination of shallow and deep quarries. The exception is the shallow site at Sæsol (trench 7) where a few millstones have been found.

The stone crosses from Myklebust are dated broadly from the 11th and 13th centuries (AD 1050-1245, AD 1175-1250 and AD 1225-1280), and may represent an influence from abroad. All the crosses are V-shaped below the arms but vary considerably in size. The smallest example has a total height of 1m and would have stood c. 50cm above the ground. The fragmentary state of some crosses makes it difficult to estimate their intended height, but they would have been much higher - probably from 1 to 3m above the ground. The crosses found seem to derive from the last and finishing production phase at the sites, and their extraction most likely dates back further. Throughout the entire period of study, the Viking Age and Middle Ages, quernstones seems to represent the main outputs from Hyllestad. This is also indicated by quernstone finds in different use contexts outside the production area.

The initial phase of Hyllestad quernstone production resulted from changes in skills and specializations, and in the use of resources and development of serial production for larger markets in the early Viking period. Hyllestad represents new usage of the outlying farm areas and a change in how the resources were exploited. The emerging urban centres in the Viking period, and the development of cargo vessels and sailing, made large-scale production and long-distance trade possible.

71 quarries have been identified at Ølve/Hatlestrand, both underground and open quarries, some of them coherent production sites extending more than 100m. It was more difficult to collect samples for C14-dating from Ølve/Hatlestrand sites than from Hyllestad. One of the investigated quarries at Fugleberg (trench 3) indicates that production began in the 11th and 12th centuries (AD 1025-1155 and AD 1045-1190), and so later than initial Hyllestad extraction. Production at the quarry is also documented in the 13th century, possibly continuing into the early modern period. A second quarry site dated at Netteland (trench 1) substantiates workings in the late 13th and 14th centuries, with a possible production also in the early modern period.

Bakestones appear to have been the main products at only one of the investigated quarries (Fugleberg, trench 3), and extraction marks of other products, such as tiles, slabs and building stone dominate in most quarries. Yet finds of bakestone fragments in the spoil heaps indicate bakestone extraction along with the manufacture of other products in all the quarries investigated. Bakestones appear in most cases to have been finished directly at the quarry sites, however three unique worksites have been identified, and the largest one, Kvitafjell, dates from the 11th to early 13th centuries (AD 1160-1220, AD 1170-1245, AD 1030-1165). During this period some quarries from Fuglebergåsen delivered unfinished bakestones to the worksite, where they received their finishing processes.

Distribution and trade

The rock types in both quarry areas seem to have been highly valued and preferred over others. The study clearly shows that production generated surpluses for exchange and trade. Both extraction areas were exploited for profit, which made it important to study the distribution of products on regional, national and international levels to see what and where were the key markets. While quernstones and millstones were used widely, bakestones represent a product of more limited use. As food preparation and cooking traditions were decisive factors, bakestone exports had a far more limited distribution abroad than quernstones and are consequently mainly found within the Norwegian cultural sphere, including the North Atlantic region.

Hyllestad quernstones and millstones are therefore the main focus for provenance study, and their distribution, both spatially and chronologically, have been investigated at sites in Scandinavia and the islands of the North Atlantic - Shetland, the Faroes and Iceland.

The widespread distribution of Hyllestad quernstones is documented from the middle of the 10th century in southern Scandinavia, even though a few finds may have an earlier date (early 9th). Quernstones from Hyllestad have been found mainly in eastern parts of medieval Denmark, Northern Jutland and the islands including Scania and Bornholm, but also towards the Baltic. The quantities of Hyllestad stones indicate organized trade in this area from the late Viking Age onwards, dominating the market in many areas. The widespread distribution to southern parts of Scandinavia indicates extended networks and routine practices of sailing long distances. Ships may have sailed more or less directly between Hyllestad and the different markets of southern Scandinavia, but some sites, such as Århus and Sebbersund, probably acted as gateways for the stones, and where Hyllestad products were reloaded for onward distribution.

Trade in Hyllestad quernstones increased from the 12th century, both geographically and quantitatively. Expanding trade to the new medieval and multi-functional Norwegian towns made them important markets for innovative products such as Hyllestad querns and bakestones. The towns had important roles as redistribution centres, with Bergen especially prominent.

Hyllestad stands out as the only quarry site distributing its quernstones widely during the Viking period; however by the medieval period in Norway these quarries were facing strong competition. From the early Middle Ages, competition from Saltdal products is indicated, and in some areas, such as Oslo and rural districts in the county of Rogaland, Saltdal stones are found in greater numbers than Hyllestad products. Stones from Saltdal were, however, never able to compete with Hyllestad stones abroad; these markets favoured Hyllestad quernstones and millstones for several hundred years. The early establishment of practices and traditions led to long-lasting trading networks and well-organized commercial trade, including exports.

Those individuals and groups involved in such transactions changed as the Viking period gave way to the Middle Ages. The extensive distribution of products throughout the period of study, especially quernstones, indicates a logistically well-established organisation and pattern of networks. It is possible that this was gradually achieved by the efforts of tenants and farmers who, over generations, managed to establish and sustain stable networks and contact areas to operate within. The refinements of logistical organisation would, however, have been easier for social elites than for the lower classes. Trade may have been undertaken by individuals or institutions from the upper strata of society, such as larger landowners, who also profited from the activities. These landowners ensured and arranged that their surplus production reached the markets, but they need not necessarily have undertaken the transportation and selling of the stones themselves. Distribution and trade was also conducted by agents; ships, crews and independent traders were hired and organized as partners, the so-called *félag*, at least in the Middle Ages.

During this same period, perhaps earlier in Hyllestad, both quarry areas studied were involved in production on demand, with the extraction of stone crosses and grave slabs from Hyllestad and building stone from Ølve/Hatlestrand. These represented commissioned products, ordered by institutions or elite agencies - with links to the Church and probably also the King. These products were transported directly from the quarries to the clients, without involving middlemen, markets or towns. Millstones may also have been commissioned products in the earliest phase, but with a development towards serial production and regular commercial exchange, at least from the 13th century onwards, with distribution through towns and marketplaces.

Control and organisation

An important aspect of the study has been to analyze the quarries of Hyllestad and Ølve/Hatlestrand in relation to those societal changes in the Viking period and the Middle Ages that may have influenced both production and distribution. Successful, large-scale quarrying and a wide-ranging distribution of products were dependent on several factors. Favourable geological conditions cannot alone define the activities. The contemporary societal structure played a decisive role, and socio-political conditions constituted an essential framework for both production and distribution.

Both quarry areas seem to have been parts of different estates in the Middle Ages, probably also in the Viking period. This seems, however, to be more articulated for Hyllestad than Ølve/Hatlestrand, where more farms appear to have been freeholds prior to the Middle Ages. Local elites, landowning farmers or magnates, may have initiated and organized the activities in both areas. In the initial stage, therefore, control and decision making to a large degree were carried out by the local communities in which the quarries were located. Significant changes in landownership and estate management occurred during the Middle Ages, where the Church (including monastic agencies) gained significant access to land and incomes. Quarrying activities would then have been effectively controlled by social, economic and political elites, in many cases by actors located far from local communities. Ruling monarchs may even have acquired land in both quarry areas, which later passed to the ecclesiastical institutions. Individuals working and living on the farms within the two production areas were therefore not necessarily those who organized the work in the quarries or who profited from the activities.

Some quarrying would still have been done by people living on the farms where the quarries were located, as part of the routine work. Such part-time or seasonal production most likely reflects the situation at most farms in Hyllestad and also Ølve/Hatlestrand. Extraction within the central production areas, such as Rønset and Myklebust in Hyllestad, and Fugleberg in Ølve, was possibly of a different scale, demanding a larger workforce and management: labourers, skilled workers, and craftsmen. Quarries in some areas would also have been rented out, as is known from Germany.

Comparing the two quarry areas, Hyllestad and Ølve/Hatlestrand both testify to serial production targeted at larger markets, and the extraction of commissioned products for agents or elite institutions. Little is known about those actually involved in production and export. The activities were controlled and run by social elites, even though it is possible that local tenants, over generations, were closely involved in production and trade. More limited production and distribution is seen at Ølve/Hatlestrand compared to Hyllestad, in terms of both duration and areas exploited. In the Middle Ages the two quarry areas operated simultaneously, but while Hyllestad at this point stands out as a well-established production area, with a tradition reaching back several hundred years, Ølve/Hatlestrand represents a new quarry complex offering a new product type not commonly used before in Norway - bakestones. Both quarry areas exhibit large-scale extraction throughout the Middle Ages, triggered by demand outside the local communities. Products were traded within the same towns, and in these centres bakestone fragments are more numerous than quernstone and millstone finds. Bakestones, however, never managed to gain the same level of importance outside the Norwegian cultural sphere as quernstones did. While organized trade in quernstones was undertaken in southern parts of Scandinavia, bakestones were only accidentally and exceptionally distributed here.

This study has attempted to provide a comprehensive and nuanced picture of the production and exploitation of resources in the Viking Age and the Middle Ages by including a contextual and societal approach to quarry operations and transactions. To this end a variety of source categories was employed. Archaeological and geological studies form the main basis, but written evidence has also played a central part as contexts for wider interpretations. The quarries were embedded in political, social and economic systems, raising both restrictions and possibilities. It was important, therefore, to reach an understanding of the interactions between the quarries and production at a micro level, and the wider local communities with aspects such as the control and rights of use at the meso level, as well as provenance studies and distribution at a macro level. This interplay between different level and scale enables a better understanding of the dynamics and complexities attached to resource utilization, production and distribution in the Viking period and Middle Ages.

Bibliography

Aaberg, Th. 1929: *Kyrkjorne i Hyllestad i gammel og ny tid.* Florø.

AB = *Aslak Bolts Jordebok,* published by Jon Gunnar Jørgensen, 1997, Oslo.

Andersen, H. Hemluth, Crabb, P. J. and Madsen, Hans J. 1971: Århus Søndervold en byarkæologisk undersøgelse. *Jysk Arkæologisk selskabs skrifter, bind IX, 1971.* København.

Andersson, Hans, Ersgård, Lars and Svensson, Eva (eds.) 1998: *Outland Use in Preindustrial Europe. Lund Studies in Medieval Archaeology 20.* Lund.

Alsvik, Elling, Sognnes, Kalle and Stalsberg, Anne 1981: *Kulturhistoriske undersøkelser ved Store Kvernfjellvatn, Selbu, Sør-Trøndelag 1978.* Det Kgl. Norske Videnskabers Selskab, Museet. Rapport Arkeologisk serie 1981:1. Trondheim.

Anderson Timothy, Villet, Damien and Serneels, Vincent 1999: La fabrication des meules en grès coquilier sur le site gallo-romain de Châbles-Les Saux. *Archäologie der Schweiz*: 182-189.

Anderson, Timothy 2006: Three Roman quern quarries in Switzerland. In: Belmont, Alain and Mangartz, Fritz (eds.) *Mühlsteinbrüche. Erforschung, Schutz und Inwersetzung eines Kulturerbes europäischer Industrie (Antike 21. Jahrhundert)*, 41-46. Mainz.

Arge, S. V. 1989: Om landnåmet på Færøerne. *Hikuin 15*, 103-128.

Barboff, Mouette, Sigaut, Francois, Griffin-Kremer, Cozette and Cremer, Robert (eds.) 2003: *Meules à Grains: Actes du Colloque Internationale de La-Ferté-sous-Jouarre 16-19 Mai 2002.* Paris.

Barndon, Randi 2002: The c*haîne op*ératoire approach, social change and modifications of the technological milieu. *Arkeologiske skrifter fra Universitetet i Bergen. No. 11*, 5-23. Bergen.

Barndon, Randi 2004: *An ethnoarchaeological Study of Iron-Smelting Practices among the Pangwa and Fipa in Tanzania.* Cambridge Monographs in African Archaeology 61. British Archaeology Reports, International Series 1308. Oxford.

Barth, Fredrik 1972: *The Role of entrepreneur in social change in Northern Norway.* Bergen.

Baug, Irene 2001: *Kvernsteinsbrota i Hyllestad. Arkeologiske punktundersøkingar i steinbrotsområdet i Hyllestad, Sogn og Fjordane.* Unpublished Master's thesis, University of Bergen.

Baug, Irene 2002: *Kvernsteinsbrota i Hyllestad. Arkeologiske punktundersøkingar i steinbrotsområdet i Hyllestad, Sogn og Fjordane.* Bergverksmuseet. Skrift nr. 22. Kongsberg.

Baug, Irene 2002b: Kvernsteinsbrota i Hyllestad - bergverksdrift frå forhistoria. *Arkeo. Nr. 2, 2002*, 12-17. Bergen.

Baug, Irene 2005a: Who owned the products? Production and exchange of quernstones, Hyllestad in Sogn, Western Norway. In: Holm, Ingunn, Innselset, Sonja and Øye, Ingvild (eds.) *UBAS International 1. University of Bergen Archaeological Series. 'Utmark'. The Outfield as Industry and Ideology in the Iron Age and the Middle Ages*, 99-108. Bergen.

Baug, Irene 2005b: Frå kvernstein til steinkross - produksjon og distribusjon. In: Førsund, Finn (ed.) *Steinkrossane på Vestlandet. Hyllestadseminaret 2004. Rapport frå fagseminar 24. april 2004*, 7-12. Førde.

Baug, Irene 2006b: Stein som handesvare. In: Førsund, Finn (ed.) *Stein som handelsvare. Hyllestadseminaret. Rapport frå fagseminar*, 5-15. Førde.

Baug, Irene 2006a: The Quarries in Hyllestad: Production of Quernstones and Millstones in Western Norway. In: Belmont, Alain and Mangartz, Fritz (eds.) *Mühlsteinbrüche. Erforschung, Schutz und Inwersetzung eines Kulturerbes europäischer Industrie (Antike-21. Jahrhundert)*, 55-59. Mainz.

Baug, Irene 2007: Prehistoric quarrying on the west coast of Norway: the production of quernstones, millstones and crosses in Hyllestad, Sogn og Fjordane. In: Klápště, Jan and Sommer, Petr (eds.) *Arts and Crafts in Medieval Rural Environments. Ruralia VI. 22nd -29th September 2005. Dobogókó, Hungary*, 219-225. Turnhout.

Baug, Irene 2011: Soapstone Finds. In: Skre, Dagfinn (ed.) *Things from the Town. Artefacts and Inhabitants in Viking-age Kaupang.* Kaupang Excavation Project. Publication Series, Volume 3. Norske Oldfunn XXIV, 311-337. Aarhus.

Baug, Irene and Løland, Torbjørn 2011: The millstone quarries in Hyllestad: an arena of research and education. In: Williams, David and Peacock, David (eds.) *Bread for the people: The Archaeology of Mills and Milling. Proceedings of a colloquium held in the British School at Rome 4th-7th November 2009.* University of Southampton Series in Archaeology, No. 3, 349-356. Oxford.

Baug, Irene and Jansen, Øystein J. 2014: Did the North Atlantic region constitute a market for quernstones from Norway during the Viking period and the Middle Ages? In: Selsing, Lotte (ed.) *Seen through a millstone.* AmS-Skrifter 24, Museum of Archaeology, University of Stavanger, 245-255. Stavanger.

Baug, Irene and Jansen, Øystein J. in prep. (a): *Collection Report: Rotary Querns in Shetland Museum.*

Baug, Irene and Jansen, Øystein J. in prep. (b): *Collection Report: Rotary Querns in the National Museum of Iceland.*

Baug, Irene and Jansen, Øystein J. in prep. (c): *Collection Report: Rotary Querns in the National Museum of the Faroe Islands.*

Baug, Irene and Jansen, Øystein J. in prep. (d): *Collection Report: Rotary Querns in the Achleswig Holsteinische Landesmuseum, Germany.*

Baug, Irene, Heldal, Tom, Jansen, Øystein J., Løland, Torbjørn and Hauken, Åsa Dahlin in prep. (a): *Collection Report: Rotary Querns in the University Museum of Bergen.*

Baug, Irene, Jansen, Øystein J., Meyer, Gurli B., Løland, Torbjørn and Grønne, Tor in prep. (b): *Collection Report: Rotary Querns in the Museum of Kroppedal, Denmark.*

Belmont, Alain 2006a: *La Pierre à pain. Les carriers de meules de moulins en France, du Moyen Âge à la révolution industrielle. Tome I.* Grenoble.

Belmont, Alain 2006b: *La Pierre à pain. Les carriers de meules de moulins en France, du Moyen Âge à la révolution industrielle. Tome II.* Grenoble.

Belmont, Alain and Hockensmith, Charles D. 2006: Milstones, Querns, and Millstone Quarry Studies in France. A bibliography. *International Molinology, No. 72, July 200*, 2-10.

Belmont, Alain and Mangartz, Fritz 2006: *Mühlsteinbrüche. Erforschung, Schutz und Inwersetzung eines Kulturerbes Europäischer Industrie (Antike-21. Jahrhundert). Internationales Kolloquium Grenoble - 22. bis 25. September 2005 - Maison des Sciences de l'homme-Alpes.* Mainz.

Beiron, Ingemar 2006: The Millstone Quarry 'Minnesfjället' in Lugnås, Sweden. In: Belmont, Alain and Fritz Mangartz (eds.) *Mühlsteinbrüche. Erforschung, Schutz und Inwersetzung eines Kulturerbes europäischer Industrie (Antike 21. Jahrhundert)*, 203-206 Mainz.

Berg, Linda 2005: *Heddal og Røldal: en komparativ analyse av to kirkesteder*. Unpublished Master's thesis, University of Bergen.

Berge, Bjørn 1979: *Vitenskapen som felleseie eller kallkverner, hjulkverner, tidevannskverner og vindmøller i Norge før i tida og i framtida?* Oslo.

Bergström, Liselotte 2007: Gräddat. Brödkultur under järnålderen i östra Mälardalen. Doctoral Thesis in Archaeological Science at Stockholm University. Stockholm.

Beuther, Anton 1540: *Bergkordnung des Löblichen newen Bergkwergs auff dem Golmsbergk im Königreich Norwegen.* Zwickau.

Bill, Jan 2008: Viking ships and the sea. In: Brink, Stefan and Price, Neil (eds.) *The Viking World*, 170-180. London and New York.

Birkedahl, Petter 2000: Sebbersund - en handelsplads med trækirke ved Limfjorden - forbindelser til Norge. In: Karmøy kommune (ed.) *Havn og Handel i 1000 år. Karmøyseminaret 1997*, 31-39. Stavanger.

Birkedahl, Petter and Johansen, Erik 2000: The eastern Limfjord in the Germanic Iron Age and the Viking Period. *Acta archaeological, Vol. 71*, 25-32.

Birkeli, Fridtjov 1973: *Norske steinkors i tidlig middelalder. Et bidrag til belysning av overgangen fra norrøn religiontil kristendom.* Skrifter utgitt av Det Norske videnskaps-akademi i Oslo II. Hist. Filos. Klasse. Ny Serie. 10. Oslo.

Birkenes, Lars Øyvind 2004: *Gard, grav og kyrkje. Ei arkeologisk analyse i Hardanger-regionen.* Unpublished Master's thesis, University of Bergen.

Bjørkvik, Halvard 1964a: Kronans finanser. Noreg. *KLNM IX*, 424-428.

Bjørkvik, Halvard 1964b: Kungsgård. Noreg. *KLNM IX*, 517-518.

Bjørkvik, Halvard 1965: Landskyld. *KLNM X*, 277-282.

Bjørkvik, Halvard 1967: Eit kyrkjeleg godskompleks i mellomalderen. Munkeliv klosters jordebok. *Heimen XIV*, 51-70. Oslo.

Bjørkvik, Halvard 1968: Bona regalia, patrimonium og kongeleg donasjonspolitikk i mellomalderen. *Nordiska historikermötet Helsingfors 1967. Plenardebatter. Sektionsmöten*, 43-55. Helsingfors.

Bjørkvik, Halvard 1970: Det norske krongodset i mellomalderen. In: Holmsen, Andreas og Simensen, Jarle (eds.) *Sverreættens komgedømme. Norske historikere i utvalg. Med bidrag fra svenske, danske og islandske forskere III*, 138-154. Oslo.

Bjørkvik, Halvard 1975: Veitsle. *KLNM XIX*, 632-634. København.

Bjørkvik, Halvard 1995a: Kva slags samfunn var det som tok imot kristendommen? In: Lidén, Hans-Emil (ed.) *Møtet mellom hedendom og kristendom i Norge*, 58-79. Oslo.

Bjørkvik, Halvard 1995b: Klostergods og klosterdrift i Noreg i mellomalderen. *Collegium Medievale: Tverrfaglig tidsskrift for middelalderforskning. Vol. 8, no. 2*, 147-163.

BK = Bergens Kalvskinn. Am 329 Fol. Norsk Riksarkiv. Published by Finn Hødnebø, 1989, Oslo.

Bl = Magnus Lagabøters bylov. Translated by Knut Robberstad, 1923. Kristiania.

Bowman, Sheridan 1990: *Interpreting the past. Radiocarbon Dating*. London.

Brattegard, Olav 1943-1946: Namni på dei gamle gardane på Bryggja i Bergen. *Historisk tidsskrift, Vol. 33*. Oslo.

Brattekværne, Cathrine 2006: *Kirkesteder i Indre Sogn. En arkeologisk analyse av religionsskiftet, med utgangspunkt i kirkesteder i Indre Sogn.* Unpublished Master's thesis, University of Bergen.

Braudel, Fernand 1980 [1969]: *On History*. London.

Brendalsmo, Jan 1990: Kirkesteder i Vestfold: Hvem bygde kirker og hvorfor - hvor ble kirkene bygd og hvorfor? Skisse til problemstillinger omkring kirkebyggeriet. In: Brendalsmo, Jan and Vea, Erik (eds.) *Vestfoldkirkene - bygning, landskap og samfunn. Arkeologiske rapporter fra Tønsberg, Nr. 3*, 37-57. Riksantikvaren, Utgravningskonteret for Tønsberg.

Brekken, Jakob 1980: Kvernberget og andre kvernsteinsbrot i Vågå og Sel. In: Engen, Arnfinn (ed.) *Årbok for Gudbrandsdalen 1980*, 78-83. Otta.

Brumfiel, Elizabeth M. and Timothy K. Earle 1987: *Specialization, exchange and complex societies*. Cambridge.

Buch, Leopold von 2011 [1813]: *Travels through Norway and Lapland during the years 1806, 1807, 1808* (Translated from the original German by John Black) London.

Buckholm, Mona Beate 1998 [1997]: *Nedlagte kirker og kirkesteder fra middelalderen i Hordaland og Sogn og Fjordane. Hvilke kirkesteder ble nedlagt, når og hvorfor*. Arkeologiske avhandlinger og rapporter fra Universitetet i Bergen. Bergen.

Buckland, P. C. and Sadler, Jon 1990: Ballast and Building Stone: A Discussion. In: Parson, David (ed.) *Stone. Quarrying and Building in England AD 43-1525*, 114-125. Sussex.

Bull, Edvard 1920: *Leding: militær- og finansforfatning i Norge i ældre tid*. Kristiania.

Buttler, Simon and Forster, Amanda K. 2009: Excavations at Clibberswick: a report on rescue work at a Norse steatite quarry in Unst, Shetland, July-August 1983. In: Forster, Amanda K. and Val E. Turner (eds.) *Kleber: Shetland's Oldest Industry. Shetland Soapstone Since Prehistory*, 85-105. Lerwick.

Bøe, Arne 1964: Kristenrettar, Noreg. *KLNM IX*, 298-304. København.

Callmer, Johan 1995: Hantverksproduktion, samhällsförändringar och bebyggelse. Iakttagelser från östra Sydskandinavien ca. 600-1100 e.Kr. In: Heid Gjøstein Resi (ed.) *Produksjon og samfunn. Om erverv, spesialisering og bosetning i Norden i 1. årtusen e.Kr. Beretning fra 2. nordiske jernaldersymposium på Granavolden Hjæstgiveri 7-10 mai 1992*, 39-72. Oslo.

Campbell, Åke 1950: *Det svenska brödet. En jämförande etnologisk-historisk undersökning*. Stockholm.

Campbell, Åke 1951: Nordvesteuropeisk brödkultur i Rigstula. In: Sahlgren, Johan (ed.) *Saga och sed: Kungl. Gustav Adolfs akademiens årsbok = Annales Academiae Regiae Gustavi Adolphi. Band 4. 1939-1943*, 5-18. Stockholm.

Carelli, Peter and Kresten, Peter 1997: Give us this day our daily bread. A study of Late Viking Age and Medievel Quernstones in South Scandinavia. *Acta Archaeologica. Vol. 68, 1997*, 109-137. Munksgaard.

Casell, Phillip 1993: *The Giddens reader*. Basingstoke.

Cederlund, Carl Olof 1964: Kvarn. *KLNM IX*, 536-538.

Clarke, Helen and Bjørn Ambrosiani 1995: *Towns in the Viking Age*. Exeter.

Christophersen, Axel 1989: Kjøpe, selge, bytte, gi. In: Andrén, Anders (ed.) *Medeltidens födelse*, 109-145. Lund.

Colic, Etel and Snæsdóttir, Mjöll 2010: *Pottery from Excavations at Stóraborg*. Nýsköpunarsjóður Námsmanna.

Costin, Cathy Lynne 1991: Craft Specialization: Issues in Defining, Documenting and Explaining the Organisation of Production. *Archaeological Method and Theory, Vol. 3*, 1-56. Tucson.

Coulter, Carolyn 2011: Of cakes and Kings: bread-making in early medieval England. In: Williams, David and Peacock, David (eds.) *Bread for the people: The Archaeology of Mills and Milling. Proceedings of a colloquium held in the British School at Rome 4th-7th November 2009*. University of Southampton Series in Archaeology, No. 3, 179-192. Oxford.

Crumlin-Pedersen, Ole 1999: Ships as indicators of trade in Northern Europe 600-1200. In: Bill, Jan and Clausen, Birthe L. (eds.) *Maritime Topography and the Medieval Town. Papers from the 5th International Conference on Waterfront Archaeology in Copenhagen, 14-16 May 1998. PNM, National Museum Studies in Archaeology and History Vol 4*, 11-20. Copenhagen.

Crumlin-Pedersen, Ole 2002: Documentation, analyses and dating. In: Crumlin-Pedersen, Ole and Olaf Olsen (eds.) *The Skuldelev Ships I. Topography, Archaeology, History, Conservation and Display*, 49-68. Roskilde.

Dahl, Gunnar 1998: *Trade, Trust and Networks. Commercial Culture in Late Medeival Italy*. Lund.

Dahlberg, Ann-Kristin 2012: *Tretten kyrkjestader I Sunnfjord. Ein arkeologisk analyse av mellomalderkyrkjestader frå kyst til innland*. Unpublished Master's thesis University of Bergen.

DN = *Diplomatarium Norvegicum*, published by C. C. A. Lange, C. R. Unger *et al.* 1847-1990. Christiania/Oslo.

Dobat, Andres S. 2013: *Kongens Borge. Rapport over undersøgelserne 2007-2010*. Jysk Arkæologisk Selskap. Højbjerg.

Dodgshon, Robert A. 1998: *Society in Time and Space. A Geographical Perspective on Change*. Cambridge.

Dommasnes, Liv Helga 1976: *Yngre jernalder i Sogn: forsøk på sosial rekonstruksjon*. Unpublished Master's thesis, University of Bergen.

Earle, Timothy K. 1982: Prehistoric Economics and the Archaeology of Exchange. In: Ericson, Jonathan E. and Earle, Timothy K. (eds.) *Context of Prehistoric Exchange*, 1-12. London.

Ek, Sven B. 1962: *Väderkvarnar och vattenmöllor. En etnologisk studie i kvarnarnas historie*. Lund.

Ekroll, Øystein 1997: *Med kleber og kalk: norsk steinbygging i mellomalderen*. Oslo.

Ekroll, Øystein 1981: Rosenkrantzgate 4. Archive report, Topografisk arkiv. Bergen, Hordaland. Bergen Museum. Bergen.

Eldjárn, Kristján and Gestson, Gisli 1952: Rannóknir á BergÞórsvoli. Árbók hins íslenzka for*nleifafélags 1951-1952*. Reykjavík.

Eldjárn, Kristján 1964: Kvarn. Island. *Kulturhistorisk leksikon for nordisk middelalder fra vikingtid til reformajsonstid 9*, 540-541. Oslo.

Elfwendahl, Magnus and Kresten, Peter 1993: Geoarkeologi innom kvarteret Bryggaren. Arkeologiska artefakter av sten från det medeltida Uppsala. *Riksantikvarieämbetet och statens historiska museer. Rapport UV 1993:5*. Uppsala.

Elvestad, Endre 2005a: Jernalderledene i Sørvest-Norge. In: Selsing, Lotte, Elvestad, Endre, Hamre, Harald,

Krøger, Jens Flemming, Midtrød, Anne and **Nærøy**, Arne Johan (eds.) *Fra Galta til Geitungen: Kystklultur i fjæresteinsarkeologi i Sørvest-Norge. AmS-Nett Nr. 7*, 8-28. Stavanger.

Elvestad, Endre 2005b: Om varder og steinkors - tidlige sjømerker i Rogaland. In: Selsing, Lotte, Elvestad, Endre, Hamre, Harald, Krøger, Jens Flemming, Midtrød, Anne and **Nærøy**, Arne Johan (eds.) *Fra Galta til Geitungen: Kystkultur og fjæresteinsarkeologi i Sørvest-Norge*. AmS-NETT Nr. 7, 81-91. Stavanger.

Eriksson-Trenter, Anna 1998: Defining Property rights. In: Anderson, Hans, Ersgård, Lars and Svensson, Eva (eds.) *Outland Use in Preindustrial Europa. Lund Studies in Medieval Archaeology 20*, 167-174, Lund.

F = *Frostatingslova*. Translated by Jan Ragnar Hagland and Jørn Sandnes, 1994, Oslo.

Fasteland, Arthur 1983: Hyllestad kommune. *Fornminne i Sogn og Fjordane rapport om topografisk-arkeologisk registrering for det økonomiske kartverket*. University of Bergen.

Fett, Eva Nissen1972: Førhistorisk tid. In: E. Vaage (ed.) *Kinnherad. Bygdesoga I*, 11-156, Bergen.

Fett, Per 1954: *Solund prestegjeld og Hyllestad prestegjeld. Førhistoriske minne i Sogn*. University of Bergen.

Fett, Per 1956: *Kvinnherad prestegjeld. Førhistoriske minne i Sunnhordland*. University of Bergen.

Feveile, Claus and Jensen, Stig (†) 2000: Ribe in the 8th and 9th century. A contribution to the Archaeological Chronology of North Western Europe. *Acta Archaeologica, Vol. 71*, 2-24. Blackwell.

Fisher, Christian 2004: *Tidlige Danske vandmøller*. Moesgård.

Forster, Amanda Kate 2004: *Shetland and the Trade of Steatite Goods in The North Atlantic Region during the Viking and Early Medieval Period.* Unpublished doctoral thesis, Department of Archaeological Sciences, University of Bradford.

Forster, Amanda K. 2009: Viking and Norse Steatite Use in Shetland. In: Forster, Amanda K. and Turner, Val E. (eds.) *Kleber: Shetland's Oldest Industry. Shetland Soapstone Since Prehistory*, 58-69. Lerwick.

Foyn, Silje Øvrebø 2008: *Rønset. En agrararkeologisk analyse av gården Rønset i Hyllestad, Sogn og Fjordane*. Unpublished Master's thesis, University of Bergen.

Friis, Peder Clausson 1632: *Norriges oc omliggende Øers sandfærdige Bescriffuelse: indholdendis huis vært er at vide, baade om Landsens oc Indbyggernis Leilighed oc Vilkor, saa vel i fordum Tid, som nu i vore Dage / korteligen tilsammen fattit aff Peder Claussøn*. Kiøbenhaffn.

Fyllingsnes, Frode 2000: *Med de tusen hjem: middelalderen*. Karmøys historie, bind 2. Kopervik.

Førsund, Finn Borgen (ed.) and Kellmer, Inger 1990: *Hyllestad bygdebok. Soga om folket. Band I*. Hyllestad kommune.

Førsund, Finn Borgen (ed.) and Kellmer, Inger 1993: *Hyllestad bygdebok. Soga om folket. Band II*. Hyllestad kommune.

Førsund, Finn Borgen 2004 (ed.): Dei første kjende hyllestadingar. *Sogeskrift frå Hyllestad 2004*, 13-17. Førde.

Førsund, Finn Borgen (ed.) 2005: *Steinkrossane på Vestlandet. Hyllestadseminaret 2004. Rapport frå fagseminar 24 April 2004*. Førde.

Førsund, Finn Borgen (ed.) 2006a: *Stein som handelsvare. Hyllestadseminaret. Rapport frå fagseminar*. Førde.

Førsund, Finn Borgen 2006b: Kvernsteinsbrota i Hyllestad. In: Førsund, Finn Borgen and Haukøy, Håkon (eds.) *Sogeskrift frå Hyllestad 2006*, 18-26. Førde.

Førsund, Finn Borgen (ed.) 2007: Ein kvernsteinshandel. In: *Sogeskrift frå Hyllestad 2007*, 42-43. Førde.

G = *Gulatingslovi*. Ttranslated by Knut Robberstad, 1952, Oslo.

Gabrielsen, Kristine Holme 2007 [2003]: Vestlandets steinkors. Monumentalisme i brytningen mellom hedendom og kristendom. In: *UBAS Universitetet i Bergen Arkeologiske Skrifter. Hovedfag/Master 2*, 115-271. Bergen.

Gardelin, Gunilla 2006: *En värld av sten. Stenhuggarnas organisation i medeltidens östergötland. Lund Studies in Historical Archaeology, Vol. 3*. Lund.

Ganerød, Morgan 2008: Alder av brynesteiner. NGUs geokronologiske laboratorium. Upublished report, NGU and Bergen Maritime Museum.

Gansum, Terje, Gro B. Jerpåsen and Christian Keller 1997: *Arkeologisk landskapsanalyse med visuelle metoder.* Stavanger.

Gestsson, Gísli 1959: Gröf í Ôræfum. *In:* Eldjárn, Kristján (ed.) Árbók hins íslenzka fornleifafélags, 5-87. Reykjavik.

Gerrard, Christopher 2005 [2003]: *Medieval archaeology. Understanding traditions and contemporary approaches*. London.

Giddens, Anthony 2004 [1984]: *The Constitution of Society. Outline of the Theory of Structuration*. Oxford.

Gollwitzer, Martin 2012: Graver, brønner og bosetningsspor fra bronsealder til middelalder på Hesby (lok. 13). In: Mjærum, Axel and Gjerpe, Lars Erik (eds.) *E18-prosjektet. Gulli-Langåker. Dyrking, bosetninger og graver i Stokke og Sandefjord*. Bind 1: 107-215. Bergen.

Granlund, John 1956: Bakning. *KLNM I*, 307-311.

Grenne, Tor, Heldal, Tom, Meyer, Gurly B. and Bloxam, Elizabeth G. 2008: From Hyllestad to Selbu: Norwegian millstone quarrying through 1300 years. *Geology of Society. Geological Survey of Norway Special Publication (Norges geologiske undersøkelse)*, 47-66. Geological Survey of Norway. Trondheim.

Grenne, Tor, Baug, Irene, Løland, Torbjørn, Jansen, Øystein J., Haack, Anne Louise Olsen and Meyer, Gurli in prep. (a): *Collection Report: Rotary Querns in the Museum of History, Nordjylland, Denmark*.

Grenne, Tor, Baug, Irene and Jansen, Øystein J. in prep. (b): *Collection Reports: Rotary Querns in the Museum of Cultural History, University of Oslo*.

Grøn, Fredrik 1939: *Brødets saga gjennem tidene*. Oslo.

Grøn, Fredrik 1942: *Om kostholdet i Norge fra omkring 1500-tallet og op til vår tid*. Skrifter utgitt av Det

Norske Videnskaps-Akademi i Oslo II. Hist.-Filos. Klasse. 1941. No. 4. Oslo.

Gullbekk, Svein H. 2006: Gjerdeskatten fra Hyllestad - en typisk og utypisk myntskatt fra sen vikingtid. In: Førsund, Finn (ed.) *Stein som handelsvare. Hyllestadseminaret. Rapport frå fagseminar*, 41-49. Førde.

Gulliksen, Steinar 1979: Hvor nøyaktig er C14-metoden? *Fortiden i søkelyset. Datering med C14-metoden gjennom25 år*, 69-80. Trondheim.

Gunnes 1987: Klosterlivet i Norge. Tilblivelse - økonomi - avvikling. I: *Foreningen til Norske fortids minnesmerkers bevaring. Årbok 1987, hefte 141*, 49-84. Oslo.

Gunnes, Erik 1996: Ordener og klostre i norskk samfunnsliv. In: Knirk, James E, (ed.) *Collegium Medievale. Interdisciplinary Journal of Medieval Research. Volume 8, 1995/2*, 131-145. Oslo.

Gustin, Ingrid 2004: *Mellan gåva och marknad. Handel, tillit og materiall kultur under vikingatid. Lund Studies in Medieval Archaeology 34*. Malmö.

Hagland, Jan Ragnar and Sandnes, Jørn 1994: Introduksjon til Frostatingslova. *Frostatingslova*. Oslo.

Hagland, Jan Ragnar 1998: Innleiing. Litt om datering på runologisk og språklig grunnlag. In: Dybdahl, Audun and Jan Ragnar Hagland (eds.) *Innskrifter og datering. Dating Inscriptions. Senter for middelalderstudier Skrifter Nr. 8*, 9-17. Trondheim.

Hamilton, J. R. C. 1956: *Excavations at Jarlshof, Shetland. Ministry of works, Archaeological Reports No. 1*. Edinburgh.

Hansen, Arild Marøy 1992: 'Kverna som maler på havets bunn'. Et kvernsteinsfunn i Alverstraumen, Lindås i Hordaland. *Sjøfartshistorisk årbok 1991*, 195-215. Bergen

Hansen, Arild Marøy 1997: Maritime perspektiv på kvernsteinsproduksjonen i Hyllestad. In: Sørheim, Helge (ed.) *'Arkeolog og kystkultur'. Foredrag på Sunnmøre Museum 25-26/10 1997*, 58-63. Ålesund.

Hansen, Arild Marøy 2013: Klosteret mitt i leia. In: Økland, Bård Gram, Jünger, Jane and Øye, Ingvild (eds.) *Halsnøy kloster. Til kongen og Augustins ære*, 185-197. Oslo.

Hansen, Gitte 1998: The Bryggen Chronology. New Light on the Dating of the Periods before Fire V. In: Øye, Ingvild (ed.) *Medieval Fires in Bergen - Revisited. The Bryggen Papers, Supplementary Series No. 6*, 81-127. Bergen.

Hansen, Gitte 2000: Bydannelse og forklaring af sociale fænomener. Individualisme, kollektivisme og Giddens strukturationsteori. In: Eriksdottir, Gunhild, Ersgård, Lars, Jonasson, Anders, Josefsson, Kristina, Mogren, Mats, Roslund, Mats and Thomasson, Joakim (eds.) *META, medeltidsarkeologisk tidskrift. 2000:4*, 2-16. Lund.

Hansen, Gitte 2005 [2004]: *Bergen c 800-c 1170. The emergence of a town. The Bryggen Papers. Main Series No 6*. Bergen.

Hatleskog, Anne-Brith 1997: Samfunnsorganisatoriske endringar eller religionsskifte i yngre jernalder. In: Selsing Lotte (ed.) *Fire fragmenter fra en forhistorisk virkelighet*. AmS-Varia 31. Arkeologisk Museum i Stavanger, 7-35. Stavanger.

Haug, Camilla 2009: *Kristningen av steder. En arkeologisk analyse av middelalderske kirkesteder på Karmaøy*. Unpublished Master's thesis, University of Bergen.

Hauken, Åsa Dahlin and Andersen, Timothy J. in press: *Collection report: Rotary Querns in the Museum of Archaeology University of Stavanger*. Forthcoming.

Haukenæs, Th. S. 1888: *Natur; Folkeliv og Folketro i Sønhordland. Natur- og Folkelivsskildringer, Eventyr, Fortællinger o.s.v. fra ældre og nyere Tid. Sjette Del: Kvinherred*. Bergen.

Hauksbók. Udgiven efter de arnamagnæanske håndskrifter no. 371, 544 og 675, 4°, samt forskellige papirhåndskrifter. 1892-96. København.

Helberg, Børn Hebba 2010: Kvernsteinsindustri i vikingtid og tidlig middelalder i Saltdal kommune, Nordland fylke. *Viking. Norsk arkeologisk årbok. Bind LXXIII-2010*,103-120. Oslo.

Helberg, Bjørn Hebba 2010/2011: *Rapport vedrørende overvåking av inngrep i skogsvei (Sypranvegen) ved kvernsteinsbruddet i Hestgarden, Saksenvik i Saltdal kommune, Nordland*. Unpublished Report, Tromsø Museum, University Museum.

Heldal, Tom and Elizaeth Bloxam 2008: *Kartlegging og karakterisering av kvernsteinsbruddene i Hyllestad. NGU Rapport 2007.079*.

Heldal, Tom and Meyer, Gurli 2011: The rise and fall of the Hyllestad millstone quarry landscape, Western Norway. In: Williams, David and Peacock, David (eds.) *Bread for the people: The Archaeology of Mills and Milling. Proceedings of a colloquium held in the British School at Rome 4th-7th November 2009*. University of Southampton Series in Archaeology, No. 3, 325-340. Oxford.

Grenne, Tor and Meyer, Gurli B. Heldal, Tom and Øystein J. Jansen 2014a: Technological development in millstone quarrying throughout the Middle Ages: the Salten quarries, Northern Norway. In: Selsing, Lotte (ed.) *Seen through a millstone*. AmS-Skrifter 24, Museum of Archaeology, University of Stavanger, 227 -244. Stavanger.

Grenne, Tor, Meyer Gurli, Heldal, Tom 2014: Unravelling the history of a complex millstone quarry landscape: Tolstadkvernberget, South Norway. In: Selsing, Lotte (ed.) *Seen through a millstone*. AmS-Skrifter 24, Museum of aarchaeology, University of Stavanger, 209-226. Stavanger.

Helland, Amund 1892: *Norsk bergret med udsigt over andre landes bergværkslovgivning*. Kristiania.

Helland, Amund 1893: *Tagskifere, hellere og vekstene*. Kristiania.

Helle, Knut 1982: *Bergen bys historie, I, Kongssete og kjøpstad. Fra opphavet til 1536*. Bergen-Oslo-Tromsø.

Helle, Knut 1998: Medieval fires in Bergen according to written sources. In: Øye, Ingvild (ed.) *The Bryggen Papers. Supplementary Series No 6*, 15-80. Bergen.

Helle, Knut 2001: *Gulatinget og Gulatingslova*. Leikanger.

Helle, Knut 2006: Del I. Fra opphavet til omkring 1500. In: Helle, Knut, Finn-Einar Eliassen, Jan Eivind Myhre and Ola Svein Stugu (eds.) *Norsk byhistorie. Urbanisering gjennom 1300 år*, 23-142. Oslo.

Henry, Francoise 1964: *Irish high crosses. Three Candles.* Dublin.

Herteig, Asbjørn 1985: *The archaeological excavations at Bryggen, the 'German warf', in Bergen, 1955-1968.* The Bryggen Papers. Main Series, Vol. 1. Bergen.

Herteig, Asbjørn 1990: *The buildings at Bryggen, their topographical and chronological development.* The Bryggen Papers. Main Series, Vol. 3, Part I. Bergen.

Hertzberg, Ebbe 1913: *Betænkning om almenningsrettens historiske oprindelse og dens udvikling: specielt i spørgsmaalet om dens sammenhæng med landbonæringen.* Hamar.

Hilberg, Volker 2008: Hedeby: an outline of its research history. In: Brink, Stefan and Price, Neil (eds.) *The Viking World*, 101-111. London and New York.

Hockensmith, Charles D. 2009: *The Millstone Industry. A summary of research on quarries and Producers in the United States, Europe and Elsewhere.* Jefferson, North Carolina.

Hodges, Richard 1989: *Dark Age Economics. The origins of towns and trade AD 600-1000.* London.

Hoff, Anne Marta and Lidén, Hans Emil 2000: *Norske minnesmerker. Norges kirker. Hordaland. Bind II.* Oslo.

Hommedal, Alf Tore 1999: Kva fortel bygningsrestane av dei norske klostera om kontinental norm og praksis innan ordenslivet? In: Mundal, Else and Øye, Ingvild (eds.) *Norm og praksis i middelaldersamfunnet*, 149-183. Bergen.

Hødnebø, Finn and Ole-Jørgen Johannessen 1989: cf. *BK*.

Hvoslef, Signe 2009: Kvernsteinfunn på Bryggen i Bergen. In: Aune, Anne and Valderhaug, Gudmund (eds). *Undervegs. Festskrift til Egil Nysæter*, 108-129. Bergen.

Håland, Randi 2004: Technology, transformation and Symbolism: Etnographic perspectives and European iron working. *Norwegian Archaeological Review 37 (1)*, 1-19. Oslo.

Håkon Håkonssóns Saga. *Norges Kongesagaer IV.* 1914-utgaven (translated byAlexander Bugge).

Håkon Håkonsson saga. Norges kongesagaer, bind 4 (translatad by Finn Hødnebø, 1979).

Hårdh, Birgitta 2008: Hacksilver and Ingots. In: Skre, Dagfinn (ed.) *Means of exchange.* Kaupang Excavation Project Publication Series, Volume 2. Norske Oldfunn XXIII, 95-118. Oslo.

Indrebø, Gustav 1936: *Fjordung. Granskingar i eldre norsk organisasjons-soge. Bergens Museums Årbok 1936. Historisk-antikvarisk rekke. Nr. 1.* Kolderup, Carl Fred. (ed.). Bergen.

Ingold, Tim 2002: *The perception of the environment. Essays in livelihood, dwelling and skill.* London.

Ingvaldsen, Lilly M. 1998: *Kirkene i søndre Sunnhordland i tidlig- og høymiddelalder : hvor ble kirkene bygget, hvorfor ble de bygget her og av hvem?* Unpublished Master's thesis, University of Bergen.

Iversen, Frode 1999 [1997]: *Var middelalderens lendmannsgårder kjerner i eldre godssamlinger? En analyse av romlig organisering av graver og eiendomsstruktur i Hordaland og Sogn og Fjordane.* Arkeologiske avhandlinger og rapporter. Bind 4. Bergen.

Iversen, Frode 2005: Royal manors and estates in Western Norway in the late Iron Age and Middle Ages. In: Iversen, Tore and Myking, John Ragnar (eds.) *Land, Lords and Peasants. Peasants' right to control land in the Middle Ages and the Early Moderen Period - Norway, Scandinavia and the Alpine region. Report from a seminar in Trondheim, November 2004. No. 52 Trondheim Studies in History*, 133-144. Trondheim.

Iversen, Frode 2008 [2004]: *Eigedom, makt og statsdannelse. Kongsgårder og gods i Hordaland i yngre jernalder og middelalder. UBAS, Nordisk, Vol. 6.* Bergen.

Iversen, Frode in press: Eit kloster på Halsnøy til heider og ære. In: Økland, Bård, Jane Jünger and Ingvild Øye (eds.) *Halsnøy kloster.* Scaninavian Academic Press. Forthcoming 2013.

Iversen, Tore 1995: Fremveksten av det norske leilendingsvesenet i middelalderen - en forklaringsskisse. *Heimen, Vol. 32*, 169-180.

Iversen, Tore and Myking, John Ragnar 2007: The workshop in Trondheim. Summary and tentative assessment. In: Iversen, Tore and John Ragnar Myking (eds.) *Land, Lords and Peasants. Peasants' right to control land in the Middle Ages and the Early Moderen Period - Norway, Scandinavia and the Alpine region. Report from a seminar in Trondheim, November 2004. No. 52 Trondheim Studies in History*, 295-299. Trondheim.

Jansen, Øystein J. 2013: Stein til klostermurane. In: Økland, Bård Gram, Jünger, Jane and Øye, Ingvild (eds.) *Halsnøy kloster*, 69-83. Oslo.

Jansen, Øystein J., Løland, Torbjørn, Meyer, Gurli, Grenne, Tor and Baug, Irene in prep.: *Collection Report: Rotary Querns in the Museum of Roskilde, Denmark.*

Jensen, Stig 1990: Handel med dagligvarer i vikingetiden. *Hikuin 16*, 119-138. Højbjerg.

Jessen, Mads Dengsø in press: Early Watermills: An archaeological genealogy of taxation.

Jobey, George 1986: Millstones and Millstone Quarries in Northumberland. *Archaeologia Aeliana 5 XIV*, 49-80.

Johnsen, Ingrid Sannes 1968: *Stuttruner i vikingtidens skrifter.* Oslo.

Kellmer, Inger 1963: *Eiendomsforhold i Hyllestad. Fra senmiddelalderen fram til omkring 1862.* Unpublished Master's thesis, University of Bergen.

Kinvig, R. H. 1975: *The Isle of Man. A social, cultural, and political history.* Liverpool.

Kleiven, Ivar 1977: *I gamle Daagaa. Forteljingo og Bygda-Minne fraa Vaagaa.* Otta.

KLNM = Kulturhistorisk leksikon for nordisk middelalder fra vikingtid til reformasjon, I-XXII, 1965-1978.

Kraft, Jens 1829: *Det vestenfjeldske Norge. Topographisk-statistisk Beskrivelse over Kongeriget Noreg. Vol. 4.* Christiania.

Kristensen, Hans Krongaard og Vellev, Jens 1999: Teknikk. In: Else Roesdahl (ed.) *Dagligliv i Danmarks middlelalder. En arkæologisk kulturhistorie*, 206-237. Copenhagen.

Kristiansen, Mette Svart 2005: *Tårnby. Gård og landsby gennem 1000 år*. Århus.

L = Magnus Lagabøtes Landslov (translated by Absalon Taranger in 1915). Oslo.

Landnámabók, after Hauksbók (translated by John Ragnar Hagland in 2002). Hafrsfjord.

Lang, James 1988: *Anglo-Saxon Sculpture. Shire Archaeology*. Aylesbury.

Langggåt, Anja Hysvær 2009: *Lyse kloster. Jordegodset som økonomisk fundament for et norsk cictercienserkloster*. Unpublished Master's thesis, University of Oslo.

Larsen, Jette Linnaa 2005a: Andet inventor. In: Kristiansen, Mette Svart (ed.) *Tårnby. Gård og landsby gennem 1000 år*, 374-383. Århus.

Larsen, Jette Linaa 2005b: Gårdens miljø - mening i materiell mangfoldighet. In: Kristiansen, Mette Svart (ed.) *Tårnby. Gård og landsby gennem 1000 år*, 405-412. Århus.

Lemonnier, Pierre 1992: *Elements of an Anthropology of Technology. Anthropological Papers, Museum of Anthropology, University of Michigan. No. 88*. Michigan.

Lemonnier, Pierre 1993: Introduction. In: Lemonnier, Pierre (ed.) *Technological choices. Transformation in material culture since the Neolithic*, 1-35. London and New York.

Leroi-Gourham, André 1993 [1964]: *Gesture and Speech*. Cambridge, Massachusetts.

Lidén, Hans-Emil 1977: Stratigrafisk-topografisk beskrivelse av feltet 'Mindets tomt'. *De arkeologiske utgravninger i Gamlebyen Oslo. Bind I. Feltet 'Mindets tomt'. Stratigrafi-topografi, daterende funngrupper*, 11-71. Oslo-Bergen-Tromsø.

Liebgott, Niels Knud 1989: *Dansk middelalderarkæologi*. København.

Liestøl, Aslak 1954: Runekrossen i Muséparken i Stavanger. *Stavanger Museum Årbok 1954*, 64-77. Stavanger.

Lillehammer, Arnvid 1968: Melding om den ark. utgravinga på tomta Skagen 3 i Stavanger sommaren 1968. Archive report, Topografisk arkiv. Stavanger, Rogaland. Arkeologisk Arkeologisk avdeling ved Stavanger Museum. Bergen.

Lisowska, Eva, Gunia, P. and Borowski, M. 2014: Rotary quernstone production in south-western Poland during the Middle Ages. In: Selsing, Lotte (ed.) *Seen through a millstone*. AmS-Skrifter 24. Museum of archaeology, University of Stavanger, 167-180. Stavanger.

Livland, Håkon 1992: *Eidsborgbryner - Eksportvare i Telemark fra vikingtid til våre dager*. Skien.

Lorentzen, Bernt 1952: *Gård og grunn i Bergen i middelalderen*. Det Hanseatiske Museums Skrifter, No. 16. Bergen.

Lossius, Siri Myrvoll 1977: *Klebermaterialet fra Borgund*. Arkeologiske avhandlinger. Historisk Museum 1, Universitetet i Bergen. Bergen.

Lund, J. B. 2001: *Tidlige danske møller. Middelalderens mølleteknologi i arkæologisk belysning*. Højbjerg.

Lundberg, Nina 2007: Øye - en arkeologisk undersøkelse av et klorittskiferbrudd. Unpublished Master's thesis, NTNU, Trondheim.

Lunden, Kåre 1972: *Korn og kaup I. Studiar over prisar og jordbruk på Vestlandet i mellomalderen*. Oslo.

Løland, Torbjørn 2007: Rapport frå skulegraving i kvernsteinsbrota i Hyllestad. Archive Report, Topografisk arkiv. University Museum of Bergen. Bergen.

Løland, Torbjørn in press: *Kvernsteinshogging i Norge. Om handverket knytt til kvernsteinsproduksjonen*. Forthcoming report NGU, Trondheim.

Magnússon, Þor 1971: Sögualdarbyggð í Hvítárholti. In: Eldjárn, Kristján (ed.) Árbók hins íslenzka fornleifafélags, 5-80. Reykjavik.

Majerowicz, Alfred and Skoczylas Janus. 1983: Petroarceologiczne badania surowców skalnych ze stanowiska 1 w Wolinie. *Materialy Zachodniopomorskie*. Wolin.

Mangartz, Fritz 1998: *Vulkanpark Forschungen. Untersuchungen zur Landschafts- und Kulturgeschichte. Band 1. Die Antiken Steinbrüche der Hohen Buche bei Andernach. Topographie tecnologie und Chronologie*. Mainz.

Mangartz, Fritz 2000: Römerzeitlicher Abbau von Basaltlava in der Osteifel: Ein bedeutender Wirtschaftszweig der Nordwestprovinzen. *Steinbruch und Berwerk. Denkmäler römischer technikgeschichte zwischen Eifel und Rhein. Vulkanpark-Forschungen. Untersuchungen zur Landschafts- und Kultugeschichte Band 2*, 6-16. Mainz.

Mangartz, Fritz 2008: *Römischer Basaltlava-Abbau zwischen Eifel und Rhein*. Mainz.

Mendels, Franklin F. 1972: The First Phase of the Industrialization Process. *The Journal of Economic History, Vol. 32, No. 1, The Tasks of Economic History*, 241-261. Cambridge.

Meyer, Gurli, Baug, Irene, Skov, Hans, Grenne, Tor, Løland, Torbjørn and Jansen, Øystein J. in prep.: *Collection Report: Rotary Querns in the Museum of Moesgård, University of Århus, Denmark*.

Miller, Heather M.-L. 2007: *Archaeological Approaches to Technology*. Amsterdam.

Molaug, Petter B. 2008: Oslo blir by - fra 1000-1200. In: Andersson, Hans, Hansen, Gitte and Øye, Ingvild (eds.) *De første 200 årene - nytt blikk på 27 skandinaviske middelalderbyer*. UBAS Nordisk Universitetet i Bergen Arkeologiske Skrifter, 73-92. Bergen.

Molland, Einar 1968: Patronatsrätt, Norge. *KLNM XIII*, 140-144.

Morcken, Roald 1964: *Den nautiske mil gjennom tusen år*. Bergen Sjøfartsmuseums Årshefte.

Moreland, John 2000a: Concepts of the early medieval economy. In: Hansen, Inge Lyse and Wickham, Chris (eds.) *The Long Eighth Century*, 1-34. Leiden, Boston, Köln.

Moreland, John 2000b: The significance of production in Eighth-century England. In: Hansen, Inge Lyse and

Wickham, Chris (eds.) *The Long Eighth Century*, 69-104. Leiden, Boston, Köln.

Morkinskinna: Norske kongesoger 1030-1157. Translated by Kåre Flokenes, 2001. Stavanger.

Mygland, Sigrid S. 2003: *Barn i byen. En arkeologisk analyse av gjenstandsmateriale knyttet til barn i Bergen, fra ca. 1100 til 1700*. Unpublished Master's thesis, University of Bergen.

Mygland, Sigrid S. 2007: *Children in Medieval Bergen: an archaeological analysis of children-related artefacts*. The Bryggen Parers, Main Series, No 7. Bergen.

Myking, John Ragnar 2005: Peasants' land control in Norway. In: Iversen, Tore and Myking, John Ragnar (eds.) *Landlords and Peasants. Peasants' rights to control land in the Middle Ages and the Early Modern Period - Norway, Scandinavia and the Alpine region. Report from a seminar in Trondheim, November 2004*, 15-28. Trondheim.

Myrvoll, Siri 1985: The trade in Eidsborg hones over Skien in the medieval period. In: Edgren, Torsten and Jugner, H. (eds.) *Proceedings of the Third Nordic Conference on the Application of Scientific Methods in Archaeology. Mariehamn,* Åland, *Finland, 8-11 October 1984. Suomen Muinaimsuistoyhdistys = Finska fornimmesföreningen. Iskos 5*, 31-47. Helsinki.

Nagel, Anne-Hilde 1986: *Staten og bergverkene. Avgjørelsesprosessen omkring bergverksprivilegier under eneveldet*. Unpublished Doctoral thesis, University of Bergen, Bergen.

Nagel, Anne-Hilde 1991: 'Wi Margrete'. 600 års norsk bergverkspolitikk - en historisk antologi. In: Berg, Bjørn Ivar (ed.) *Bergverksmuseet. Skrift Nr. 6*, 41-45. Kongsberg.

Naterstad, Johan 1984: Den geologiske bakgrunn for bakstehelleindustrien ved Kvitebergvatnet i Hardanger. *Viking Bind XLVII, 1983*, 161-164.

Naterstad, Johan 1989: Stein brukt i baksteheller. Muligheter for lokalisering av bruddsteder. *Fortiden i Trondheim bygrynn: Folkebibliotekstomten. Meddelelser nr. 15. Baksteheller - en handelsvare*, 23-32. Trondheim.

Nedkvitne, Arnved 1983: *Utenrikshandelen fra det vestafjelske Norge 1100-1600*. Bergen.

NG = Norske Gaardnavne opplysninger samlede til brug ved matrikkelens revision I-XIX. Kristiania.

NgL = *Norges gamle Love indtil 1387 1-5* (published by Gustav Storm and Ebbe Hertzberg). Christiania.

NIyR IV = *Norges innskrifter med de yngre tuner. Bind IV*. Olsen, Magns and Aslak Liestøl 1957. Oslo.

Nordeide, Sæbjørg Walaker and Gulliksen, Steinar 2007: First generation Christians, second generation radiocarbon dates: the cemetery at St. Clement's in Oslo. *Norwegian Archaeological Review, Volume 40, No. 1*, 1-25. Oslo.

Nordeide, Sæbjørg Walaker 2009: Cross Monuments in North-Western Europe. *ZAM Zeitschrift für Archäologie des Mittelalters, Jahrgang 37*, 163-178. Bonn.

Nordeide, Sæbjørg Walaker 2011: Steinkors og korssteiner i nordvestre Europa. *Collegium Medievale, colum 24, 2011*, 127-148. Oslo.

Normann, Einar 1936: *Mottakarverk til korn, turkor, reinske-, male- og sliteverk, andre mylnemaskinar. Norske mylnor og kvernar: utgreidingar frå Statens Kornforretning, 1, Statistikk yver mylnor og kvernar i 1927-1929. Bind II*. Oslo.

Norske middelalderdokumenter. Bagge, Sverre, Synnøve Holstad Smedsdal and Knut Helle (1973). Bergen

Norstrand, Ingemar 2007: Mangt og kuriøst om kvernstein. In: Førsund, Finn Borgen (ed.) *Sogeskrift frå Hyllestad 2007*, 26-33. Førde.

NRJ = *Norske Regnskaber og Jordebøger fra det 16. de Aarhundrede 1-4*, utgitt for Det Norske historiske kildeskriftfond ved H.J. Huitfeldt-Kaas (1887-1906). Christiania.

NIyR IV = Norges innskrifter med de yngre runer. Bind IV. Olsen, Magnus and Aslak Liestøl 1957. Oslo.

Nymoen, Pål 2011: Kun for den smarte skippers regning? Skipsvrak med omsettelig ballast: på sporet av kvernstein, kleber og brynehandelen fra Norge ca 800-1800. *Nicolay. Arkeologisk tidsskrift, no. 114*, 65-75. Oslo.

Nymoen, Pål 2012: Til bunns for 1000 år siden. *Norsk Maritimt Museum. Årbok 2011*, 171-184. Oslo.

Nysæter, Egil 2013: Jordegods og andre verdslege gode i klostertida. In: Økland, Bård, Jünger, Jane and Øye, Ingvild (eds.) *Halsnøy kloster. Til kongen og Augustins ære*, 143-171. Oslo.

Nøttveit, Ole-Magne 2010: *Rapport fra marinarkeologisk forundersøkelse i Vågen, Bergen. Delraundersøkelse: Dokumentasjonsdykking*. Unpublished Report, Bergen Maritime Musem, Bergen.

Oesterwind, Bernd C. 2000: Mayen - Zentrum der Muhlsteinherstellung in der Römerzeit. *Steinbruch und Berwerk. Denkmäler römischer technikgeschichte zwischen Eifel und Rhein. Vulkanpark-Forschungen. Untersuchungen zur Landschafts- und Kultugeschichte Band 2*, 33-57. Mainz.

OE = Olav Engelbrektssøns jordebog. Register paa St. Olavs jorder forfattet under erkebiskop Olav Engelbrektssøn. Anhang: Erebiskop Gautes jordebøger. Published by Norges Rigsarkiv. Brinchmann, Chr. and Agerholt, Johan (ed.). 1926. Oslo.

Olaus Magnus = Olaus Magnus 2010 [1555]: *Historia om de nordiska folken*. Vilnius.

Olsen, Bjørnar 1997: *Fra ting til tekst. Teoretiske perspektiv i arkeologisk forsking*. Oslo.

Ommundsen, Åslaug 2010: Nonneseter i Bergen - eit benediktinarkloster. *Historisk tidsskrift, hefte 89 (2010) No. 4*, 547-571. Oslo.

Parkhouse, Jonathan 1976: The Dorestad quernstones. *Berichten van de Rijkdienst voor het Oudheidkundig = Proceedings of the State Service for Archaeological Investigations in the Netherlands*, 181-188. Amersfoort.

Parkhouse, Jonathan 1997: The Distribution and Exchange of Mayen Lava Quernsones in Early Medieval Northwestern Europe. In: Boe, Guy De and Frans Verhaeghe (eds.). *Exchange and Trade in Medieval*

Europe. Papers of the 'Medieval Europe Brugge 1997' Conference. Volume 3, 97-106. Zellik.

Pendergast, D. 2000: The problems raised by small charcoal samples for radiocarbon analysis. *Jounal of Field Archaeology, Vol. 27, No. 2*, 237-239. Boston.

PN = Pavelige Nuntiers Regnskabs- og Dagböker, førte under Tiende Opkrævningen i Norden 1282-1334. Udgivne efter Offentlig Foranstaltning ved P.A. Munch: med et Anhang av Diplomer. 1864, Christiania.

Petersen, Jan 1919: *De norske vikingesverd: en typologisk-kronologisk studie vikingetidens vaaben*. Kristiania.

Petersen, Jan 1933: *Fra forhistorisk tid og middelalder*. Gamle gårdsanlegg i Rogaland. Bind/hefte nr. 1. Instituttet for sammenlignende kulturforskning. Serie B, Skrifter; 23, 31. Oslo.

Petersen, Jan 1951: *Vikingtidens redskaper*. Skrifter (Det Norske videnskaps-akademi i Oslo) / 2 Historisk-filosofisk klasse. Bind/hefte nr. 1951:4. Oslo.

Pettersen, Gunnar I. 2000: Handel og priser i Norge i middelalderen. *Collegium medievale, Vol. 13*, 203-228. Oslo.

Pettersson, Täpp John-Erik 1977: *Kvernstenshuggning. En ineldande översikt på skandinavisk grund med utgångspunkt från Malun socken*. Uppsats i etnologi. Påbyggnadskurs C I, ved Uppsala Universitet.

Pilø, Lars 2007: The settlement: Artefacts and Site Periods. In: Skre, Dagfinn (ed.) *Kaupang in Skiringssal*. Kaupang Excavation Project. Publication Series, Volume 1. Norske Oldfunn XXII, 161 -178. Aarhus.

Placat og Jorde-bog over Den Hans Kongel. Majest. Tilhørende og udi Norge beliggende Halsnøklosters Gaard og Gods samt herunder liggende Leedinger, Tiender og Grundelejer, med videre, som, med de herefter til Slutning avførte Conditioner i følge hans Kongl. Majetets. Wernaadigte Refolution af 2. julii 1744 paa de Sider og Stæder, som, bedhr. Stiftbefalnings-mand à Möinochen, fornd bliver besiendiorte, ved offentlig action skal bortsælges. 1744, Kjøbenhavn.

Pfaffenberg, Bryan 1992: Social Anthropology of Technology. In: *Annual Review of Anthropology, Vol. 21 (1992)*, 491-516. California.

Pohl, Meinrad 2011: Querns as markers for the determination of Medieval northern European trade spheres. In: Williams, David and David Peacock (eds.) *Bread for the People: The Archaeology of Mills and Milling. Proceedings of a Colloquium Held in the British School at Rome 4th-7th November 2009*, University of Southampton Series in Archaeology, No. 3, 169-177. Oxford.

Pohl, Meinrad 2012 [2011]: *Steinreich - Mühlsteine, Tuff und Trass aus der östlichen Vulkaneifel und ihr Markt. Eine vergleichende Analyse vorindustrieller Produktions- und Handelsstrukturen*. Bochum.

Pred, Alan 1985: The Social Becomes the Spatial, the Spatial Becomes the Social: Enclosures, Social Change and the Becoming of Places in Skåne. In: Gregory, Derek and Urry, John (eds.) *Social Relations and Spatial Structures*, 337-365. New York.

Prøsch-Danielsen, Lisbeth and Soltvedt, Eli-Christine 2011: From saddle to rotary hand querns in South-Western Norway and the corresponding crop plant assemblages. *Acta Archaeologica, Vol. 82*, 129-162. Oxford.

Prøsch-Danielsen, Lisbeth and Heldal, Tom 2014: A local millstone quarry at Nord-Talbje in Southwestern Norway: mapping, interpretation and age estimate In: Selsing, Lotte (ed.) *Seen through a millstone*. AmS-Skrifter 24. Museum of Archaeology, University of Stavanger, 193-208. Stavanger.

Ramstad, Morten, Halvorsen, Lene Synnøve and Olsen, Asle Bruen 2011: Bjørkum: feasting, craft production and specialisation on a Viking Age rural site in Norway. *Antiquity*, Vol. 0855, Issue 328 June 2011.

Ramstad, Morten in prep.: Excavation report from Bjørkum. Preliminary Report, Bergen University Museum.

Reiersen, Marit 1999: *Mat i yngre jernalder og middelalder. Mattilberedning og m*åltid langs kysten av nordre Nordland *og Troms*. Unpublished Master's thesis, University of Trømsø.

Resi, Heid Gjøstein 2008: Whetstones and grindstones used in everyday life at Kaupang. *Whetstones and grindstones in the settlement area: the 1956-1974 excavations. The Kaupang finds, Volume III C*. Oslo.

Resi, Heid Gjøstein 2011: Whetstones, Grindstones, Touchstones and Smoothers. In: Skre, Dagfinn (ed.) *Things from the Town. Artefacts and Inhabitants in Viking-age Kaupang*. Kaupang Excavation Project. Publication Series, Vol. 3. Norske Oldfunn XXIV, 373-393. Aarhus.

Rijal, Suman 1998: *The Traditional System of Iron Working: Technology, Social Context and Rituals of Transformation - an Ethnoarcaeological Study from Eastern Nepal*. Unpublished Master's thesis, University of Bergen.

Ringstad, Bjørn 1992: Økonomsike og politiske sentra på Vestlandet ca. 400-1000 e. Kr. In: Mikkelsen, Egil and Larsen, Jan Henning (eds.) Økonomiske og p*olitiske sentra i Norden ca 400-1000 e. Kr.* Åkerseminaret, Hamar 1990. Universitetets Oldsamlings skrifter. Ny rekke 13, 120-130. Oslo.

RN = *Regesta Norvegica I* (1989).

Robberstad, Knut 1952: Jordeiga til Holsnaklosteret. *Sunnhordland tidsskrift. Band XXXIII*, 42-54. Stord.

Robberstad, Knut 1967: Odelsrett. *KLNM VII*, 94-106.

Rollseth, P.O. 1947: *Kvernfjellet*. Oslo.

Rynning, Lorentz 1934: *Bidrag til norsk almenningsrett I*. Oslo.

Rønneseth, Ottar 1968: Das Zentrum der ältesten Mühlsteinindustrie in Norwegen. In: Martin, Claus, Haarnagel, Werner and Raddatz, Klaus (eds.) *Studien zur europäischen Vor- und Frühgeschichte*, 241-252. Neunmünster.

Rønneseth, Ottar 1977: Kvernsteinsbrota ved Åfjorden. *Sogeskrift frå Hyllestad. Hefte 2*, 5-28. Førde.

Sabo, Katalin Schmidt 2005: *Den medeltida byns sociala dimensioner. Riksantikvarieämbetet arkeologiske undersökningar. Skrifter 67. Lund studies in historical archaeology 1*. Stockholm.

Salzman, L. F. 1967: *Building in England down to 1540*. Oxford.

Sandberg, Per-Øyvind 1986: *Gørvel Fadersdatters regneskap over Giske og Giskegodset 1563. Norsk Historisk kjeldeskrit-institutt*. Oslo.

Sandnes, Jørn and Stemshaug, Ola 1976: *Norsk stadnamnleksikon*. Oslo

Sanbdberg, Per-Øyvind 1979: *Jordeboka over Hans Sigurdssons norske godskompleks. Et forsøk på rekonstruksjon av Sudrheimættens jordegods i Norge ca. 1500*. Unpublished Master's thesis, University of Oslo.

Schön, Volkmar 1995: *Die Mühlsteine von Haithabu und Schleswig. Ein Beitrag zur Entwicklungsgeschichte des mittelalterlichen Mühlenwesens in Nordwesteuropa*. Neunmünster.

Selsing, Lotte 2014 (ed.): *Seen through a millstone*. AmS-Skrifter 24. Museum of Archaeology, University of Stavanger. Stavanger.

Sindbæk, Søren Michael 2005: *Ruter og rutinisering. Vikingetidens fjernhandel i Nordeuropa*. København.

Sindbæk, Søren 2007: The Small World of the Vikings: Networks in Early Medieval Communication and Exchange. *Norwegian Archaeological Review, Vol. 40, No. 1, 2007*, 59-74. London.

Sindbøk, Søren 2008: Local and long-distance exchange. In: Brink, Stefan and Price, Neil (eds.) *The Viking World*, 150-158. London and New York.

Skaare, Kolbjørn 1976: *Coins and coinage in Viking Age Norway*. Oslo.

Skattematrikkelen 1647 XI Hordaland fylke. Fladby, Rolf and Winge, Harald (eds.) 1995. Oslo.

Skattematrikkelen 1647 XII Sogn og Fjordane fylke. Fladby, Rolf and Marthinsen, Liv (eds.) 1977. Oslo.

Skeised, Angunn 2007: *Seks kyrkjestader i Rogaland. Ein arkeologisk analyse av kyrkjestader i ytre og indre delar av Rogaland*. Unpublished Master's thesis, University of Bergen.

Skjølsvold, Arne 1961: *Klebersteinsindustrien i vikingetiden*. Oslo.

Skov, Hans 1999: Archaeological evidence of trade in Århus, Denmark, from the 10th to the 17th centuries. In: Dunckel, Regina, Gläser, Manfred and Oltmanns. Ulrike (eds.) *Lübecker Kolloquium zur Stadtharchäologie im Hansaraum II. Der Handel*, 603-611. Lübeck.

Skov, Hans 2008: Det ældste Århus, ca. 770-1200. In: Andersson, Hans, Hansen, Gitte and Øye, Ingvild (eds.) *De første 200 årene - nytt blikk på 27 skandinaviske middelalderbyer*. UBAS Nordisk. Universitetet i Bergen Arkeologiske Skrifter 5, 215-226. Bergen.

Skre, Dagfinn 1988: *Gård og kirke, bygd og sogn. Organiseringsmodeller og organiseringsenheter i middelalderens kirkebygging i Sør-Gudbrandsdalen. Riksantikvarens rapporter 16*. Øvre Ervik.

Skre, Dagfinn 1997: Haug og grav. Hva betyr gravhaugene? In: Christensson, Ann, Mundal, Else and Øye, Ingvild (eds.) *Middelalderens symboler*, 37-52. Bergen.

Skre, Dagfinn 1998: *Herredømmet. Bosetning og besittelse på Romerike 200-1350 e.Kr*. Acta Humaniora. Oslo.

Skre, Dagfinn 2007: Towns and Markets, Kings and Central Places in South-western Scandinavia c. AD 800-950. In: Skre, Dagfinn (ed.) *Kaupang in Skiringssal*. Kaupang Excavation Project Publication Series, Vol. 1. Norske Oldfunn XXII, 431-444. Oslo.

Skre, Dagfinn 2008a: Dealing with Silver: Economic Agency in South-Western Scaninavia AD 600-1000. In: Skre, Dagfinn (ed.) *Means of exchange*. Kaupang Excavation Project Publication Series, Vol. 2. Norske Oldfunn XXIII, 343-355. Aarhus.

Skre, Dagfinn 2008b: Post-Substantivist Towns and Trade AD 600-1000. In: Dagfinn Skre (ed.) *Means of exchange*. Kaupang Excavation Project Publication Series, Vol. 2. Norske Oldfunn XXIII, 327-341. Aarhus.

Skre, Dagfinn 2010: From Dorestad to Kaupang. Frankish Traders and Settlers in the 9th-century Scandinavian Town. In: Willemsen, Annemarieke and Kik, Hanneke (eds.) *Dorestad in an international framework: new research on centres of trade and coinage in Carolingian times; proceedings of the first 'Dorestad congress' held at the National Museum of Antiquities Leiden, The Netherlands, June 24-2*, 131-135. Turnhout.

Skre, Dagfinn 2011: Commodity Money, Silver and Coinage in Viking-Age Scandinavia. In: Graham-Campbell, James, Sindbæk, Sørenm and Williams, Gareth (eds.) *Silver Economies, Monetisation and Society in Scandinavia, AD 800-1100*, 67-91. Aarhus.

Skyllberg, Eva 2001: *Södermanlands medeltida bergsbruk - en feudal angelägenhet*. Sørmländska handlingar nr 50. Doctoral thesis, University of Stockholm.

Smith, A. N. 1999: Steatite: vessels, bakestones and other objects. In: Crawford, B. E. and Ballin-Smith, B. (eds.) *The Biggings, Papa Stour, Shetland; the history and archaeology of a royal Norwegian farm*. Society and Antiquaries of Scotland and Det Norske Videnskaps-Akademi, Society of Antiquaries of Scotland Monograph Series, Number 15, 129-143. Edinburgh.

Solberg, Bergljot 1984: *Norwegian spear-heads from the Merovingian and Viking periods*. Unpublished Doctoral thesis, University of Bergen.

Solberg, Bergljot 1985; Social Status in the Merovingian and Viking Periods in Norway from Archaeological and Historical Sources. *Norwegian Archaeological Review, Vol. 18, Nos. 1 -2, 1985*, 62-76. London.

Solem, Jon 2003: Den norske allmenningsinstitusjonen i middelalderen. In: Lillehammer, Arnvid, Tysdal, Olav and Østrem, Nils Olav (eds.) *Heimen. Lokalhistorisk tidsskrift, Vol. 40, 4, 2003*, 243-265.

Specification over Halsnøe-Klosters Hoved gaard med underliggende Gods samt Leeding, Tiender og Grundelejer udi Bergen. 1757, Kjøbenhavn.

Steinnes, Asgaut 1982: Mål, vekt og verderekning i Noreg. *Gammel mål og vekt i Norge*, 16-86. Oslo.

Stemshaug, Ola 1976: *Namn i Noreg: ei innføring i norsk stadnamngransking*. Oslo.

Steuer, Heiko 1987: Der Handel der Wikingerzeit zwischen Nord- und Westeuropa aurgrund archäelogischer Zeugnisse. In: Düwel, Klaus (ed.) *Unterzuchungen zu Handel und Verkehr der vor- und frühgeschichtlichen Zeit in Mittel- und Nordeuropa*, 113-197. Göttingen.

Stigum, Hilmar 1964: Kvarn. Norge. *KLNM IX*. Oslo.

Stuland, Arne 1924: *Kvinnherad Naturtilhøve og gardssoga*. Norheimsund.

Stumman, Steffen Hansen and Larsen, Anne-Christine 2000: Miniature quern- and millstones from Shetland's Scandinavian past. In: Stumman Hansen, Steffen and Randsborg, Klavs (eds.) *Vikings in the west*. Acta Archaeological, Vol. 71-2000, 105-121. København.

Stylegard, Frans-Arne and Torbjøn Landmark 1999: Bergverksforskning - et 'nytt' tema i norsk arkeologi. *Viking, No. 42*, 81-99. Oslo.

Sætre, Anja Nordvik 2008: *Sæsol. Jordbruk i kvernsteinslandskap. En arkeologisk punktundersøkelse av en gård i drift. Hyllestad, Sogn og Fjordane*. Unpublished Master's thesis, University of Bergen.

Søyland, Eirik Herdlevær 2009: *Elleve kirkesteder i Hordaland: en arkeologisk analyse av kirkesteder på Osterøy og i Lindås*. Unpublished Master's thesis, University of Bergen.

Taranger, Absalon 1915: *Magnus Lagabøtes Landslov* (translated by Absalon Taranger 1915). Oslo/Bergen/Tromsø.

Teigum, Ivar 2011: Tolstadkvernberget - frå kongeleg pant i Amsterdam til bondegods i Vågå. In: Nordstrand, Leiv, Hutchinson, Alan and Aas, Steinar (eds.) *Heimen, Lokal og regional historie, Bind 48*, 47-58. Stavanger.

Teigum, Ivar 2001: *Bygdebok for Vågå og Sel, Band 1: Frå dei eldste tider til 1600*. Otta.

Tengesdal, Sigrun Solbakken 2010: Å steike! En kontekstuell materialanalyse av steikeheller funnet i et bygårdskompleks i middelalderens Bergen. Unpublished Master's thesis, University of Bergen.

Tilley, Christopher 1994: A *phenomenology of landscape. Places, paths and monuments*. Oxford.

Titland, H. P. 2003: Kvernsteinsbruddene i Saltdal. *Saltdalsboka 2002/2003*, 35-43. Saltdal.

Thue, Johs. B. 2000: *Livets steinar: produksjon og eksport av kvernstein frå Hyllestad i mellomalderen*. Leikanger.

Trones, J. I. 2003: Om bergverksdrift frå middelalderen og muligens tidligere. *Saltdalsboka 2992/2003*, 44-47. Saltdal.

Tryti, Anna Elisa 1994: Bergen som sentrum i bispedømmet. In: Øye, Ingvild (ed.) *By og land i middelalderen*. Onsdagskvelder i Bryggens Museum, IX, 23-43. Bergen.

Tucker, Gordon 1982: Millstones north and south of the Scottish border. *Industrial Archaeology Review. Vol. VI, Number 3*, 186-193.

Tucker, Gordon 1984: Millstone making in Scotland. *Proceedings of the Society of Antiquaries of Scotland*, 539-556.

Turner, Val 2009: Kleber: Shetland's first industry. In: Forster, Amanda K. and Val E. Turner (eds.) *Kleber: Shetland's Oldest Industry. Shetland Soapstone Since Prehistory*, 1-3. Lerwick.

Turner, Val, Paul Sharman and Carter, Stephen 2009: Excavations at Catpund, Cunningsburgh. In: Forster, Amanda K. and Turner, Val E. (eds.) *Kleber: Shetland's Oldest Industry. Shetland Soapstone since Prehistory*, 70-84. Lerwick.

Tveitane, Mattias 1981: Kviteseid - Eidsborg. Om namn, egedomsforhold og samferdsel I Vest-Telemar i elder tid. *Heimen no. 2 1981. Bind XVIII, hefte 10*, 635-650.

Tveiten, Ole 2010: Kven rådde over vinna? Jarnutvinning og råderett i utmark i mellomalderen. *Heimen, bind 47*, 243-262.

Tveiten, Ole 2012: *Mellom aust og vest. Ein arkeologisk analyse av jarnvinna kring Langfjella i yngre jarnalder og mellomalder*. Unpublished Doctoral thesis, University of Bergen.

Ulriksen, Eli 2008: Tønsberg - bebyggelse og beboere fra 1000-tall til 1200-tall. In: Andersson, Hans, Hansen, Gitte and Øye, Ingvild (eds.) *De første 200 årene - nytt blikk på 27 skandinaviske middelalderbyer*. UBAS Nordisk Universitetet i Bergen Arkeologiske Skrifter, 93-108. Bergen.

Vaage, Erling 1972: *Kvinnherad. Bygdesoga I*. Bergen.

Vaage, Erling 1987: *Kvinnherad. Gards- og ættesoga. Band 2, soknene Ølve Hatlestrand Ænes*. Bergen.

Waage, Astrid 2005: Kvernsteinsdrifta I Hyllestad - eit nyoppdaga kapittel av norsk bergverkssoge. *Fortidsvern, 30*, 28 -29.

Watts, Martin 2002: *The Archaeology of Mills and Milling*. Gloucestershire.

Watts, Susan 2006: *Rotary querns c. 700-1700. Datasheet 38*. http://www.findsresearchgroup700-1700.org.uk.

Weber, Birthe 1984: 'I Hardanger er Qvernberg og Helleberg… og Hellerne, det er tyndhugne Steene, bruger man til at bage det tynde Brød Fladbrød på…' *Viking Bind XLVII, 1983*, 149-160. Oslo.

Weber, Birthe 1989: *Fortiden i Trondheims bygrunn: Folkebibliotekstomten. Meddelelser nr. 15. Baksteheller - en handelsvare*. Trondheim.

Weber, Birthe 1990: Tregjenstander. In: Molaug, Erik and Petter B. Molaug (eds.) *De arkeologiske utgravninger i Gamlebyen, Oslo. Bind 7. Dagliglivets gjenstander, del I*, 11-180. Oslo.

Weber, Birthe 1999: Bakestones. In: Crawford, Barbara E. and Beverley Ballin Smith (eds.) *The Biggings, Papa Stour, Shetland. The history and excavation of a royal Norwegian farm. Society of Antiquaries of Scotland Monograph Series, Number 15*, 134-139. Edinburgh.

Welin, Sanfrid 1942: *Kvernstensbrytningen vid Lugnås under äldre tider*. Mariestad.

Williams, David and Peacock, David (eds.) 2011: *Bread for the people: The archaeology of mills and milling. Proceedings of a conference held in the British School at Rome, 4th-7th November 2009*. University of Southampton Series in Archaeology, No. 3. Oxford.

Wright, M. Elizabeth 1982: Beehive quern manufacture in the south-east Pennines. *Scottish Archaeological Review*. Glasgow.

Zachrisson, Torun 1994: The Odal and its manifestations in the Landscape. *Current Swedish Archaeology 2*, 219-238. Stockholm.

Zachrisson, Torun 2004: The holiness of Helgö. *Excavations at Helgö XVI. Exotic and Sacral Finds from Helgö*, 143-175. Stockholm.

Zachrisson, Torun 2014: Rotary querns and bread - A social history of Iron Age Sweden. In: Selsing, Lotte (ed.)

Seen through a millstone. AmS-Skrifter 24, Museum of Archaeology, University of Stavanger, 181-191. Stavanger.

Zehetner, Janicke Larsen 2006: *Fra sted til gård: en agrararkeologisk analyse av Indre Matre i Kvinnherad, Hordaland*. Unpublished Doctoral thesis, University of Bergen.

Ødegaard, Marie 2010: Graver og grenser - territoriell inndleing av jernalderens jordbrukslandskap I Vestfold. *Primitive tider 2010, 12, Årgang*, 27-40.

Østerås, Bodil 2002: *Slipsteinsberget I Sparbu. Kva undersøkingar av eit klebersteinsbrot kan fortelje om gamle steinhoggartradisjonar*. Unpublished Master's thesis, Vitenskapsmuseet, NTNU. Trondheim.

Øye, Ingvild 1989: Kaupangen i Sogn i komparativ belysning. *Viking 1989*, 144-165.

Øye, Ingvild 2002: Landbruk under press 800-1350. *Jorda blir levevei: 4000 f.Kr.-1350 e. Kr. Norges Landbrukshistorie I*, 215-414. Oslo.

Øye, Ingvild 2004: Farming systems and rural societies ca. 800-1350. In: Almås, Reidar (ed.) *Norwegian Agricultural History*, 80-143. Trondheim.

Øye, Ingvild 2005: Introduction. In: Holm, Ingunn, Innselset, Sonja and Øye, Ingvild (eds.) *'Utmark'. The Outfield as Industry and Ideology in the Iron Age and the Middle Ages*. UBAS International 1. University of Bergen Archaeological Series, 9-20. Bergen.

Øye, Ingvild 2008: Kontakten mellom Bergen og Færøyene i middelalderen i arkeologisk og historisk lys. In: Paulsen, Caroline and Helgi D. Michelsen (eds.) *Símunarbók. Heiðursrit til Símun V. Arga á 60 ára degnum. 5. September 2008*, 243-252. Tórshavn.

Øye, Ingvild 2009: Food and technology - Cooking utensils and food processing in medieval Norway. In: Klápŝtê, Jan and Sommer, Petr (eds.) *Processing, Storage, Distribution of Food. Food in the Medieval Rural Environment. Ruralia VIII. 7th-12th September 2009. Lorca, Spain*, 225-234. Turnhout.

Øye, Ingvild 2011: Settlements and Agrarian Landscapes. Chronological Issues and Archaeological Challenges. In: Sigmundsson, Svavar (ed.) *Viking Settlements and Viking Society. Papers from the Proceedings of the Sixteenth Viking Congress, Reykjvík and Reykjholt, 16th-23rd August 2009*, 494-506. Reykjavik.

URL (accessed December 2014)

http://askeladden.ra.no
http://www.millstone.no
http://www.unimus.no

Correspondence and personal communication

Arild Marøy Hansen - Bergen Maritime Museum
Tom Heldal - NGU
Conny Johansson Hervén - Head of Cultural Heritage Departement, Kulturhistoriske föreningen för sødra Sverige
Øystein J. Jansen - University Museum of Bergen
Gurli Meyer - NGU
Jenny Murray - National Museum in Shetland
Knut Johan Nerhus - Hatlestrand
Gunnar Nygård - Hyllestad
Audun Oppedal - Hyllestad
Asbjørn Rønneseth - Hyllestad
Hans Skov - Mosgård Museum
Lars Stenvik - NTNU
Ole Tveiten - Oppland County Council